PERSPECTIVES IN LATTICE QCD

PERSPECTIVES IN LATTICE QCD

Proceedings of the Workshop

Nara International Seminar House
Nara, Japan

31 October – 11 November 2005

Edited by

Yoshinobu Kuramashi
University of Tsukuba, Japan

World Scientific

NEW JERSEY • LONDON • SINGAPORE • BEIJING • SHANGHAI • HONG KONG • TAIPEI • CHENNAI

Published by

World Scientific Publishing Co. Pte. Ltd.

5 Toh Tuck Link, Singapore 596224

USA office: 27 Warren Street, Suite 401-402, Hackensack, NJ 07601

UK office: 57 Shelton Street, Covent Garden, London WC2H 9HE

British Library Cataloguing-in-Publication Data
A catalogue record for this book is available from the British Library.

PERSPECTIVES IN LATTICE QCD
Proceedings of the Workshop

ISBN-13 978-981-270-000-1
ISBN-10 981-270-000-5

Printed in Singapore by World Scientific Printers (S) Pte Ltd

Preface

A school-type workshop entitled "Perspectives in Lattice QCD" was held from 31 October to 11 November 2005 at Nara International Seminar House, Nara, Japan. This is a part of activities of ILFT (International Lattice Field Theory) Network, which is a two-year research program sponsored by the Japan Society for Promotion of Science aiming at building and expanding a cooperative international framework in the lattice field theories among universities and research institutions in Japan, the US, Germany, Italy and the UK.

In the past decade the progress of simulation algorithms and the availability of more affluent computational resources have enabled us to investigate two- and three-flavor dynamical quark effects with lighter quark masses. A full QCD simulation with the physical up, down and strange quark masses are now getting into our view. On the other hand, theoretical understanding of the chiral fermion on the lattice has also developed remarkably since the early 90s. Although the dynamical quark simulation with the exact chiral symmetry is extremely demanding, some exploratory studies have already started. Under these circumstances, the workshop was intended to provide us an opportunity of systemaic understanding of the recent progress in lattice QCD and basic grounds for thinking of the future direction. The lectures and seminars covers a wide range of subjects including forefront topics:

- Renormalization group and fixed point action,
- Algorithms in lattice QCD simulation,
- Lattice QCD with the domain decomposition algorithm,
- Chiral fermions on the lattice,
- Numerical simulation with the overlap fermion,
- On-shell improvement and nonperturbative renormalization,
- Twisted mass lattice QCD,
- Phenomenology with lattice QCD,
- Chiral perturbation theory.

They are delivered at an advanced level based on the lecturer's expertise in the subject. We would like to express our sincere gratitude to all the speakers for their excellent lectures and seminars. The written version of the lectures are collected in this volume, which is indebted to the efforts of the contributers.

We gratefully acknowledge financial support by the Japan Society for Promotion of Science, the Ministry of Education, Culture, Sports, Science and Technology, the Center for Computational Sciences (CCS) in the University of Tsukuba. We thank young physicists at the University of Tsukuba, KEK and the University of Tokyo for their devoted assistance in preparing the workshop. Special thanks are due to Tomomi Ishikawa, Naoto Tsutsui and Naoya Ukita who prepared a first draft to aid the lecturers in the preparation of their manuscript.

We hope this kind of school will be succeeded and contribute to the progress of the field in future.

Editor

Yoshinobu Kuramashi

CONTENTS

FIXED POINT ACTIONS, SYMMETRIES AND SYMMETRY TRANSFORMATIONS ON THE LATTICE *

P. HASENFRATZ

Institute for Theoretical Physics,
University of Bern,
Sidlerstrasse 5, CH-3012 Bern, Switzerland

Unwanted symmetry breaking by the lattice regularization will produce cut-off effects and distort the results. Symmetries are realized on the lattice frequently in an unusual way. Fixed point actions preserve all the classical symmetries of the theory and help to abstract not only the symmetry conditions, but the form of the lattice symmetry transformations also.

1. Introduction

Defining the action of a field theory on a hypercubic lattice is the first step towards a non-perturbative treatment of the corresponding quantum field theory. It seems to be obvious that certain symmetries of the continuum action (like space-time independent, 'internal' symmetries) will be trivially respected, while certain space time symmetries (like infinitesimal translations, rotations) unavoidably will be broken by the lattice action. This is, however not true. The internal chiral symmetry in its well known continuum formulation can not be kept on the lattice without violating some basic principles [1]. On the other hand, simple renormalization group considerations suggest that there exist lattice *actions* which inherit all the features of translation and rotation symmetries of the continuum. Even more, they inherit all the features of all the continuum symmetries. Since the action defines the classical field theory, these lattice formulations are classically perfect [2]. Refering to their role in renormalization group theory we shall call these actions here 'fixed point (FP) actions'.

These theoretical considerations are directly connected to earth bound

*This summary is based on parts of the lectures given at the ILFTN workshop at the Nara International Seminar House, 31. Oct.-11. Nov. 2005.

numerical experiments. Unwanted symmetry breaking by the lattice regularization will produce cut-off effects and distort the results.

On the basis of the accumulated experience on stochastic calculations since the first primitive computers entered the scene more than fifty years ago, it is not very probable that we shall see a great breakthrough in full QCD calculations in the near future. It is hard to expect miracles: good scaling, good chiral properties, theoretical safety and expenses will remain in balance. Probably, we shall see a plethora of full QCD simulations adapted to the physical problem investigated. In some interesting, but difficult situations like the ϵ-, or δ-regime the approximated fixed point action might be a competitive choice.

2. Cut-off effects: a numerical experiment on the running coupling

We shall start with an example. Fig 1. shows the stochastic values of a physical quantity as the function of the lattice resolution a^3. The results are plotted against a^2, since the quantity is expected to approach the continuum limit with a^2 corrections if a^2 is small. Actually, the figure refers to a $d = 2$ quantum field theory, the two dimensional $O(3)$ non-linear σ-model. This model has many analogies with a $d = 4$ Yang-Mills theory: it is asymptotically free, has dimensional transmutation (i.e. it has massive excitations although the classical theory is scale invariant) and has exact classical solutions with topology (instantons).

The continuum extrapolation does not seem to be easy: the onset of the a^2 behaviour is delayed. One might perform the continuum extrapolation with an Ansatz including a^2 powers and logs and obtain an estimate for the continuum limit. One might also perform further simulations on larger lattices with smaller a and make the extrapolation more reliable.

An alternative procedure is to simulate a better action in order to make the extrapolation easier. For example, one might consider a theoretical construction for a local action on the lattice which has no cut-off effects at all in the classical field theory. Such local actions, called fixed point (FP) actions, exist and, as we shall discuss, are defined by classical equations. These equations can be solved approximately and a local action can be constructed which is approximately classically perfect. Simulating this action produces the full triangles in Fig.2. An excellent global behaviour is observed. Of course, these results should be weighted with the expenses of the numerical procedure. Constructing a good approximation to the FP action requests analytic and numerical work. The simulation is more expensive than that

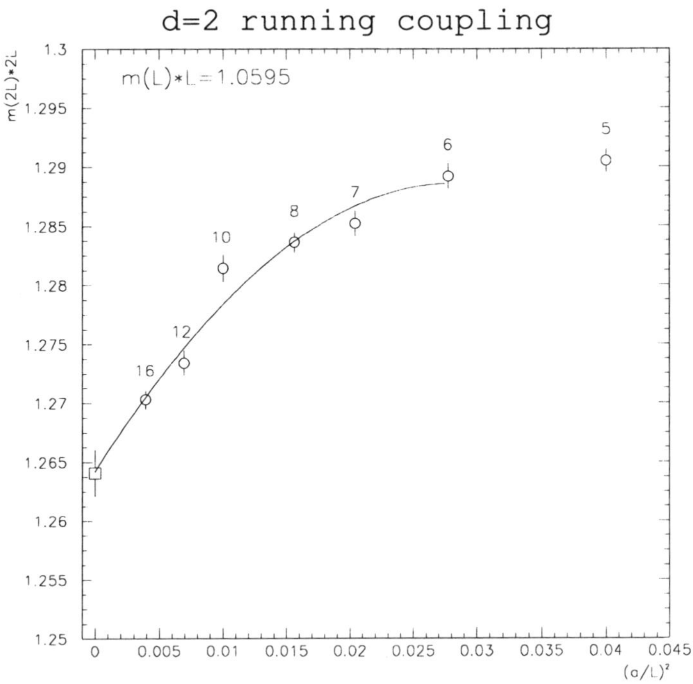

Fig. 1. The cutoff effects and the continuum extrapolation of $m(2L)2L$ if $m(L)L = 1.0595$ [3], where L is the spatial size of the lattice and $m(L)$ is the mass gap in this box.

of the simplest discretizations. The gain might also depend on the physical quantity measured. In this example the FP action seems to be a very useful idea.

The physical quantity considered above is a specially defined running coupling in this model. Since an analogous quantity plays an important role in QCD, let us discuss it in this simpler model.

The continuum action of the $d = 2, O(3)$ non-linear σ-model in Euclidean space reads

$$\mathcal{A} = \frac{1}{2g}\int d^2x \partial_\mu \mathbf{S} \partial_\mu \mathbf{S}, \qquad \mu = 0,1\,, \qquad \mathbf{S} = (S^1, S^2, S^3), \qquad \mathbf{S}(x)^2 = 1\,, \tag{1}$$

where the bare coupling g is dimensionless. Consider this system in a finite, periodic box of size L. Take a small box $L \ll 1/m$, where m is the infinite volume mass-gap. From the point of view of low energy excitations the sector with non-zero momenta, and so the x^1-dependence of the field

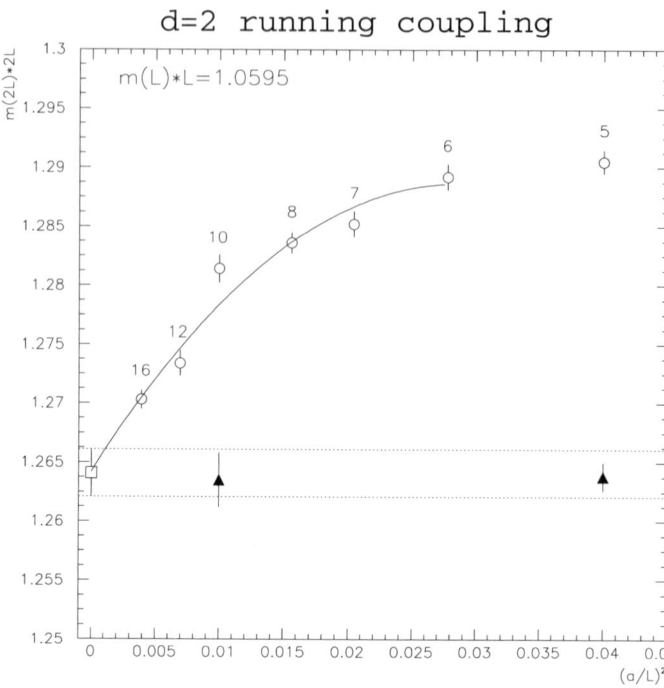

Fig. 2. The same as Fig.1 but adding data obtained with an approximation of the fixed point action[2].

$\mathbf{S}(x^0, x^1)$, can be suppressed in this case

$$\mathbf{S}(x^0, x^1) \sim \mathbf{S}(x^0), \qquad \mathcal{A} \sim \frac{L}{2g} \int dx^0 \partial_0 \mathbf{S} \partial_0 \mathbf{S} \,. \tag{2}$$

Here $\mathbf{S}$ is a unit vector with three components which depend on x^0, so the action in Eq. (2) describes the quantum mechanics of a free particle moving on the surface of a unit sphere: we have a quantum rotator. The energy spectrum is $E_l = l(l+1)/2\Theta$ with $\Theta = g/L$. The lowest excitation in this box (the mass-gap $m(L)$) is

$$m(L) = E_1 - E_0 = g/L \,. \tag{3}$$

Since the bare coupling $g \to 0$ in the continuum limit, the rotator excitations are indeed much lighter than those with non-zero momenta. The dimensionless quantity $Lm(L)$ is a valid definition of the scale dependent ('running') coupling constant $g(L)$. In a small box $L \ll \Lambda_{O(3)}^{-1}$ ($\Lambda_{O(3)}$ is the Lambda-parameter of the theory) i.e. in the ultraviolet region, it is equal to the bare coupling g.

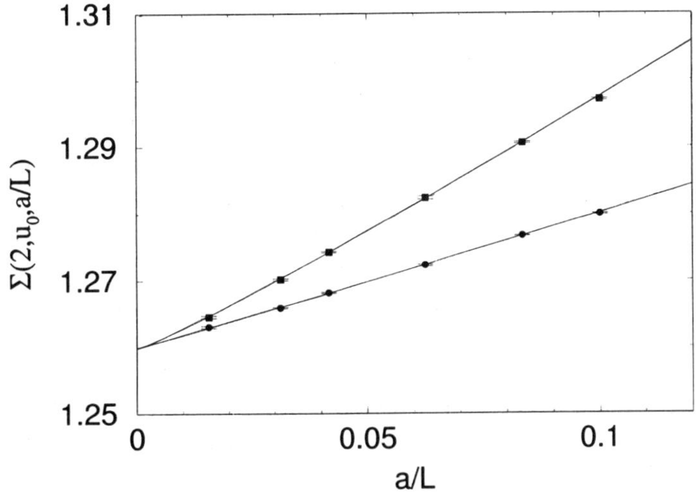

Fig. 3. The cutoff dependence of high precision data obtained with two different actions[4] are shown in this figure. The starting coupling is $g(L) = Lm(L) = 1.0595$ as previously. The cutoff dependent running coupling in a box of $2L$ is denoted here by Σ. In the continuum limit $\Sigma \to g(2L)$.

Assume that at some value of L the running coupling takes the value $g(L) = r$, where r is some positive real number. The dynamics of the $O(3)$ σ-model defines then the value of $g(2L)$ in the continuum limit. Since the theory is asymptotically free, we expect that $g(2L) > g(L)$. At finite lattice constant a this number depends on the resolution. This cutoff dependence is shown in Fig. 1 for the arbitrarily taken case with $r = 1.0595$.

The data in Fig. 1 are more than ten years old. They have large statistical errors and do not go deep in the continuum limit. Fig. 3 shows some more recent results with two simple actions [4†]. Universality predicts that the two sets of data hit the same point in the continuum limit. This makes the extrapolation easier and leads to the prediction $g(2L) = 1.2614(3)$. A few years later the exact continuum number became known also [5]: $1.261208\ldots$. It is satisfying to see that using special cluster algorithms and estimating the errors carefully high precision results can be obtained in this $d = 2$ model. For much more difficult problems, like QCD, one would like to see more theoretical inputs, however.

---

†Note that the data are plotted against a/L in this figure.

3. Renormalization group and the fixed point action

A QFT is defined over a large span of scales from low physical scales up to the cut-off which goes to infinity in the continuum limit. Although field variables associated with very high scales do influence the physical predictions through a complicated cascade process, no physical question involves them directly. Their presence and indirect influence makes it difficult to establish an intuitive connection between the form of the interaction and the final predictions. The presence of a large number of degrees of freedom makes the problem technically difficult also. It is, therefore a natural idea to partially integrate them out in the path integral. This process, which reduces the number of degrees of freedom, taking into account their effect on the remaining variables exactly, is called a renormalization group transformation [6,7] (RGT). A technically simple introduction to lattice regularization and Wilson's RG can be found in Sect.2-3 of Ref. 8.

Consider some lattice regularization of the $SU(N)$ Yang-Mills gauge action

$$\mathcal{A} = \frac{\beta}{8N}\int F^a_{\mu\nu}(x)F^a_{\mu\nu}(x), \qquad a = 1, ..., N^2 - 1\,, \tag{4}$$

where $\beta = 2N/g^2$ and g is very small. Imagine performing repeatedly RG transformations by some gauge invariant averaging. Tree level perturbation theory predicts that all but one of the possible gauge invariant operators die out rapidly under this repeated averaging. The surviving operator is a special discretization of Eq. (4) whose detailed form depends on the averaging procedure. This special operator (action) is called the fixed point of the RG transformation since it is reproduced by the transformation. The form of the fixed point action is determined by classical saddle-point equations[10]. The theoretical properties of the fixed point action and the construction of approximate solutions to be used in simulations are based on these saddle-point equations which will be our starting point in the next section. For further details, which will not be needed here, we refer to the literature[10].

4. Saddle point equation for the fixed point action in QCD

The saddle point equation and its solution, the fixed point action, depend on the averaging procedure. Let $U(n)_\mu, \psi_n, \bar{\psi}_n$ denote the gauge and fermion fields on a $d = 4$ Euclidean ('fine') lattice whose points are indexed by n. The averaged ('blocked') gauge and fermion fields will be denoted by $V(n_B)_\mu, \chi_{n_B}, \bar{\chi}_{n_B}$. This lattice ('coarse') is indexed by n_B. The gauge fields are $SU(N)$ matrices, the fermion fields are Grassmann variables.

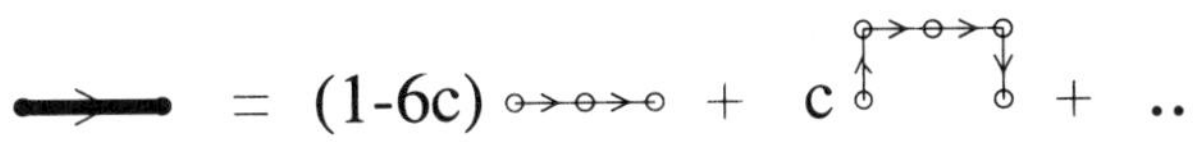

Fig. 4. A simple example for a gauge invariant blocking

The averaging procedure for the gauge variables is coded by an $N \times N$ complex matrix $Q_\mu(n_B)$. It is a gauge covariant average of the fine link variables U in the neighbourhood of the coarse link (n_B, μ), i.e. it is a weighted sum of paths (each path being a product of U matrices) between the points n_B and $n_B + \hat{\mu}$ of the coarse lattice. A simple example for a scale 2 blocking is (see Fig. 4):

$$\begin{aligned} Q_\mu(n_B) = (1-6c)U_\mu(n)U_\mu(n+\hat{\mu}) + \\ c\sum_{\nu\neq\mu}\big[U_\nu(n)U_\mu(n+\hat{\nu})U_\mu(n+\hat{\mu}+\hat{\nu})U_\nu^\dagger(n+2\hat{\mu}) + \\ U_\nu^\dagger(n-\hat{\nu})U_\mu(n-\hat{\nu})U_\mu(n+\hat{\mu}-\hat{\nu})U_\nu(n+2\hat{\mu}-\hat{\nu})\big] \, , \end{aligned} \tag{5}$$

where c, the relative weight of the staples versus the central link, is a parameter.

The fixed point action has the standard structure

$$A^{FP}(\overline{\chi}, \chi, V) = \mathcal{A}_g^{FP}(V) + \sum_{n_B, n'_B} \overline{\chi}_{n_B} D^{FP}(n_B, n'_B)\chi_{n'_B} \, , \tag{6}$$

where $\mathcal{A}_g^{FP}(V)$ is the gauge action and D^{FP} is the fixed point Dirac operator. The fixed point gauge action is determined by the equation[10,11]

$$\mathcal{A}_g^{FP}(V) = \min_{\{U\}}\left(\mathcal{A}_g^{FP}(U) - \frac{\kappa_g}{N}\sum_{n_B,\mu}\left[ReTr(V_\mu(n_B)Q_\mu^\dagger(n_B)) - f(Q_\mu(n_B))\right]\right) \, , \tag{7}$$

where

$$f(Q) = \max_W\left[ReTr(WQ^\dagger)\right], \qquad W \in SU(N) \, . \tag{8}$$

Given an arbitrary gauge field configuration $\{V\}$ on the coarse lattice one is looking for the value of the FP action (a real number) on this configuration. This number is given by the minimum on the r.h.s. of Eq. (7) with respect to the gauge field configuration U on the fine lattice. The FP action should be local, only such solutions are acceptable. The parameters κ_g and c in Eq. (7) and Eq. (6) can be used to optimise the action being not only local but as compact as possible.

The FP Dirac operator satisfies the equation

$$D^{FP}(V)^{-1}_{n_B,n'_B} = \frac{1}{\kappa_f}\delta_{n_B,n'_B} + b_f^2 \sum_{n,n'} \omega(U_{\min})_{n_B,n} D^{FP}(U_{\min})^{-1}_{n,n'} \omega(U_{\min})^{\dagger}_{n',n'_B} \tag{9}$$

where $U_{\min}$ is the minimising field in Eq. (7), $\omega(U)$ defines the gauge covariant averaging of fermions, while κ_f and b_f are parameters[‡]. Eqs. (7,9) determine the FP action of QCD.

5. Perfect classical lattice theories

The FP Yang-Mills action and the FP Dirac operator have amazing properties: they are perfect on any physical question in the classical limit even on coarse lattices. So, for example, the classical (Euler-Lagrange) equation obtained from $A_g^{FP}(V)$ has exact, scale invariant instanton solutions on the lattice. The value of the instanton action is the same as in the continuum. This is true even if the instanton size is a few lattice unit only. Another example is the free FP Dirac operator: $D^{FP}(V=1)$. The energy-impulse dispersion relation $E = E(\mathbf{p}) = |\mathbf{p}|$ is exact and $E \in (0,\infty)$ like in the continuum.

For illustration let us prove the statement on instantons. We show first that if the lattice gauge configuration V satisfies the FP classical (Euler-Lagrange) equations $\delta\mathcal{A}_g^{FP}(V)/\delta V = 0$ and V is a local minimum of $\mathcal{A}_g^{FP}(V)$ then the configuration $U_{\min}$ which minimises the r.h.s. of the FP equation Eq. (7) satisfies the FP classical equations as well. In addition, the value of the action is the same: $\mathcal{A}^{FP}(U_{\min}(V)) = \mathcal{A}^{FP}(V)$.

The argument is as follows. In the equation $\delta\mathcal{A}_g^{FP}(V)/\delta V = 0$ the FP action can be replaced by the r.h.s. of the FP equation Eq. (7). The r.h.s. of Eq. (7) depends on V explicitly and also implicitly through the V dependence of $U_{\min}$. This last contribution is zero since $U_{\min}$ is the minimum. The explicit dependence comes from the middle term in the r.h.s. of Eq. (7), therefore this term takes its maximum and so the last terms cancel each other. Therefore $\mathcal{A}_g^{FP}(U)$ takes its minimum at $U_{\min}$, i.e. $U_{\min}$ is a solution of the Euler-Lagrange equation of the action $\mathcal{A}_g^{FP}(U)$ on the fine lattice and

$$\mathcal{A}_g^{FP}(V) = \mathcal{A}_g^{FP}(U_{\min}) \,. \tag{10}$$

[‡]For simplicity we quoted the FP equation for the inverse of the Dirac operator, which can be used only if D^{FP} has no zero modes.

This result implies that to any solution V with a characteristic size ρ there exists another solution $U_{\rm min}(V)$ of size 2ρ [§] with the same value of the action, i.e. these are scale invariant solutions. Repeating the argument there exist solutions with scale $2^2\rho,\ldots,2^k\rho,\ldots$. The very large solutions become arbitrarily smooth and the value of the action is equal to the continuum value. Since this value is independent of ρ, all the solutions have the continuum value.

6. The FP Dirac operator satisfies the Ginsparg-Wilson relation

Nielsen and Ninomiya[1] demonstrated that if the lattice Dirac operator satisfies some basic conditions (locality and massless fermions without doublers), than the standard chiral symmetry relation $D\gamma_5+\gamma_5 D=0$ is unavoidably violated. In a following paper Ginsparg and Wilson[12] argued that the correct chiral symmetry relation on the lattice is $D\gamma_5+\gamma_5 D = D\gamma_5 R D$, where R is an arbitrary local operator, trivial in Dirac space. No Dirac operators were around that time which satisfied this non-linear relation and so this work and the message remained largely unnoticed. More than fifteen years later the FP [13] and soon after the overlap [14] Dirac operators were identified as solutions of the Ginsparg-Wilson relation.

Using the FP equation Eq. (7) it is easy to demonstrate that D^{FP} satisfies the GW relation. Eq. (7) refers to a fine and a coarse lattice whose lattice units differ by a factor of 2^k. Start from a very fine lattice, take k very large and consider the anticommutator of γ_5 with the inverse of FP Dirac operator on the coarse lattice:

$$\{\gamma_5, D^{FP}_{n_B,n'_B}(V)^{-1}\}\,. \tag{11}$$

On the r.h.s. of Eq. (7) the first term gives $2/\kappa_f\delta_{n_B,n'_B}$. In the second term D^{FP} lives on the very fine (for $k\to\infty$, infinitely fine, i.e. continuum) gauge configuration $U_{\rm min}(V)$. Any legitim Dirac operator goes over the continuum (massless) Dirac operator in this limit. We get then

$$\{\gamma_5, D^{FP}_{n_B,n'_B}(V)^{-1}\} = \frac{2}{\kappa_f}\,, \tag{12}$$

or equivalently

$$\{\gamma_5, D^{FP}(V)\} = \frac{2}{\kappa_f}D^{FP}(V)\gamma_5 D^{FP}(V)\,. \tag{13}$$

[§] We consider a scale=2 RG transformation

Since a RG transformation does not change the physical content of the theory, the lattice action (which is actually the fixed point action) inherits all the symmetries of the starting continuum action. We give now a simple procedure to find the lattice symmetry transformations.

The path integral describing this block transformation is Gaussian which is equivalent to a formal minimization problem:

$$\overline{\chi}\mathcal{D}\chi = \min_{\overline{\psi},\psi}\left\{\overline{\psi}D\psi + \left(\overline{\chi} - \overline{\psi}\omega^{\dagger}\right)\left(\chi - \omega\psi\right)\right\}, \tag{17}$$

where the fermion fields ψ_x and χ_n live in the continuum and on the lattice respectively, ω_{nx} is the blocking matrix, $D_{xx'} = (\gamma^{\mu}\partial_{\mu})_{xx'}$ and $\mathcal{D}_{nn'}$ are the continuum and lattice Dirac operators. For the blocking we take a flat, non-overlapping averaging

$$\omega_{nx} = \begin{cases} 1 & \text{if} \quad x \in \text{block}\, n\,, \\ 0 & \text{otherwise}\,. \end{cases} \tag{18}$$

With this choice one has $\sum_x \omega_{nx}\omega^{\dagger}_{xn'} = \delta_{nn'}$, i.e. $\omega\omega^{\dagger} = 1$.

The minimising fields $\psi_0 = \psi_0(\chi)$ and $\overline{\psi}_0 = \overline{\psi}_0(\overline{\chi})$ from Eq. (17) are given by

$$\begin{aligned} \psi_0(\chi) &= A^{-1}\omega^{\dagger}\chi, \\ \overline{\psi}_0(\overline{\chi}) &= \overline{\chi}\omega A^{-1}, \end{aligned} \tag{19}$$

where

$$A = D + \omega^{\dagger}\omega. \tag{20}$$

Inserting Eq. (19) into Eq. (17) gives the lattice Dirac operator

$$\mathcal{D} = 1 - \omega A^{-1}\omega^{\dagger}\,. \tag{21}$$

From the equations above it is easy to derive the following useful relations

$$\begin{aligned} \omega\psi_0(\chi) &= (1-\mathcal{D})\chi\,, & \overline{\psi}_0(\overline{\chi})\omega^{\dagger} &= \overline{\chi}(1-\mathcal{D})\,, \\ D\psi_0(\chi) &= \omega^{\dagger}\mathcal{D}\chi\,, & \overline{\psi}_0(\overline{\chi})D &= \overline{\chi}\mathcal{D}\omega\,. \end{aligned} \tag{22}$$

The Ginsparg-Wilson relation can be obtained then from Eq. (21) by using $\{D, \gamma_5\} = 0$ and the relations above¶:

$$\{\mathcal{D}, \gamma_5\} = 2\mathcal{D}\gamma_5\mathcal{D}\,. \tag{23}$$

¶Note that with our choice of the coefficients in Eq. (17) the factor 2 appears in the GW relation. This is, of course, just a convention.

We formulate now a general procedure to find the lattice symmetry *transformation* to any infinitesimal symmetry transformation in the continuum[17].
Let $\delta\psi$ and $\delta\overline{\psi}$ be the change of the corresponding continuum fields under an infinitesimal symmetry transformation which leaves the continuum action $\overline{\psi}D\psi$ invariant.

Define the infinitesimal change of the lattice fields by

$$\delta\chi = \omega\delta\psi_0(\chi)\,, \quad \delta\overline{\chi} = \delta\overline{\psi}_0(\overline{\chi})\omega^\dagger\,. \tag{24}$$

Then the lattice action $\overline{\chi}\mathcal{D}\chi$ is invariant under this infinitesimal transformations.

Rather than detailing the proof[17], let us consider a few examples.

U(1) axial transformation

The standard infinitesimal axial rotation in the continuum reads

$$\delta\psi_0(\chi) = \mathrm{i}\epsilon\gamma_5\psi_0(\chi), \quad \delta\overline{\psi}_0(\overline{\chi}) = \mathrm{i}\epsilon\overline{\psi}_0(\overline{\chi})\gamma_5\,. \tag{25}$$

The corresponding lattice transformation has the form

$$\begin{aligned}\delta\chi &= \mathrm{i}\epsilon\gamma_5\omega\psi_0(\chi) = \mathrm{i}\epsilon\gamma_5(1-\mathcal{D})\chi\,,\\ \delta\overline{\chi} &= \mathrm{i}\epsilon\overline{\psi}_0(\overline{\chi})\gamma_5\omega^\dagger = \mathrm{i}\epsilon\overline{\chi}(1-\mathcal{D})\gamma_5\,,\end{aligned} \tag{26}$$

where we used Eq. (24) and Eq. (22). These transformations have the well known form found by Lüscher [15]. Notice, however that the axial transformation in the continuum is not unique and so the lattice transformation is not unique either.

Infinitesimal translation

In the continuum we have $\delta\psi_0(\chi) = \epsilon\hat{\partial}_\mu\psi_0(\chi)$, $\delta\overline{\psi}_0(\overline{\chi}) = \epsilon\overline{\psi}_0(\overline{\chi})\hat{\partial}^\dagger_\mu$, where $\left(\hat{\partial}_\mu\right)_{xy} = \partial^x_\mu\delta(x-y)$. Our general procedure leads to the lattice transformations

$$\begin{aligned}\delta\chi &= \epsilon\omega\hat{\partial}_\mu\psi_0(\chi)\,,\\ \delta\overline{\chi} &= \epsilon\overline{\psi}_0(\overline{\chi})\hat{\partial}^\dagger_\mu\omega^\dagger\,.\end{aligned} \tag{27}$$

Using $[D,\hat{\partial}_\mu] = 0$ it is a simple exercise to show explicitly that the lattice action is invariant under this infinitesimal translation.

Further examples, also in the presence of interactions, can be found in Ref. 17.

Acknowledgements

I thank the organisers for the invitation and hospitality and many of the participants for the discussions at these meetings. This work was supported by the Schweizerischer Nationalfonds.

References

1. N. B. Nielsen and M. Ninomiya, *Nucl. Phys.* **B185**, 20 (1981).
2. P. Hasenfratz and F. Niedermayer, *Nucl. Phys.* **B414**, 785 (1994) [arXiv:hep-lat/9308004].
3. M. Lüscher, P. Weisz and U. Wolff, *Nucl. Phys.* **B359**, 221 (1991).
4. M. Hasenbusch, P. Hasenfratz, F. Niedermayer, B. Seefeld and U. Wolff, *Nucl. Phys. Proc. Suppl.* **106** 911 (2002) [arXiv:hep-lat/0110202].
5. J. Balog and A. Hegedus, *J. Phys.* **A37** 1881 (2004) [arXiv:hep-th/0309009].
6. K. G. Wilson and J. B. Kogut, *Phys. Rep.* **C12** 75 (1974).
7. K. G. Wilson, *Rev. Mod. Phys.* **47** 773 (1975).
8. P. Hasenfratz, *Prog. Theor. Phys. Suppl.* **No.131** 189 (1998).
9. K. G. Wilson, *Phys. Rev.* **D10**, 2445 (1974).
10. T. DeGrand, A. Hasenfratz, P. Hasenfratz and F. Niedermayer, *Nucl. Phys.* **B454**, 587 (1995) [arXiv:hep-lat/9506030]; W. Bietenholz and U.-J. Wiese, *Nucl. Phys.* **B464**, 319 (1996) [arXiv:hep-lat/9510025]; T. DeGrand, A. Hasenfratz, D. Zhu, *Nucl. Phys.* **B475**, 321 (1996) [arXiv:hep-lat/9603015].
11. M. Blatter and F. Niedermayer, *Nucl. Phys.* **B482**, 286 (1996) [arXiv:hep-lat/9605017].
12. P. H. Ginsparg and K. G. Wilson, *Phys. Rev.* **D25**, 2549 (1982).
13. P. Hasenfratz, *Nucl. Phys. Proc. Suppl.* **63A-C**, 53 (1998) [arXiv:hep-lat/9709110].
14. H. Neuberger, *Phys. Lett.* **B427**, 353 (1998) [arXiv:hep-lat/9801031].
15. M. Lüscher, *Phys. Lett.* **B428**, 342 (1998) [arXiv:hep-lat/9802011].
16. P. Hasenfratz, *Nucl. Phys.* **B525**, 401 (1998) [arXiv:hep-lat/9802007].
17. P. Hasenfratz, F. Niedermayer and R. von Allmen, [arXiv:hep-lat/0606021].
18. U.-J. Wiese, *Phys. Lett.* **B315**, 417 (1993) [arXiv:hep-lat/9306003].

ALGORITHMS FOR DYNAMICAL FERMIONS

A. D. KENNEDY

School of Physics, University of Edinburgh,
King's Buildings, Mayfield Road,
Edinburgh EH9 3JZ, Scotland
E-mail: adk@ph.ed.ac.uk

This is the write-up of three lectures on algorithms for dynamical fermions that were given at the ILFTN workshop 'Perspectives in Lattice QCD' in Nara during November 2005. The first lecture is on the fundamentals of Markov Chain Monte Carlo methods and introduces the Hybrid Monte Carlo (HMC) algorithm and symplectic integrators; the second lecture covers topics in approximation theory and thereby introduces the Rational Hybrid Monte Carlo (RHMC) algorithm and ways of evading integrator instabilities by means of multiple pseudofermion fields; the third lecture introduces on-shell chiral (Ginsparg–Wilson) lattice fermions and discusses five-dimensional formulations for computing fermion propagators for such fermions.

1. Introduction

This is a written version of a set of lectures on algorithms for dynamical fermions. The organization of these lecture notes is as follows.

In the first lecture (§2) we introduce some 'building blocks' from which we can construct Monte Carlo algorithms for the evaluation of the functional integrals that describe quantum field theories on the lattice, and in particular for such computations including the dynamical effects of fermions. After introducing the basic idea of Monte Carlo integration and proving the Central Limit theorem we introduce Markov chains (§2.3), and prove the basic theorem on their convergence (§2.4). We explain how detailed balance and the Metropolis algorithm provide a simple way of constructing Markov steps with some specified fixed point, and then how we can construct composite Markov steps that are likely to be ergodic.

Next we briefly introduce perfect sampling by the method of 'Coupling from the Past' that in some cases lets us generate completely uncorrelated samples from the exact fixed point distribution of a Markov chain. Following this we consider the effects of autocorrelations, and introduce the Hybrid

Monte Carlo (HMC) algorithm as a way of reducing them without having to resort to approximating the equilibrium distribution. The requirement HMC has for reversible and area preserving integrators for Hamiltonian systems leads us to analyze symplectic integrators (§2.13) using the Baker–Campbell–Hausdorff (BCH) formula (§2.12). The amplification of floating point arithmetic rounding errors by such integrators is considered in §2.16, where we also see the effects of the instabilities (§2.13.1) that must occur in symplectic integrators when the step size becomes too large. The use of multiple timescale integration schemes to avoid these instabilities is discussed in §2.14. Finally we discuss the use of pseudofermions within the HMC algorithm to handle dynamical fermions (§2.15).

We begin the second lecture by introducing the theory of optimal polynomial approximation and establishing Chebyshev's criterion. After elucidating the rôle of Chebyshev polynomials (§3.3) we consider Chebyshev optimal rational function approximation in §3.4. A discussion of the significance of rational approximations for functions of matrices is given in §3.6. The use of multiple pseudofermion fields (§3.8) to reduce the fluctuations in the force exerted by the pseudofermions on the gauge fields was introduced by Hasenbusch (§3.8.1), and the way in which this may be implemented using RHMC follows.

The results of numerical comparison of RHMC with the R algorithm are given both for finite temperature QCD (§3.11.1) and for domain wall fermions (§3.11.2). Data showing the efficacy of using multiple timescale integrators for multiple pseudofermions follows, as does data showing how these methods succeed in 'bringing down the Berlin Wall' for Wilson fermions.

The third lecture (§4) is concerned with five-dimensional algorithms for Ginsparg–Wilson (GW) fermions. We introduce the concept of chiral lattice fermions in what we believe to be a logical (rather than historical) approach, starting with Lüscher's on-shell chiral transformation, deriving the GW identity from it, and then showing that Neuberger's operator is essentially the unique solution (up to the choice of kernel, and assuming γ_5 hermiticity).

We then turn to a class of five-dimensional algorithms to invert Neuberger's operator: in these the Schur complement (§4.5) of a matrix plays a central rôle. The algorithms are characterized by four independent choices: the kernel of the Neuberger operator (§4.4); whether we introduce five-dimensional pseudofermions as dynamical fields in a Markov process or just view them as constraints in the computation of the inverse of the

Schur complement (§4.6); the choice of rational approximation of the sgn (signum) function (§4.7); and the choice of the five dimensional matrix used to linearize the approximation to the Neuberger operator (§4.8). The different choices of five dimensional matrices correspond to different ways of representing a rational functions; as a continued fraction, a partial fraction, or as a Euclidean Cayley transform. The latter lead directly to the domain wall fermions formulation and its generalizations (§4.8.3). Finally we consider the characterization of chiral symmetry breaking in these approaches, and look at the results of preliminary numerical studies.

2. Building Blocks for Monte Carlo Methods

We start by reviewing the basic ideas of Monte Carlo integration and Markov processes.[1]

2.1. *Monte Carlo Methods*

The basic idea of Monte Carlo integration is the mathematical identification of probabilities with integration measures: we evaluate an integral by sampling the integrand at points selected at random from a probability distribution proportional to the integration measure.

Of course, there are much better methods for carrying out numerical integration (quadrature) in low dimensional spaces; however, all these methods become hopelessly expensive for high dimensional integrals. Since in lattice quantum field theory there is one integration per degree of freedom, the only practical way to carry out such integrations is to use Monte Carlo methods.

The fundamental objective of (Euclidean) quantum field theory is to compute the *expectation value* of some operator $\Omega(\phi)$ that depends on the field ϕ

$$\langle\Omega\rangle = \frac{1}{Z}\int d\phi\, e^{-S(\phi)}\,\Omega(\phi),$$

where the action is $S(\phi)$, the measure is $d\phi$, and the partition function Z is introduced to impose the normalisation condition $\langle 1\rangle = 1$.

In order to calculate this expectation value, we generate a sequence of field configurations $(\phi_1, \phi_2, \ldots, \phi_t, \ldots, \phi_N)$ chosen from the probability distribution

$$P(\phi_t)\, d\phi_t = \frac{1}{Z}\, e^{-S(\phi_t)};$$

how this may be done will be explained later (§2.3). We then measure the value of the operator Ω on each configuration, and compute its average over all the configurations

$$\bar{\Omega} \equiv \frac{1}{N} \sum_{t=1}^{N} \Omega(\phi_t).$$

This *sample average*, which we denote by writing a bar over the quantity averaged, is to be contrasted with the expectation value which is denoted by enclosing the quantity in angle brackets. The *law of large numbers* then tells us that the configuration average $\bar{\Omega}$ tends to the expectation value $\langle\Omega\rangle$ as N, the number of configurations sampled, tends to infinity,

$$\langle\Omega\rangle = \lim_{N\to\infty} \bar{\Omega}.$$

2.2. *Central Limit Theorem*

The Laplace–DeMoivre *central limit theorem* establishes the stronger result that under very general conditions the sample average tends to become Gaussian distributed with the expectation value $\langle\Omega\rangle$ as its mean and with a standard deviation which falls as $1/\sqrt{N}$,

$$P(\bar{\Omega}) \sim \text{constant} \times \exp\left[-\frac{\left(\bar{\Omega} - \langle\Omega\rangle\right)^2}{2C_2/N}\right],$$

where $C_2 \equiv \left\langle \left(\Omega - \langle\Omega\rangle\right)^2 \right\rangle$ is the *variance* of the distribution of Ω. Note that the central limit theorem is an asymptotic expansion in $1/\sqrt{N}$ for the probability distribution of $\bar{\Omega}$.

In order to prove the central limit theorem let us consider the distribution of the configuration average: the distribution of the values of a single sample, $\omega = \Omega(\phi)$, is

$$P_\Omega(\omega) \equiv \int d\phi\, P(\phi)\, \delta\big(\omega - \Omega(\phi)\big) = \Big\langle \delta\big(\omega - \Omega(\phi)\big) \Big\rangle.$$

If we take the logarithm of the Fourier transform of this we obtain the generating function for the connected moments (cumulants) of Ω, namely

$$\begin{aligned} W_\Omega(k) &\equiv \ln \int d\omega\, P_\Omega(\omega)\, e^{ik\omega} = \ln \int d\phi\, P(\phi) e^{ik\Omega(\phi)} \\ &= \ln\langle e^{ik\Omega}\rangle \equiv \sum_{n=0}^{\infty} \frac{(ik)^n}{n!}\, C_n, \end{aligned}$$

where the first few cumulants[a] are

$$C_0 = 0, \qquad C_3 = \langle(\Omega - \langle\Omega\rangle)^3\rangle,$$
$$C_1 = \langle\Omega\rangle, \qquad C_4 = \langle(\Omega - \langle\Omega\rangle)^4\rangle - 3C_2^2.$$
$$C_2 = \langle(\Omega - \langle\Omega\rangle)^2\rangle,$$

Next we consider the distribution of the value of the average of N samples,

$$P_{\bar{\Omega}}(\bar{\omega}) \equiv \int d\phi_1 \cdots d\phi_N \, P(\phi_1)\cdots P(\phi_N)\, \delta\Big(\bar{\omega} - \frac{1}{N}\sum_{t=1}^{N}\Omega(\phi_t)\Big),$$

and we construct its connected generating function

$$\begin{aligned}
W_{\bar{\Omega}}(k) &\equiv \ln \int d\bar{\omega}\, P_{\bar{\Omega}}(\bar{\omega}) e^{ik\bar{\omega}} \\
&= \ln \int d\phi_1 \cdots d\phi_N \, P(\phi_1)\cdots P(\phi_N)\, \exp\Big(\frac{ik}{N}\sum_{t=1}^{N}\Omega(\phi_t)\Big) \\
&= \ln\Big[\int d\phi\, P(\phi) e^{ik\Omega(\phi)/N}\Big]^N = N \ln \Big\langle e^{ik\Omega/N}\Big\rangle \\
&= N W_{\Omega}(k/N) = \sum_{n=1}^{\infty} \frac{(ik)^n}{n!}\frac{C_n}{N^{n-1}}.
\end{aligned}$$

We may take the inverse Fourier transform of $W_{\bar{\Omega}}$ to obtain an explicit expression for the distribution $P_{\bar{\Omega}}$,

$$\begin{aligned}
P_{\bar{\Omega}}(\bar{\omega}) &= \frac{1}{2\pi}\int dk\, e^{W_{\bar{\Omega}}(k)} e^{-ik\bar{\omega}} \\
&\sim \exp\left[\sum_{n=3}^{\infty} \frac{(-1)^n C_n}{n! N^{n-1}}\left(\frac{d}{d\bar{\omega}}\right)^n\right] \int \frac{dk}{2\pi} \exp\left[ik\langle\Omega\rangle + (ik)^2 \frac{C_2}{2N} - ik\bar{\omega}\right] \\
&= \exp\left[-\frac{C_3}{3! N^2}\left(\frac{d}{d\bar{\omega}}\right)^3 + \frac{C_4}{4! N^3}\left(\frac{d}{d\bar{\omega}}\right)^4 - \cdots\right] \frac{\exp\left[-\frac{(\bar{\omega}-\langle\Omega\rangle)^2}{2C_2/N}\right]}{\sqrt{2\pi C_2/N}}.
\end{aligned}$$

This is an asymptotic expansion because the cumulant expansion in general only converges for sufficiently small values of $|k|$, whereas the integral is over all values of k. It can be shown that this leads to corrections of $O(e^{-\alpha N})$ for some constant $\alpha > 0$ that, for any given value of N, will exceed the $1/N^{\ell}$ term in the series for some ℓ.[2]

---

[a] C_1 and C_2 are the mean and variance, $\sigma \equiv \sqrt{C_2}$ is the standard deviation, and C_3/σ^3 and C_4/σ^4 are called the skewness and kurtosis.

The distribution $P_{\bar{\Omega}}$ tends to a δ function as $N \to \infty$, and in order to see its Gaussian nature it is useful to rescale its argument to $\xi \equiv \sqrt{N}(\bar{\omega} - \langle\Omega\rangle)$, in terms of which

$$P_{\bar{\Omega}}(\bar{\omega}) = F(\xi)\frac{d\xi}{d\bar{\omega}}$$

with

$$F(\xi) = \left[1 + \frac{C_3\xi(\xi^2 - 3C_2)}{6C_2^3\sqrt{N}} + \cdots\right]\frac{e^{-\xi^2/2C_2}}{\sqrt{2\pi C_2}}.$$

Figure 1 illustrates the central limit theorem by showing how the scaled probability distribution $F(\xi)$ approaches a Gaussian distribution for the case where a single sample x is chosen uniformly in $-\frac{1}{2} \leq x \leq \frac{1}{2}$.

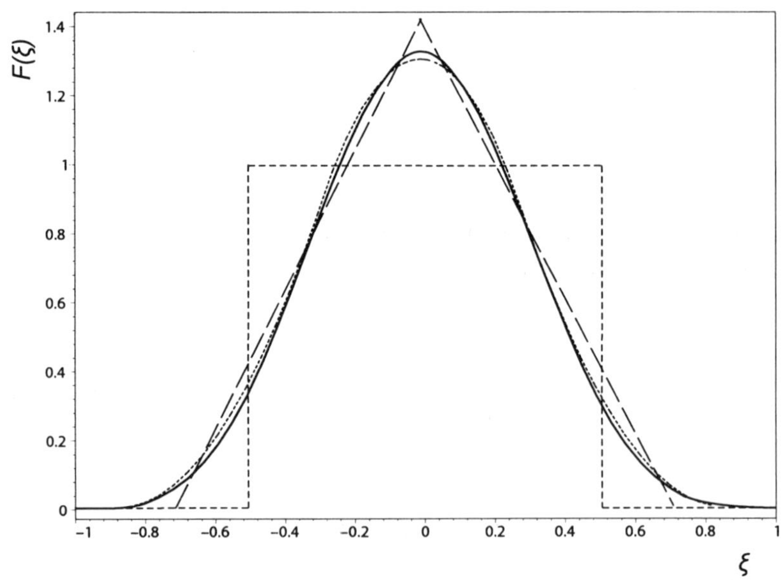

Figure 1. A simple illustration of how the probability distribution of the resçaled sample mean approaches a Gaussian. The four curves show the functions $F_N(\xi)$ of the scaled variable $\xi = x\sqrt{N}$ for $N = 1, \ldots, 4$. These functions are defined by $F_N(\xi)d\xi = P_N(x)dx$, where the unscaled probability distributions are $P_N(x) \equiv \int dx_1 \cdots dx_N\, P_1(x_1)\cdots P_1(x_N)\,\delta\Big(x - \frac{1}{N}\sum_{i=1}^N x_i\Big)$, starting with the very simple (and very non-Gaussian) distribution $P_1(x) = \theta(\frac{1}{2} + x)\theta(\frac{1}{2} - x)$.

2.3. *Markov Chains*

In order to implement the Monte Carlo integration procedure outlined in the previous section we need to generate a sequence of configurations distributed with probability proportional to e^{-S}. The only practicable way that we can do this is by utilising Markov chains, a procedure that mathematicians call *Markov Chain Monte Carlo* (MCMC).

Let Ω be the state space: for example in the present case each point in this space is gauge field configuration. We consider an *ergodic* stochastic mapping $P : \Omega \to \Omega$; by 'stochastic' we mean that $P(j \leftarrow i)$ gives the *probability* that state i will be mapped to state j, but that the actual state that i is taken to is selected at random with this probability. By 'ergodic' we mean that the probability getting from any state to any other state is greater than zero. The key to analyzing the properties of a Markov chain is to think of it as a deterministic mapping of probability distributions on the state space rather than as a stochastic mapping on the state space *per se.* A probability distribution is a mapping $Q : \Omega \to \mathbb{R}$ which is positive and normalized, $Q(x) > 0 \ \forall x \in \Omega$ and $\int_\Omega dx\, Q(x) = 1$. If we call the space of all such mappings[b] Q_Ω then the Markov process P induces a map $P : Q_\Omega \to Q_\Omega$.

2.4. *Convergence of Markov Chains*

The principal result in the theory of Markov chains is that whatever distribution of states we start with (for example we could start with a very specfic state, such as a totally ordered gauge configuration or 'cold start', or we could choose our starting state entirely at random, a 'hot start'), the sequence of distributions generated by repeatedly applying the Markov mapping P converges to unique fixed point distribution $\bar{Q}$. The purpose of this section is to establish this result.

To this end we introduce a metric on the space Q_Ω of probability distributions by defining the distance[c] between two probability distributions

---

[b]Strictly speaking we define Q_Ω to be the space of equivalence classes of such mappings, as two probability distributions are to be considered equal if their difference vanishes almost everywhere: that is, if their difference vanishes except on a set of measure zero.
[c]It is readily verified that this satisfies the axioms $d(x, x) = 0$, $d(x, y) = d(y, x)$, and the *triangle inequality* $d(x, y) \le d(x, z) + d(z, y) \ \forall x, y, z \in Q_\Omega$; indeed, it just corresponds to the L_1 norm.

Q_1 and Q_2

$$d(Q_1, Q_2) \equiv \int dx \big| Q_1(x) - Q_2(x) \big|. \tag{1}$$

Our approach to the proof is that the Markov process is a *contraction mapping* with respect to this distance: when we apply P to two probability distributions the distance between the resulting distributions is smaller than the distance between the two original probability distributions

$$d(PQ_1, PQ_2) \leq (1-\alpha) d(Q_1, Q_2),$$

where $0 < \alpha \leq 1$.

Once we have established this, we may argue that P is a contraction mapping with respect to the metric of Eq. (1), so the sequence $(Q, PQ, P^2Q, P^3Q, \ldots)$ is Cauchy, namely that for any $\varepsilon > 0$ there is an integer N such that for all $n \geq m \geq N$ we have $d(P^mQ, P^nQ) < \varepsilon$. This is just the *Banach fixed-point* theorem, which is easily proved as follows:

$$d(P^mQ, P^nQ) \leq \sum_{j=0}^{n-m-1} d(P^{m+j}Q, P^{m+j+1}Q)$$

by repeated application of the triangle inequality,

$$\leq \sum_{j=0}^{n-m-1} (1-\alpha)^j d(P^mQ, P^{m+1}Q) \leq \sum_{j=0}^{\infty} (1-\alpha)^j d(P^mQ, P^{m+1}Q)$$

$$= \frac{1}{\alpha} d(P^mQ, P^{m+1}Q) \leq \frac{(1-\alpha)^m}{\alpha} d(Q, PQ) \leq \frac{(1-\alpha)^N}{\alpha} d(Q, PQ) < \varepsilon,$$

where the final inequality holds provided that

$$N > \frac{\ln\left(\dfrac{\varepsilon\alpha}{d(Q, PQ)}\right)}{\ln(1-\alpha)}$$

or $d(Q, PQ) = 0$. Note that the condition $d(PQ_1, PQ_2) < d(Q_1, Q_2)$ is not strong enough to prove this; indeed, this weaker inequality holds without requiring the assumption of ergodicity which is in general necessary for the result to hold. As the space of probability distributions is complete we can conclude that the sequence converges to a unique fixed point $\bar{Q} = \lim_{n\to\infty} P^nQ$.

In order to show that the Markov process is a contraction mapping, we proceed as follows:

$$\begin{aligned} d(PQ_1, PQ_2) &= \int dx \left| PQ_1(x) - PQ_2(x) \right| \\ &= \int dx \left| \int dy P(x \leftarrow y) Q_1(y) - \int dy P(x \leftarrow y) Q_2(y) \right| \\ &= \int dx \left| \int dy P(x \leftarrow y) \Delta Q(y) \right| \end{aligned}$$

where we have defined the quantity $\Delta Q(y) \equiv Q_1(y) - Q_2(y)$,

$$= \int dx \left| \int dy P(x \leftarrow y) \Delta Q(y) \Big[\theta(\Delta Q(y)) + \theta(-\Delta Q(y)) \Big] \right|$$

upon inserting the trivial step function identity $\theta(y) + \theta(-y) = 1$. Next we make use of the identity $\big||a| - |b|\big| = |a| + |b| - 2\min\big(|a|, |b|\big)$, which is readily established by considering the cases $|a| \geq |b|$ and $|a| < |b|$ separately,

$$\begin{aligned} &= \int dx \int dy\, P(x \leftarrow y) |\Delta Q(y)| \\ &\quad -2 \int dx \min_{\pm} \left| \int dy\, P(x \leftarrow y) \Delta Q(y) \theta(\pm \Delta Q(y)) \right|. \end{aligned}$$

The first term simplifies if we note that $\int dx\, P(x \leftarrow y) = 1$ (the probability of going from state y to somewhere must be unity), so

$$\begin{aligned} & d(PQ_1, PQ_2) \\ &= \int dy\, |\Delta Q(y)| - 2 \int dx \min_{\pm} \left| \int dy\, P(x \leftarrow y) \Delta Q(y) \theta(\pm \Delta Q(y)) \right| \\ &\leq \int dy |\Delta Q(y)| - 2 \int dx \inf_y P(x \leftarrow y) \min_{\pm} \left| \int dy\, \Delta Q(y) \theta(\pm \Delta Q(y)) \right|. \end{aligned}$$

We now observe that

$$\begin{aligned} & \int dy\, \Delta Q(y) \theta(\Delta Q(y)) + \int dy\, \Delta Q(y) \theta(-\Delta Q(y)) \\ &\qquad = \int dy\, \Delta Q(y) = \int dy\, Q_1(y) - \int dy\, Q_2(y) = 1 - 1 = 0, \end{aligned}$$

hence

$$\left| \int dy\, \Delta Q(y) \theta(\pm \Delta Q(y)) \right| = \tfrac{1}{2} \int dy |\Delta Q(y)|.$$

We thus finally reach the desired result

$$\begin{aligned} d(PQ_1, PQ_2) &\leq \int dy |\Delta Q(y)| - \int dx \inf_y P(x \leftarrow y) \int dy |\Delta Q(y)| \\ &\leq (1 - \alpha) d(Q_1, Q_2), \end{aligned}$$

with $0 < \alpha \equiv \int dx\, \inf_y P(x \leftarrow y) \leq 1$.

Suppose that we can construct an ergodic Markov process P that has some desired distribution $\bar{Q}$ as its fixed point. We then start with an arbitrary state ('field configuration') and iterate the Markov process until it has converged ('thermalized'). Thereafter, successive configurations will be distributed according to $\bar{Q}$, but in general they will correlated.

An important point is that we only need the *relative probabilities* of states $Q(i)/Q(j)$ to construct P: we do not need to know the absolute normalization of Q. Conversely, suppose we want to evaluate $\int_\Omega dx\, w(x)$ with $w(x) \geq 0 \quad \forall x \in \Omega$ by Monte Carlo using a Markov chain to generate suitably distributed samples of $x \in \Omega$. The Markov chain samples $x \in \Omega$ with probability proportional to $w(x)$, but gives us no hint as to what the absolute probability is, so we are unable to find the value of the integral. In other words, we cannot use Markov chains to compute integrals directly, but only ratios of integrals of the form

$$\frac{\int_\Omega dx\, w(x) f(x)}{\int_\Omega dx\, w(x)}.$$

Fortunately this is usually what we want in quantum field theory where we are not interested in the value of the partition function *per se.*

2.5. *Detailed Balance and the Metropolis Algorithm*

We now consider how to construct a Markov process with a specified fixed point

$$\bar{Q}(x) = \int dy P(x \leftarrow y)\bar{Q}(y).$$

A sufficient (but not necessary) condition is to make it satisfy *detailed balance*

$$P(y \leftarrow x)\bar{Q}(x) = P(x \leftarrow y)\bar{Q}(y) :$$

we can easily show that this implies the fixed point condition by integrating both sides with respect to y. One simple way of implementing detailed balance is the *Metropolis algorithm* where we select a candidate state $x \in \Omega$

at random[d] and then accept it with probability

$$P(x \leftarrow y) = \min\left(1, \frac{\bar{Q}(x)}{\bar{Q}(y)}\right) \tag{2}$$

or otherwise keep the initial state y as the next state in the Markov chain. We can show that this implies detailed balance (and hence has Q as its fixed point) by considering the cases $\bar{Q}(x) > \bar{Q}(y)$ and $\bar{Q}(x) \leq \bar{Q}(y)$ separately.[e]

The particular form of the acceptance probability of Eq. (2) is not unique: other choices are possible, e.g.,

$$P(x \leftarrow y) = \frac{\bar{Q}(x)}{\bar{Q}(x) + \bar{Q}(y)},$$

but they have lower acceptance.

2.6. *Composition of Markov Steps*

The reason why we can construct bespoke Markov processes from a toolkit of methods is that we can combine different Markov steps together. Let P_1 and P_2 be two Markov steps that both have the desired fixed point distribution, but are not necessarily ergodic. Then the composition of the two steps $P_2 \circ P_1$ is a Markov step that also has the desired fixed point, and it may be ergodic. This trivially generalizes to any (fixed) number of steps. For the case where P_1 is not ergodic but P_1^n is the terminology *weakly* and *strongly* ergodic is sometimes used.

This result justifies 'sweeping' through a lattice performing single site Metropolis updates. Each individual single site update has the desired fixed point because it satisfies detailed balance; the entire sweep therefore has the desired fixed point, and furthermore is ergodic. On the other hand, the entire sweep does not in general satisfy detailed balance; 'undoing' the single site updates would correspond to sweeping through the lattice in the reverse order. Of course it would satisfy detailed balance if the sites were updated in a random order, but this is neither necessary nor desirable (because it puts too much randomness into the system).

---

[d]It is not necessary to choose the candidate entirely at random; it suffices that the probability of choosing x when starting from y is the same as the probability of choosing y when starting from x.

[e]The case $x = y$ is special as it must 'mop up' all the rejections.

2.7. *Coupling from the Past*

It would appear that two of the fundamental limitations of MCMC are that the distribution of states generated only converges to the desired fixed point distribution and never exactly reaches it, and that successive states are necessarily correlated to some extent. Before we turn to methods to alleviate these problems it is interesting to note that there is a way of sampling a sequence of completely independent states from the exact fixed point distribution.

The method of *coupling from the past* or *perfect sampling* was introduced by Propp and Wilson.[3] Imagine that we have some ergodic Markov chain, where the stochasticity of each step is implemented by using a book of random numbers. For example, if the system is in state i at step k then we select the nth random number $r_k \in \mathbb{R}$ from our book, where $0 \leq r_k \leq 1$, and set the new state at step $k+1$ to be j where

$$\sum_{\ell=0}^{j-1} P(\ell \leftarrow i) < r_k \leq \sum_{\ell=0}^{j} P(\ell \leftarrow i).$$

We now ask what state f will the system be in at step 0 if it was in state i at step $-N$? (The use of negative step numbers is just a notational convenience). This has a well-defined answer that depends on the random numbers $r_{-N}, r_{-N+1}, \ldots, r_{-1}$ in our book. If N is large enough then with probability one f will be independent of i, because there is a positive probability at each step that any two states will map to the same state, and thereafter they will continue through the same sequence of states since we are using the same sequence of random numbers from our book (I like to call this the 'flypaper principle,' once two states have coalesced they stay together forever, and ergodicity guarantees that all the states must coalesce eventually).

Coupling from the Past therefore just consists of finding this state f; it will be sampled from the exact fixed point distribution because it is the state that the Markov chain would reach at step 0 if it were started at step $-\infty$. We can then repeat the entire procedure using a different book of random numbers to get a completely independent sample.

So far the algorithm is entirely impractical, as it requires following the sequence of states visited starting with each state i at step $-N$, and in cases of interest the number of states is extremely large. What makes it practicable in some situations is the existence of a partial ordering of the states with a largest and a smallest state (this is what mathematicians call a *lattice*) that is preserved by each Markov step. In other words we have an ordering[f] such that $i \succeq j \Rightarrow i' \succeq j'$ where the Markov step takes the unprimed states into the corresponding primed ones, using the same random number(s) in both cases. The ordering need only be a partial ordering, so for any pair of states i and j we can have $i \succeq j$, $j \succeq i$, both, or neither. Nevertheless the ordering is a lattice, so there is an $s_{\min}$ and an $s_{\max}$ satisifying $s_{\max} \succeq i \succeq s_{\min}\ \forall i$. With such a partial ordering we just need to see if the sequence of states starting with $s_{\min}$ and $s_{\max}$ coalesce: if they do then all states must coalesce and end at the same state f at step 0; if they do not then we need to increase N and repeat the calculation (using the same book of random numbers, of course).

An interesting non-trivial example where this is practicable is provided by the Ising model, where a state consists of an array s of spins each taking values ± 1, and the desired fixed point distribution is $Q(s) \propto \exp\left[\beta \sum_{\langle ij \rangle} s_i s_j\right]$ where the sum is over all pairs of nearest-neighbour spins and $\beta > 0$. The Markov update step consists of sweeping through the lattice updating each spin in turn from a heatbath,[g] specifically the new value s'_i of the spin at site i is chosen, independently of the old value s_i, by taking a random number $0 \leq r \leq 1$ and setting

$$s'_i = \begin{cases} +1 & \text{if } r \leq \exp[\beta \sum_{\langle ij \rangle} s_j]/Z \\ -1 & \text{otherwise} \end{cases}$$

with $Z \equiv \exp[\beta \sum_{\langle ij \rangle} s_j] + \exp[-\beta \sum_{\langle ij \rangle} s_j]$. The partial ordering is that $s \succeq t$ if $s_i \geq t_j\ \forall i$, so for example $101101 \succeq 000101$ but 101101 and 011001 are not comparable. $s_{\max}$ is the state with all spins $+1$, and $s_{\min}$ that with all spins -1. We just verify that the heatbath update preserves this ordering: if the nearest neighbours of the spin s_i being updated are the same then this is trivial as the new value, $s'_i = t'_i$, is the same for both states. If the some of the neighbours differ then $s \succeq t \Rightarrow \sum_{\langle ij \rangle} s_j \geq \sum_{\langle ij \rangle} t_j$, and thus $t'_i = -1 \Rightarrow s'_i = -1$ so $s' \succeq t'$.[h]

---

[f] An *order relation* satisfies the axioms $x \succeq y \wedge y \succeq x \Leftrightarrow x = y$, and $x \succeq y \wedge y \succeq z \Rightarrow x \succeq z\ \forall x, y, z$.

[g] In general a *heatbath* algorithm is one that directly samples the desired distrution. This is straightforward for a single-site update in the Ising model, and with more effort can also be done for single-link updates in SU(2) gauge theories.

[h] Of course $s'_i = -1 \not\Rightarrow t'_i = -1$.

2.8. *Autocorrelations*

Successive states in a Markov chain are correlated in general. There are two different ways to measure this autocorrection between states: the first is the *exponential autocorrelation* (§2.8.1), which is an intrinsic property of the Markov process itself; and the second is the *integrated autocorrelation* (§2.8.2) for some observable Ω, which is more useful insofar as it is directly related to the statistical error of its measured estimator $\bar{\Omega}$.

2.8.1. *Exponential Autocorrelation*

In §2.3 we proved that an ergodic Markov process converges to a unique fixed point. In terms of the transition matrix $P_{ij} \equiv P(i \leftarrow j)$ this corresponds to P having a unique eigenvector with eigenvalue one and all its other eigenvalues lying strictly within the unit circle in the complex plane. In particular, the magnitude of the largest subleading eigenvalue must be smaller than 1,[i] $|\lambda_{\max}| < 1$. Any initial state vector may be expanded in a basis of normalized eigenvectors of P, $v = \sum_i \alpha_i u_i$ with $Pu_i = \lambda_i u_i$ and $\lambda_1 = 1 > \lambda_2 = \lambda_{\max} \geq \lambda_3 \geq \cdots$,

$$P^N u = \sum_i \lambda_i^N \alpha_i u_i = \alpha_1 u_1 + \alpha_2 \lambda_{\max}^N u_2 + \cdots ,$$

so the leading deviation from the equilibrium state represented by u_1 is of magnitude $|\lambda_{\max}^N| = e^{N \ln|\lambda_{\max}|} = e^{-N/N_{\exp}}$, which falls off exponentially with the number of Markov steps with a characteristic scale or exponential autocorrelation time of

$$N_{\exp} \equiv -\frac{1}{\ln|\lambda_{\max}|} > 0.$$

2.8.2. *Integrated Autocorrelation*

Consider the autocorrelation of some operator Ω measured on a sequence of successive configurations from a Markov chain. Without loss of generality we may assume $\langle\Omega\rangle = 0$. The variance of the estimator (sample average)

[i]Indeed, from the proof of §2.3, $|\lambda_{\max}| \leq 1 - \alpha$ with $\alpha = \sum_i \min_j P_{ij}$.

$\bar{\Omega}$ is

$$\begin{aligned}\langle\bar{\Omega}^2\rangle &= \frac{1}{N^2}\sum_{t=1}^{N}\sum_{t'=1}^{N}\langle\Omega(\phi_t)\Omega(\phi_{t'})\rangle \\ &= \frac{1}{N^2}\Big\{\sum_{t=1}^{N}\langle\Omega(\phi_t)^2\rangle + 2\sum_{t=1}^{N-1}\sum_{t'=t+1}^{N}\langle\Omega(\phi_t)\Omega(\phi_{t'})\rangle\Big\};\end{aligned}$$

and if we introduced the *autocorrelation function* (which is independent of t)

$$C_\Omega(\ell) \equiv \frac{\langle\Omega(\phi_{t+\ell})\Omega(\phi_t)\rangle}{\langle\Omega(\phi)^2\rangle}$$

this becomes

$$\langle\bar{\Omega}^2\rangle = \frac{1}{N}\Big\{\langle\Omega^2\rangle + \frac{2}{N}\sum_{\ell=1}^{N-1}(N-\ell)C_\Omega(\ell)\langle\Omega^2\rangle\Big\}.$$

The autocorrelation function must fall faster than the exponential autocorrelation

$$\big|C_\Omega(\ell)\big| \le \lambda_{\max}^{\ell} = e^{-\ell/N_{\exp}},$$

so, for a sufficiently large number of samples $N \gg N_{\exp}$,

$$\begin{aligned}\langle\bar{\Omega}^2\rangle &= \Big\{1 + 2\sum_{\ell=1}^{\infty}C_\Omega(\ell)\Big\}\frac{\langle\Omega^2\rangle}{N}\Big[1 + O\Big(\frac{N_{\exp}}{N}\Big)\Big] \\ &= (1 + 2A_\Omega)\frac{\langle\Omega^2\rangle}{N}\Big[1 + O\Big(\frac{N_{\exp}}{N}\Big)\Big],\end{aligned}$$

where we have defined the *integrated autocorrelation* function

$$A_\Omega \equiv \sum_{\ell=1}^{\infty}C_\Omega(\ell).$$

This result tells us that on average $1 + 2A_\Omega$ correlated measurements are needed to reduce the variance of $\bar{\Omega}$ by the same amount as a single truly independent measurement.

2.9. *Hybrid Monte Carlo*

In order to carry out Monte Carlo computations that include fermion dynamics we would like to find an algorithm which has following features:

- it updates the fields globally, because single link updates are not cheap if the action is not local;

- it takes large steps through configuration space, because small-step methods carry out a random walk which leads to critical slowing down with a *dynamical critical exponent*[j] $z = 2$.
- it does not introduce any systematic errors.

A useful class of algorithms with these properties is the (Generalized) Hybrid Monte Carlo (HMC) method.[4] In the HMC algorithm, we introduce a 'fictitious' momentum p for each dynamical degree of freedom q, and we construct a Markov chain with fixed point $e^{-H(q,p)}$ where H is the fictitious Hamiltonian $\frac{1}{2}p^2 + S(q)$: here the action S of the underlying quantum field theory plays the rôle of the potential in the fictitious classical mechanical system. This generates the desired distribution $e^{-S(q)}$ if we ignore the momenta p (statisticians call the distribution of q ignoring p a *marginal distribution*).

The HMC Markov chain alternates two Markov steps: the first step is *Molecular Dynamics Monte Carlo* (MDMC) (§2.10), and the second is (Partial) Momentum Refreshment (§2.11). Both have the desired fixed point, and their composite is clearly[k] ergodic.

2.10. *MDMC*

If we could integrate Hamilton's equations exactly we would follow a trajectory $(q, p) \to (q', p')$ of constant fictitious energy, $H(q, p) = H(q', p')$, for fictitious time τ: this corresponds to traversing a set of equiprobable fictitious phase space configurations. Liouville's theorem tells us that this trajectory preserves the measure[l] $dq \wedge dp = dq' \wedge dp'$, and reversing the momenta at the end of the trajectory ensures that it is reversible, $(q', -p') \to (q, -p)$, so such an update satisfies detailed balance and therefore would provide the desired Markov step.

Of course in general we cannot integrate Hamilton's equations exactly, but if we can find an approximate integration scheme that is exactly reversible and area-preserving (q.v., §2.13) then it may be used to suggest configurations to a Metropolis test with acceptance probability $\min(1, e^{-\delta H})$, where $\delta H \equiv H(q', p') - H(q, p)$ is the amount by which our integrator fails

---

[j] z relates an autocorrelation to a correlation length of the system, $A_\Omega \propto \xi^z$. The correlation length ξ is a characteristic length-scale of the physical system that diverges as the system approaches a continuous phase transition such as the continuum limit in the case of lattice quantum field theory.

[k] But it is not clear that this can be proved rigorously for any but the simplest systems.

[l] This is required for detailed balance to be satisfied for a continuous system.

to conserve fictitious energy. This too gives a Markov step that satisfies detailed balance.

We therefore build the MDMC Markov step out of three parts:

(1) *Molecular Dynamics* (MD), an approximate integrator $U(\tau)$: $(q,p) \mapsto (q',p')$ that is exactly area-preserving

$$\det U_* = \det \left[\frac{\partial(q',p')}{\partial(q,p)} \right] = 1,$$

and reversible

$$F \circ U(\tau) \circ F \circ U(\tau) = 1;$$

(2) a momentum reveral $F : p \mapsto -p$;
(3) a Metropolis test.

The composition of these, implementing the Metropolis test using a uniform random number $0 \le r \le 1$, gives

$$\begin{pmatrix} q \\ p \end{pmatrix} \to \begin{pmatrix} q' \\ p' \end{pmatrix} = \left[F \circ U(\tau)\, \theta(e^{-\delta H} - r) + \mathbb{I}\, \theta(r - e^{-\delta H}) \right] \begin{pmatrix} q \\ p \end{pmatrix}.$$

2.11. *Partial Momentum Refreshment*

The MDMC steps enables us to find acceptable candidate points in phase space far from the starting place, but it is far from ergodic because it almost stays on a constant fictitious energy surface. We remedy this by alternating it with a momentum refreshement which updates the momenta p from a Gaussian heatbath without touching the q: while this is also manifestly not ergodic it can easily make large changes in the fictitious energy.

Partial momentum refreshment is a minor generalization of this: it mixes the old Gaussian distributed momenta p with Gaussian noise ξ,

$$\begin{pmatrix} p' \\ \xi' \end{pmatrix} = \begin{pmatrix} \cos\theta & \sin\theta \\ -\sin\theta & \cos\theta \end{pmatrix} \circ F \begin{pmatrix} p \\ \xi \end{pmatrix}.$$

The Gaussian distribution of p is invariant under F. The extra momentum reversal F ensures that for small θ the momenta are reversed after a rejection rather than after an acceptance. For $\theta = \pi/2$, which is what is chosen for the standard HMC algorithm, all momentum reversals are irrelevant.

The reason for introducing partial momentum refreshment in GHMC (also known as second-order Langevin or Kramer's algorithm) was that for small θ and τ it introduces a small amount of noise frequently into the classical dynamics, rather than introducing a lot of noise occasionally as in HMC, where $\theta = \pi/2$ and $\tau \approx \xi$. Unfortunately any benefits this may have are negated by the fact that the momentum has to reversed after each Metropolis rejection, leading to 'zitterbewegung' back and forth along the trajectory.

2.12. *Baker–Campbell–Hausdorff (BCH) formula*

Fortunately there is a large class of reversible and area-preserving integrators: symmetric symplectic integrators (§2.13). A useful tool to analyze these is the BCH formula: if A and B belong to an associative algebra then $e^A e^B = e^{A+B+\delta}$, where δ is constructed from commutators of A and B, i.e., is in the *Free Lie Algebra* generated by the set $\{A, B\}$. More precisely, $\ln(e^A e^B) = \sum_{n\geq 1} c_n$, where $c_1 = A + B$, and

$$c_{n+1} = \frac{1}{n+1}\Bigg\{ -\tfrac{1}{2}\,\mathrm{ad}_{c_n}(A - B) + \sum_{m=0}^{\lfloor n/2 \rfloor} \frac{B_{2m}}{(2m)!} \sum_{\substack{k_1,\ldots,k_{2m}\geq 1 \\ k_1+\cdots+k_{2m}=n}} \mathrm{ad}_{c_{k_1}} \cdots \mathrm{ad}_{c_{k_{2m}}}(A + B)\Bigg\},$$

where $\mathrm{ad}_X Y \equiv [X, Y]$, and the B_n are Bernoulli numbers. Explicitly, the first few terms are

$$\begin{aligned}
\ln(e^A e^B) = {} & \{A + B\} + \tfrac{1}{2}[A, B] + \tfrac{1}{12}\Big\{[A, [A, B]] - [B, [A, B]]\Big\} \\
& -\tfrac{1}{24}\big[B, [A, [A, B]]\big] \\
& +\tfrac{1}{720}\Big\{ -4\,\big[B, \big[A, [A, [A, B]]\big]\big] - 6\,\big[[A, B], [A, [A, B]]\big] \\
& \qquad + 4\,\big[B, \big[B, [A, [A, B]]\big]\big] - 2\,\big[[A, B], [B, [A, B]]\big] \\
& \qquad - \big[A, \big[A, [A, [A, B]]\big]\big] + \big[B, \big[B, [B, [A, B]]\big]\big]\Big\} + \cdots;
\end{aligned}$$

from this we easily obtain the formula for a symmetric product

$$\begin{aligned}
\ln(e^{\frac{1}{2}A} e^B e^{\frac{1}{2}A}) = {} & \{A + B\} - \tfrac{1}{24}\Big\{2[B, [A, B]] + [A, [A, B]]\Big\} \\
& + \tfrac{1}{5760}\Big\{ 32\,\big[B, \big[B, [A, [A, B]]\big]\big] - 16\,\big[[A, B], [B, [A, B]]\big] \\
& \qquad + 28\,\big[B, \big[A, [A, [A, B]]\big]\big] + 12\,\big[[A, B], [A, [A, B]]\big] \\
& \qquad + 8\,\big[B, \big[B, [B, [A, B]]\big]\big] + 7\,\big[A, \big[A, [A, [A, B]]\big]\big]\Big\} + \cdots.
\end{aligned} \tag{3}$$

2.13. *Symplectic Integrators*

We are interested in finding the classical trajectory in phase space of a system described by the Hamiltonian

$$H(q,p) = T(p) + S(q) = \tfrac{1}{2}p^2 + S(q).$$

The basic idea of symplectic integrator is to write the time evolution operator as

$$\begin{aligned}\exp\left(\tau\frac{d}{dt}\right) &\equiv \exp\left(\tau\left\{\frac{dp}{dt}\frac{\partial}{\partial p} + \frac{dq}{dt}\frac{\partial}{\partial q}\right\}\right)\\ &= \exp\left(\tau\left\{-\frac{\partial H}{\partial q}\frac{\partial}{\partial p} + \frac{\partial H}{\partial p}\frac{\partial}{\partial q}\right\}\right)\\ &= \exp\left(\tau\left\{-S'(q)\frac{\partial}{\partial p} + T'(p)\frac{\partial}{\partial q}\right\}\right) \equiv \exp(\tau h),\end{aligned}$$

which relates a time derivative on the left to a linear differential operator on phase space, h, on the the right. In differential geometry such an operator is a *vector field*, and in the particular case where it is derived from a Hamiltonian function,

$$h = \iota(H) \equiv \frac{\partial H}{\partial p}\frac{\partial}{\partial q} - \frac{\partial H}{\partial q}\frac{\partial}{\partial p},$$

it is called a *Hamiltonian vector field.* Let us define

$$Q \equiv \iota(T) = T'(p)\frac{\partial}{\partial q}, \qquad P \equiv \iota(S) = -S'(q)\frac{\partial}{\partial p},$$

so that $h = P + Q$. Since the kinetic energy T is a function only of p and the potential energy S is a function only of q, it follows that the action of $e^{\tau P}$ and $e^{\tau Q}$ may be evaluated trivially: by Taylor's theorem

$$\begin{aligned}e^{\tau Q} &: f(q,p) \mapsto f\big(q + \tau T'(p), p\big),\\ e^{\tau P} &: f(q,p) \mapsto f\big(q, p - \tau S'(q)\big).\end{aligned}$$

It also means that these maps are area-preserving,

$$\frac{\partial\big(e^{\tau Q}(q,p)\big)}{\partial(q,p)} = \det\begin{pmatrix}1 & \tau T''(p)\\ 0 & 1\end{pmatrix} = 1,$$

$$\frac{\partial\big(e^{\tau P}(q,p)\big)}{\partial(q,p)} = \det\begin{pmatrix}1 & 0\\ -\tau S''(q) & 1\end{pmatrix} = 1.$$

From the BCH formula (3) we find that the PQP symmetric symplectic integrator is given by

$$\begin{aligned}
U_0(\delta\tau)^{\tau/\delta\tau} &\equiv \left(e^{\frac{1}{2}\delta\tau P}e^{\delta\tau Q}e^{\frac{1}{2}\delta\tau P}\right)^{\tau/\delta\tau} \\
&= \left(\exp\left\{(P+Q)\delta\tau - \tfrac{1}{24}\left([P,[P,Q]] + 2[Q,[P,Q]]\right)\delta\tau^3 + O(\delta\tau^5)\right\}\right)^{\tau/\delta\tau} \\
&= \exp\left\{\tau\left((P+Q) - \tfrac{1}{24}\left([P,[P,Q]] + 2[Q,[P,Q]]\right)\delta\tau^2 + O(\delta\tau^4)\right)\right\} \\
&= e^{\tau h'} = e^{\tau(P+Q)} + O(\delta\tau^2).
\end{aligned}$$

Such symmetric symplectic integrators are manifestly area-preserving and reversible in addition to conserving energy to $O(\delta\tau^2)$. What is even more remarkable is that since the vector field h' is built out of commutators of Hamiltonian vector fields it is itself a Hamiltonian vector field:

$$h' = \iota(H') \equiv \frac{\partial H'}{\partial p}\frac{\partial}{\partial q} - \frac{\partial H'}{\partial q}\frac{\partial}{\partial p}.$$

This means that for each symplectic integrator there exists a Hamiltonian H' that is exactly conserved.

In fact we can calculate H' quite easily. If $h_1 = \iota(H_1)$ and $h_2 = \iota(H_2)$ are two Hamiltonian vector fields derived from Hamiltonians H_1 and H_2 respectively then their commutator $h_3 \equiv [h_1, h_2]$ is a Hamiltonian vector field derived from the Hamiltonian that is the *Poisson bracket* of the Hamiltonians,

$$h_3 = [h_1, h_2] = \iota(H_3) \quad \text{where} \quad H_3 = \{H_1, H_2\} \equiv \frac{\partial H_1}{\partial p}\frac{\partial H_2}{\partial q} - \frac{\partial H_1}{\partial q}\frac{\partial H_2}{\partial p}.$$

Poisson brackets satisfy the Jacobi relation $\{A,\{B,C\}\} + \{B,\{C,A\}\} + \{C,\{A,B\}\} = 0$, so they endow the space of Hamiltonian functions with a Lie algebra structure. Indeed, the exactly conserved Hamiltonian for our PQP integrator may be written using the BCH formula (3) with the substitution of Poisson brackets for the corresponding commutators, $[Q,P] = [\iota(T), \iota(S)] \mapsto \{T, S\}$. With this technology it is easy to see that

$$\begin{aligned}
H' = H &+ \tfrac{1}{24}(2p^2 S' - S'^2)\delta\tau^2 \\
&+ \tfrac{1}{720}\left(-p^4 S^{(4)} + 6p^2(S'S''' + 2S''^2) - 3S'^2 S''\right)\delta\tau^4 + O(\delta\tau^6).
\end{aligned}$$

This expansion converges for small values of $\delta\tau$, and presumably up to the value of $\delta\tau$ for which the integrator becomes unstable (q.v., §2.13.1).

Note that H' cannot be written as the sum of a p-dependent kinetic term and a q-dependent potential term, so we cannot make use of this to find an integrator that exactly conserves H. Moreover, as H' is conserved, $\delta H = H(q',p') - H(q,p) = [H(q',p') - H'(q',p')] - [H(q,p) - H'(q,p)]$ is of $O(\delta\tau^2)$ for trajectories of arbitrary length even if $\tau = O(\delta\tau^{-k})$ with $k > 1$.

2.13.1. *Integrator Instability*

What happens if we take the integration step size $\delta\tau$ to be large? Clearly the Metropolis acceptance rate will fall as $\delta\tau$ increases, but this behaviour undergoes a sudden change at some value of $\delta\tau$ where the integrator goes unstable. We can see this for our PQP integrator even for the simple case where $S = \frac{1}{2}q^2$;[5] in this case the update $U_0(\delta\tau)$ is a linear mapping

$$\begin{pmatrix} q \\ p \end{pmatrix} \mapsto \begin{pmatrix} q' \\ p' \end{pmatrix} = \begin{pmatrix} 1 - \frac{1}{2}\delta\tau^2 & dt \\ -\delta\tau + \frac{1}{4}\delta\tau^3 & 1 - \frac{1}{2}\delta\tau^2 \end{pmatrix} \begin{pmatrix} q \\ p \end{pmatrix}.$$

The determinant of this matrix is unity because of area-preservation, and its characteristic polynomial has discriminant $\delta\tau^2(\delta\tau^2 - 4)$. When $\delta\tau = 2$ the system changes from having two complex conjugate eigenvalues of magnitude one, $e^{\pm i\phi}(\phi \in \mathbb{R})$, to having two real eigenvalues, $e^{\pm\nu}(\nu \in \mathbb{R})$; for $\delta\tau > 2$ this means that the errors increase exponentially with characteristic Liapunov exponent ν.

2.14. *Multiple Timescales*

We are not restricted to using simple symmetric symplectic integrators such as those described so far.[6] Suppose that the Hamiltonian is split into pieces

$$H(q,p) = T(p) + S_1(q) + S_2(q),$$

then we can define

$$Q \equiv \iota(T) = T'(p)\frac{\partial}{\partial q}, \qquad P_i \equiv \iota(S_i) = -S_i'(q)\frac{\partial}{\partial p},$$

so that $h = P_1 + P_2 + Q$. We may introduce a symmetric symplectic integrator of the form

$$U_{\mathrm{SW}}(\delta\tau)^{\tau/\delta\tau} =$$
$$\left\{\left[\exp\left(\frac{\delta\tau}{2n}P_2\right)\exp\left(\frac{\delta\tau}{2n}Q\right)\right]^n e^{\delta\tau P_1}\left[\exp\left(\frac{\delta\tau}{2n}Q\right)\exp\left(\frac{\delta\tau}{2n}P_2\right)\right]^n\right\}^{\tau/\delta\tau}.$$

We have a lot of freedom: all we need do is assemble the pieces symmetrically and ensure the the leading term in the BCH expansion is H. The remaining freedom can be used to reduce or eliminate higher-order errors in δH, or to make the step size small for a particular force term so as to avoid instabilities. For instance, if $2n\|P_1\|_\infty \approx \|P_2\|_\infty$, then the instability in the integrator is tickled equally by each sub-step. This helps if the most expensive force computation does not correspond to the largest force.

2.15. *Dynamical Fermions*

The direct simulation of Grassmann fields is not feasible: the problem is not that of manipulating anticommuting Grassmann variables in a computer, but that $e^{-S_F} = e^{-\bar{\psi}\mathcal{M}\psi}$ is not positive definite and this leads to poor importance sampling and thus a huge variance in measured quantities.

We therefore integrate out the quadratic fermion fields to obtain the fermion determinant

$$\int d\bar{\psi}\, d\psi\, e^{-\bar{\psi}\mathcal{M}\psi} \propto \det M.$$

The overall sign of the exponent is unimportant.

Any operator $\Omega(\phi, \bar{\psi}, \psi)$ can be expressed solely in terms of the bosonic field by Schwinger's technique of adding source terms $\bar{\psi}\eta + \bar{\eta}\psi$ to S_F before integrating over the fermion fields:

$$\Omega'(\phi) = \Omega\left(\phi, \frac{\partial}{\partial\eta}, \frac{\partial}{\partial\bar{\eta}}\right) e^{\bar{\eta}\mathcal{M}(\phi)^{-1}\eta}\Big|_{\eta=\bar{\eta}=0};$$

e.g., the fermion propagator is

$$G(x, y) = \langle\psi(x)\bar{\psi}(y)\rangle = \mathcal{M}^{-1}(x, y).$$

One obvious way of proceeding would be to include the determinant as part of the observable to be measured and computing a ratio of functional integrals

$$\langle\Omega\rangle = \frac{\langle\det\mathcal{M}(\phi)\Omega(\phi)\rangle_{S_B}}{\langle\det\mathcal{M}(\phi)\rangle_{S_B}}$$

with $S_B(\phi)$ being the bosonic (gauge field) part of the action; but this is not feasible because the determinant is extensive in the lattice volume, and we get hopelessly poor importance sampling.

We therefore proceed by representing the fermion determinant as a bosonic Gaussian integral with a non-local kernel

$$\det\mathcal{M}(\phi) \propto \int d\bar{\chi} d\chi \, \exp\left[-\bar{\chi}\mathcal{M}^{-1}(\phi)\chi\right].$$

The fermion kernel must be positive definite (all its eigenvalues must have positive real parts) as otherwise the bosonic integral will not converge. The new bosonic fields are called *pseudofermions.*

It is usually convenient to introduce two flavours of fermion and to write

$$\left(\det \mathcal{M}(\phi)\right)^2 = \det\left(\mathcal{M}(\phi)\mathcal{M}^\dagger(\phi)\right) \propto \int d\bar{\chi}d\chi \, \exp\left[-\bar{\chi}(\mathcal{M}^\dagger\mathcal{M})^{-1}\chi\right].$$

This not only guarantees positivity, but also allows us to generate the pseudofermions from a global heatbath by applying $\mathcal{M}^\dagger$ to a random Gaussian distributed field.

The equations for motion for the boson (gauge) fields are

$$\begin{aligned}\dot{\phi} &= \pi, \\ \dot{\pi} &= -\frac{\partial S_{\mathrm{B}}}{\partial \phi} - \chi^\dagger \frac{\partial (\mathcal{M}^\dagger\mathcal{M})^{-1}}{\partial \phi}\chi \\ &= -\frac{\partial S_{\mathrm{B}}}{\partial \phi} + \left[(\mathcal{M}^\dagger\mathcal{M})^{-1}\chi\right]^\dagger \frac{\partial (\mathcal{M}^\dagger\mathcal{M})}{\partial \phi}\left[(\mathcal{M}^\dagger\mathcal{M})^{-1}\chi\right]. \qquad (4)\end{aligned}$$

The evaluation of the pseudofermion action and the corresponding force then requires us to find the solution of a (large) set of linear equations $\mathcal{M}^\dagger\mathcal{M}\psi = \chi$.

It is not necessary to carry out the inversions required for the equations of motion exactly, there is a trade-off between the cost of computing the force and the acceptance rate of the Metropolis MDMC step. The inversions required to compute the pseudofermion action for the Metropolis accept/reject step do need to be computed exactly, however. We usually take 'exactly' to by synonymous with 'to machine precision'.

2.16. *Reversibility*

We now want to address the question as to whether HMC trajectories are reversible and area-preserving in practice.[5] The only source of irreversibility is the rounding errors caused by finite precision floating point arithmetic, so the fundamental reason for irreversibility is just that floating point arithmetic is not associative. What we are really studying is how much the MD evolution amplifies this noise.

For fermionic systems we can also introduce irreversibility by choosing the starting vector for the iterative linear equation solver time-asymmetrically, as we do if we use a chronological inverter, which takes some extrapolation of the previous solutions as the starting vector. Of course we do not have to use chronological inverter, we can start with a zero vector; alternatively we can find the solution sufficiently accurately that is does not depend on the initial guess.

A way that we may study the irreversibility is to follow a trajectory for time τ, then reverse the momenta and follow it back again; in other words we compute $U \circ F \circ U \circ F$, which in exact arithmetic should take us back exactly to where we started. We then measure the distance Δ in phase space from our starting point, using some suitable norm. What we observe is that the rounding errors are amplified exponentially with the trajectory length, $\Delta \propto \tau^{\nu}$, we call the exponent ν the *Liapunov exponent.* In practice if we work with parameters such that the Liapunov exponent is small then the resulting irreversibility is not a big problem, because the 'seed' rounding errors are of order 10^{-7} for 32-bit floating point arithmetic and 10^{-15} for 64-bit precision: rounding errors fall exponentially with the number of bits we use to represent floating point numbers.

In Figure 2 we show data for pure SU(3) gauge theory and full QCD (both on tiny lattices) where the Liapunov exponent is plotted as a function of the integration step size $\delta\tau$.

The fact that for small step size the Liapunov exponent $\nu \neq 0$ appears to tell us that the underlying continuous-time equations of motion for gauge field fictitious dynamics are chaotic. In QCD the Liapunov exponent appears to scale with β as the system approaches the continuum limit, that is $\lim_{\beta\to\infty} \nu\xi \approx$ constant where ξ is the correlation length as before. This can be interpreted as saying that the Liapunov exponent characterizes the chaotic nature of the continuum classical equations of motion, and is not a lattice artefact. If this is so we should not have to worry about reversibility breaking down as we approach the continuum limit, as the amplification factor for a trajectory of length $\tau \approx \xi$ stays fixed. However, beware that this is based on data from tiny lattices, and is not at all conclusive.

More significantly, at some particular value of the step size, $\delta\tau \approx 0.6$ for the top two graphs and $\delta\tau \approx 0.1$ for the bottom one, the Liapunov exponent start to increase, perhaps linearly. This behaviour is what we expect when the integrator has become unstable. For the top two graphs this is not very interesting, as the change in energy along a trajectory of length $\tau \approx 1$ at this step size is very large, $\delta H \gg 1$, so the acceptance rate is essentially zero anyhow. For the bottom graph — the case with light fermions — it is the instability rather than the 'bulk' behaviour of δH that limits the step size we can get away with. We shall investigate on way to circumvent this problem in §3.8.

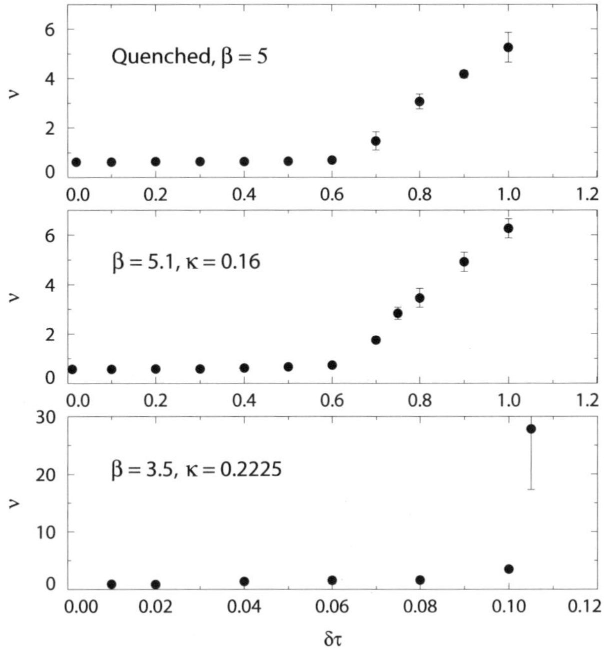

Figure 2. The Liapunov exponent ν is shown as a function of the integration step size $\delta\tau$. The top graph is for pure $SU(3)$ gauge theory, the middle one is for QCD with heavy dynamical Wilson quarks, and the bottom one is for QCD with light dynamical Wilson quarks. Note that the scale for the light quark case is quite different from the other two. The error bars show the standard deviation of measurements made on three independent configurations. All the data is from 4^4 lattices.

3. The RHMC Algorithm

In this second lecture, we will discuss the Rational Hybrid Monte Carlo (RHMC) algorithm.[7] For this purpose we shall first take a brief look at approximation theory.

3.1. *Polynomial Approximation*

We start by consider the problem of approximating a continuous function by a polynomial. It turns out that the rational approximation theory is simple generalization of this.

We want to address the question of what is the best polynomial approximation $p(x)$ to a continuous function $f(x)$ over the unit interval $[0, 1]$. To address the question we have to specify an appropriate norm on the space of functions. The most important class of norms are the L_n norms, defined by

$$\|p - f\|_n = \left(\int_0^1 dx\, \big|p(x) - f(x)\big|^n\right)^{1/n},$$

which satisfies the required axioms[m] provided $n \geq 1$. The case $n = 2$, for example, is usual Euclidean norm. The case $n = 1$ is the L_1 norm which was used in the proof of convergence of Markov processes in §2.4. The *minimax norm* is defined by taking the limit $n \to \infty$

$$\|p - f\|_\infty = \max_{0 \leq x \leq 1} \big|p(x) - f(x)\big|, \tag{5}$$

because when $n \to \infty$ the integral is dominated by the point at which the error $p(x) - f(x)$ is maximal. What we want to do in the following is to find optimal polynomial approximations with respect to the L_∞ norm, since this guarantees a bound on the pointwise error. We can generalize these definitions by including a positive weight function, but we will not consider this further in this introduction.

The fundamental theorem on the approximation of continuous functions by polynomials was given by Weierstrass, who proved that any continuous function can be arbitrarily well approximated over the unit interval by a polynomial. This result is very important in functional analysis, where it is the essential ingredient in many proofs, although it is of less significance in finding approximations for practical use. The most elegant proof of Weierstrass' theorem was given by Bernstein, who showed that the *Bernstein polynomials*

$$p_n(x) \equiv \sum_{k=0}^{n} f\left(\frac{k}{n}\right) \binom{n}{k} x^n (1 - x)^{n-k}$$

can arbitrarily well approximate f over the unit interval by taking n to be sufficiently large, that is $\lim_{n\to\infty} \|p_n - f\|_\infty = 0$.

3.2. *Chebyshev's theorem*

If we restrict ourselves to considering polynomials of some fixed degree then the Bernstein polynomial has no reason to be the best approximation to f. What we really want is a solution to the 'minimax' problem: that is to find a polynomial p of fixed degree which gives the minimum L_∞ error

$$\Delta = \min_p \|p - f\|_\infty = \min_p \max_{0 \leq x \leq 1} |p(x) - f(x)|.$$

---

[m]These are that $\|f\| \geq 0$, $\|f + g\| \leq \|f\| + \|g\|$, $\|\alpha f\| = |\alpha| \|f\|$ $\forall f, g$, where α is a scalar, and $\|f\| = 0$ iff $f = 0$ almost everywhere.

A surprisingly elegant solution to the problem of finding the best approximation of fixed degree was given by Chebyshev, who proved that for any degree d there is always a unique polynomial p that minimizes the L_1 norm (5), and it is characterized by the criterion that the error $p(x) - f(x)$ attains its maximum absolute value at exactly $d + 2$ points on the unit interval, and the sign of the error alternates between successive maxima.

We shall prove Chebyshev's theorem in two steps; we will first establish the necessity of the criterion stated above, and then prove its sufficiency.

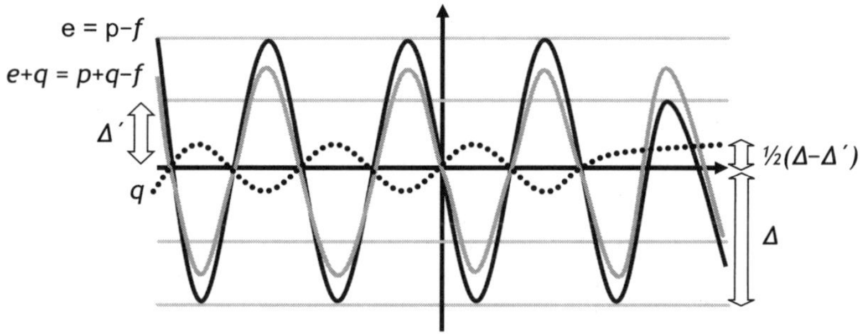

Figure 3. An illustration of the proof of necessity of Chebyshev's condition. p is a polynomial of degree $d = 7$ that approximates a continuous function f. The heavy line shows the error $e = p - f$ that attains its maximum value $\Delta = \|p - f\|_\infty$ at only $d+1 = 8$ points (including one end point of the interval). The next largest maximum of $|e(x)|$ has a value of Δ', as shown. We therefore construct the polynomial q that passes through a zero of e between each of the points where $e(x) = \pm\Delta$; there are $d = 7$ such points, so q is also a polynomial of degree $d = 7$. It is chosen to have the opposite sign to e at each of the points where $|e(x)| = \Delta$, and a maximum value of $\frac{1}{2}(\Delta - \Delta')$ over the interval. It is shown as the dotted line on the graph. The polynomial $p + q$ is then a better approximation than p, as shown by its error $p + q - f = e + q$ which is the light gray line.

Suppose that p is an optimal polynomial of degree d for which the error $e(x) \equiv p(x) - f(x)$ has less than $d + 2$ alternating extrema of equal magnitude; then at most $d + 1$ of its maxima can exceed some magnitude Δ'. This defines a 'gap', $\Delta - \Delta'$. As e is continuous it must have at least one zero between two successive maxima; let us suppose these occur at $x_i\,(i = 1, \ldots, d)$. We can construct a polynomial q of degree d which has the opposite sign to e at each of the $d + 1$ maxima, $q(x) = k \prod_{i=1}^{d+1}(x - x_i)$, and whose magnitude is smaller than the 'gap' by a suitable choice of the magnitude and sign of the constant k. But this contradicts the assumption, for the polynomial $p + q$ is a better approximation than p to f.

Sufficiency is shown as following. Suppose that p satisfies Chebyshev's

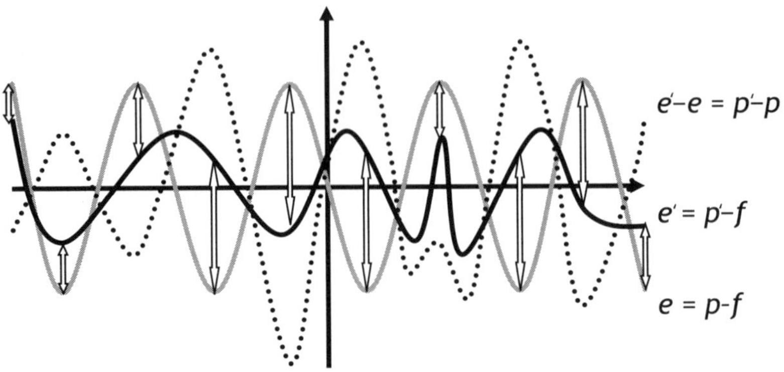

Figure 4. An illustration of the sufficiency of Chebyshev's criterion. The polynomial p of degree $d = 8$ satisfies Chebyshev's criterion, as indicated by the graph of its error $e = p - f$ (light gray line) which has $d + 2 = 10$ alternating extrema of equal magnitude on the interval (including the end points). p' is another polynomial also of degree $d = 8$ with a smaller error, $\|e'\|_\infty \leq \|e\|_\infty$, as shown on the graph by the solid line. The difference $p' - p = (p' - f) - (p - f) = e' - e$ shown by the dotted line then must have $d + 1 = 9$ zeros on the interval since $e(x) - e'(x)$ has an alternating sign at each of the extrema of e (as shown by the arrows), because otherwise the maximum error of $|e'|$ would not be smaller than that of e. Since $p - p'$ is a polynomial of degree $d = 8$ this means it must vanish identically by the fundamental theorem of algebra.

criterion, and there is a better polynomial approximation p' than p, i.e., $\|p' - f\|_\infty \leq \|p - f\|_\infty$; then $|p'(x_i) - f(x_i)| \leq |p(x_i) - f(x_i)|$ at each of the $d + 2$ extrema. But then $p' - p$ must have $d + 1$ zeros on the unit interval, thus $p' - p = 0$ as it is a polynomial of degree d.

3.3. *Chebyshev Polynomials*

The rôle of Chebyshev polynomials in approximation theory causes some confusion. The expansion of a function f in Chebyshev polynomials up to degree d does not, in general, give the best approximation of that degree. What the Chebyshev polynomials do give us it the best approximation of degree $d - 1$ to the function x^d over the compact interval $[-1, 1]$; this is given by

$$p_{d-1}(x) \equiv x^d - (\tfrac{1}{2})^{d-1} T_d(x), \tag{6}$$

where the Chebyshev polynomials are defined by

$$T_d(x) \equiv \cos\bigl(d \cos^{-1}(x)\bigr).$$

The use of the letter T is from an old transliteration of Chebyshev's name. Perhaps the most surprising thing about Chebyshev polynomials is that they are polynomials, despite being defined in terms of transcendental functions. This is shown by expanding

$$\begin{aligned}\cos(d\xi) &= \operatorname{Re} e^{id\xi} = \operatorname{Re}(\cos\xi + i\sin\xi)^d = \operatorname{Re}\sum_{j=0}^{d}\binom{d}{j}(\cos\xi)^{d-j}(i\sin\xi)^j \\ &= \sum_{k=0}^{\lfloor d/2\rfloor}\binom{d}{2k}(\cos\xi)^{d-2k}(-1)^k\left[1-(\cos\xi)^2\right]^k,\end{aligned}$$

from which we immediately see the polynomial nature of $T_d(x)$: indeed writing $x = \cos\xi$ we may proceed to obtain an explicit form

$$T_d(x) = \sum_{k=0}^{\lfloor d/2\rfloor}\sum_{\ell=0}^{k}\binom{d}{2k}\binom{k}{\ell}(-1)^{k+\ell}x^{d-2k+2\ell}.$$

From this we see that the leading monomial is

$$\begin{aligned}\sum_{k=0}^{\lfloor d/2\rfloor}\binom{d}{2k}x^d &= \sum_{j=0}^{d}\binom{d}{j}\left[1+(-1)^j\right]\frac{x^d}{2} = \left\{\sum_{j=0}^{d}\binom{d}{j} + \sum_{j=0}^{d}\binom{d}{j}(-1)^j\right\}\frac{x^d}{2} \\ &= \left\{(1+1)^d + (1-1)^d\right\}\frac{x^d}{2} = 2^{d-1}x^d.\end{aligned}$$

The error in the approximation of Eq. (6) is

$$\left\|x^d - p_{d-1}(x)\right\|_\infty = \left(\tfrac{1}{2}\right)^{d-1}\left\|T_d(x)\right\|_\infty = 2e^{-d\ln 2},$$

which obviously satisfies Chebyshev's criterion because $\cos d\xi$ takes the values ± 1 exactly $d+1$ times on the interval $0 \le \xi \le \pi$.

In this case the error falls exponentially with the degree d because the coefficient of the leading term in the Chebyshev polynomials grows as 2^{d-1}, but in general such exponential convergence does not occur.

3.4. *Chebyshev Optimal Rational Approximation*

Chebyshev's theorem is easily extended to rational functions (ratios of polynomials); the criterion that a rational approximation of degree (n, d) is optimal is that its error must attain its maximum magnitude at $n + d + 2$ alternating points on the interval.

Rational functions with nearly equal degree numerator and denominator are usually the best choice. An approximation with $n < d$ cannot be better than one with $n = d$ unless Chebyshev's criterion would lead to the leading $d - n$ coefficients of the latter vanishing; this usually only happens when there is an obvious reason for it, such as trying to approximate a rational function by one of higher degree, or trying to approximate an odd function in which case the optimal approximation will generally be of degree $(d \pm 1, d)$.

An empirical observation is that convergence is often more often exponential for optimal rational approximations than for polynomial ones, and that for a given degree rational functions usually give a much better approximation, which probably is not surprising.

There are only a few special cases where it is possible to compute optimal polynomial or rational approximations in closed form: we have seen one in §3.3 and we will come across another in §4.7.2. In general we have to find the optimal approximations numerically. A simple (but somewhat slow) numerical algorithm for finding the optimal Chebyshev rational approximation was given by Remez. The Remez algorithm is basically just an way of implemeting Chebyshev's proof; it alternates (i) a phase of searching for an alternating set of $n + d + 2$ points at which the global maximum of the error occurs, and (ii) a phase of adjusting the $n + d + 1$ coefficients of the rational function to make the magnitude of the error take the same value on the set of points found in the previous phase. A little though will explain why it is difficult to find the optimal approximation to a function like $\sin \frac{1}{x}$.

A realistic example of a rational approximation to the function $1/\sqrt{x}$ is

$$\frac{1}{\sqrt{x}} \approx 0.3904603901 \frac{(x+2.3475661045)(x+0.1048344600)(x+0.0073063814)}{(x+0.4105999719)(x+0.0286165446)(x+0.0012779193)}.$$

This is accurate to within almost 0.1% over the range $[0.003, 1]$. Using a partial fraction expansion of such rational functions allows us to use a multishift linear equation solver, thus reducing the cost significantly. The partial fraction expansion of the rational function above is

$$\frac{1}{\sqrt{x}} \approx 0.3904603901 + \frac{0.0511093775}{x+0.0012779193} + \frac{0.1408286237}{x+0.0286165446} + \frac{0.5964845033}{x+0.4105999719}.$$

This is numerically stable because all the terms are positive and the roots of the denominators are all real. The fact that this desirable property holds for optimal rational approximations to x^α for $-1 \leq \alpha \leq 1$ is one of the principle reasons for the efficacy of the RHMC algorithm, but why it does is not at all clear.

It is interesting to compare the optimal rational and polynomial approximations for $1/\sqrt{x}$, as this is one of the few cases where we can analyze their convergence analytically. The optimal rational approximation of degree $d \pm 1, d$ over the interval $\varepsilon \leq |x| \leq 1$ with a weight function $w(x) = \sqrt{x}$, i.e., where we minimise the maximum relative error, was found analytically by Zolotarev (§4.7.2) and has an L_∞ error that falls as $\Delta \sim \exp(d/\ln\varepsilon)$. The optimal L_2 approximation with respect to the weight function $1/\sqrt{1-x^2}$ is given by the Fourier series

$$\sum_{j=0}^{d} \frac{(-1)^j 4}{(2j+1)\pi} T_{2j+1}(x),$$

which has L_2 error that falls as $O(1/d)$. The optimal L_∞ approximation cannot be too much better, or it would lead to a better L_2 approximation, so it certainly only converges as $1/d$. Note that we are not being entirely fair in this comparison, as the weights, as well as the approximation intervals, are different in the two cases.

It is also worth noting in comparing rational and polynomial approximations that using optimal polynomial approximation to $1/x$ to compute the inverse of a matrix (§3.6) is essentially the same as using a Jacobi iteration scheme, and this is well-known to be inferior to Gauss-Seidel iteration or Krylov space methods.

3.5. *Non-Linearity of CG Solver*

We have now completed our survey of approximation theory,[8] and we turn to considering its application in the RHMC algorithm. To motivate the need for this let us consider the following suggestion of a method to accelerate matrix inversion.

Suppose we want to solve a system of linear equations $A^2x = b$ where A is a positive Hermitian matrix using the conjugate gradient (CG) method. It is well known that it is better to solve $Ay = b$ and $Ax = y$ successively, because the condition number[n] is reduced from $\kappa(A^2) = \kappa(A)^2$ to $2\kappa(A)$.

Suppose now that we want to solve $Ax = b$. Why don't we solve $\sqrt{A}y = b$ and $\sqrt{A}x = y$ successively? The square root of A is uniquely defined as $A > 0$, and we can even choose it to be positive too. Indeed, for application to quantum field theory most of our time is spent finding $(\mathcal{M}^\dagger\mathcal{M})^{-1}\chi$, and the fermion kernel is manifestly positive, $\mathcal{M}^\dagger\mathcal{M} > 0$. All this generalizes trivially to n^{th} roots, so it seems that we have the opportunity of reducing the cost even further.

[n] The *condition number* of a matrix is defined to be the ratio of its largest and smallest eigenvalues. To a first approximation the cost of solving a system of linear equations is proportional to the condition number.

The question that needs to be addressed is how do we apply the square root of a matrix?

3.6. *Rational Matrix Approximation*

Before we can investigate how to evaluate functions of matrices we need to define what we mean by them. For our purposes we only need to define a function of a Hermitian matrix H, and this can be diagonalized by unitary transformation $H = UDU^{-1}$. We define the function $f(H)$ just by changing to the basis where H is diagonal, applying f to the diagonal elements (eigenvalues), and then transforming back to the original basis,

$$f(H) = f(UDU^{-1}) \equiv Uf(D)U^{-1}.$$

The great advantage of rational functions of matrices is that they do not require this diagonalisation to be carried out explicitly, since by linearity

$$\alpha H^m + \beta H^n = U(\alpha D^m + \beta D^n)U^{-1} \quad \text{and} \quad H^{-1} = UD^{-1}U^{-1}.$$

It suffices to compute the rational function using the operations of matrix addition, matrix multiplication, and matrix inversion in place of their scalar analogues. Furthermore, if we only want the result of applying the matrix function to a vector, $f(H)v$, then we can use the appropriate matrix-vector operations rather than the more costly matrix-matrix ones; in particular we can use a linear equation solver rather than finding a complete matrix inverse.

3.7. *'No Free Lunch' Theorem*

We now have an efficient way of computing matrix functions by finding a good rational approximation to the function and applying it using matrix operations. Let us return to our problem of solving $Ax = b$ by solving n systems of the form $A^{1/n}x = y$ successively: for each of these we use a rational approximation to compute[o] $x = A^{-1/n}y$. To do this, we must apply the rational approximation of the nth root $r(A) \approx A^{1/n}$. This is done efficiently by expanding the rational function in partial fractions and then applying all the terms at once using a multishift CG solver. The condition number for each term in the partial fraction expansion is approximately that of the original matrix $\kappa(A)$, so the cost of applying $A^{1/n}$ is proportional to

[o] We save a lot of work here by noticing that it is just as easy to approximate $A^{-1/n}$ as it is to approximate $A^{1/n}$, so there is no need for nested CG solvers.

$\kappa(A)$. We have therefore found out why our suggested inversion method fails: even though the condition number $\kappa(A^{1/n}) = \kappa(A)^{1/n}$ and $\kappa\big(r(A)\big) = \kappa(A)^{1/n}$ the cost of applying $r(A)$ is $\kappa(A)$, so unfortunately we don't win anything.

3.8. *Multiple Pseudofermions*

Let us return to the topic we were considering in the previous lecture (§2.15), where we introduced pseudofermions as a means of representing the fermion determinant. Recall that we rewrote the determinant as a bosonic functional integral over a pseudofermion field ϕ with kernel $\mathcal{M}^{-1}$

$$\det \mathcal{M} \propto \int d\phi^\dagger d\phi \, \exp\left[-\phi^\dagger \mathcal{M}^{-1} \phi\right].$$

What we are doing is to evaluate functional integrals that include the fermion determinant $\det \mathcal{M}$ by using a stochastic estimate of the determinant, namely we are approximating the integral over the pseudofermion fields by a single configuration — in other words evaluating the integrand at only one point — yet this produces a Markov process with exactly the correct fixed point. We again seem to be getting something for nothing, so perhaps we should be suspicious that we are paying a hidden price somewhere.

The hidden price is that we are introducing extra noise into the system by using a single pseudofermion field to sample this functional integral. This noise manifests itself as fluctuations in the force exerted by the pseudofermions on the gauge fields, and in turn

- this increases the maximum fermion force;
- which triggers the integrator instability;
- which requires decreasing the integration step size.

A better estimate is to write the fermion determinant as

$$\det \mathcal{M} = \left[\det \mathcal{M}^{1/n}\right]^n,$$

and introduce a separate pseudofermion field for each factor

$$\det \mathcal{M} = \left[\det \mathcal{M}^{1/n}\right] \propto \prod_{j=1}^{n} \int d\phi_j^\dagger d\phi_j \, \exp\left[-\phi_j^\dagger \mathcal{M}^{-1/n} \phi_j\right].$$

3.8.1. *The Hasenbusch Method*

Before we go further into the RHMC implementation of multiple pseudofermions we ought to note that the idea of using multiple pseudofermions was originally due to Hasenbusch.[9] He considered a theory with the Wilson fermion action $\mathcal{M}(\kappa)$ where κ is the *hopping parameter* which controls the fermion mass. He introduced a heavy fermion kernel $\mathcal{M}' = \mathcal{M}(\kappa')$, and made use of the trivial identity following from the associativity of matrix multiplication $\mathcal{M} = \mathcal{M}'(\mathcal{M}'^{-1}\mathcal{M})$ to write the fermion determinant as $\det \mathcal{M} = \det \mathcal{M}' \det(\mathcal{M}'^{-1}\mathcal{M})$. He then introduced separate pseudofermions for each determinant, and tuned κ' to minimise the cost of the overall computation. This can be all be easily generalized to more than two pseudofermions and to the clover-improved Wilson action.

3.9. *Violation of NFL theorem*

Let us now return to using our n^{th} root trick to implement multiple pseudofermions. As before we observe that the condition number of the nth root $\kappa\big(r(\mathcal{M})\big) = \kappa(\mathcal{M})^{1/n}$, and we may use the simple model that the largest contribution to the force acting on the gauge fields due to one of our pseudofermion fields is inversely proportional to the smallest eigenvalue of the fermion kernel, at least when the fermion mass is sufficiently small. This follows by considering Eq. (4), and recalling that we used $\mathcal{M}^\dagger\mathcal{M}$ where we now have $\mathcal{M}^{1/n}$. In fact we might even expect the force to grow faster than this, but let us stick with our simple model. As the largest eigenvalue of $\mathcal{M}$ is more-or-less fixed this force is proportional to $\kappa(\mathcal{M})^{1/n}$; we have n pseudofermions each contributing to the force, and if we are conservative and assume that these contributions add coherently we get a total force proportional to $n\kappa(\mathcal{M})^{1/n}$. This is to be compared to the original single pseudofermionic force of $\kappa(\mathcal{M})$, so we expect that the maximum force is reduced by a factor of $n\kappa(\mathcal{M})^{\frac{1}{n}-1}$. This is a good approximation if the condition number is dominated by a few isolated tiny eigenvalues, which is just the what happens in the cases of interest. If the force is reduced by this factor the according to our previous considerations (§2.16) we can increase the step size by the reciprocal of this factor, and all other things being equal, the overall cost is reduced by a factor of $n\kappa(\mathcal{M})^{\frac{1}{n}-1}$.

If we take this model seriously then we can easily calculate the optimal number of pseudofermions by minimising the cost, and we find that the optimal value is $n_{\text{opt}} \approx \ln\big(\kappa(M)\big)$, and the corresponding optimal cost reduction is $e \ln\kappa/\kappa$.

So, by introducing a small number of pseudofermion fields (of order $\ln\kappa(\mathcal{M})$) we expect to get a cost reduction of order $1/\kappa(\mathcal{M})$, and this works in practice — at least there is a significant cost reduction, the exact scaling laws of our simple model are undoubtedly only followed very approximately at best — thereby violating the infamous 'no free lunch' theorem.

The advantage that this method has over the Hasenbusch method is that it is trivially applicable to any fermion kernel and that no parameter tuning is required: the condition number is automatically equipartitioned between the pseudofermion fields. On the other hand, the Hasenbusch method has the advantage that one of the pseudofermions may be cheap to apply because it is heavy.

3.10. *Rational Hybrid Monte Carlo*

Let us now go through the details of the Rational Hybrid Monte Carlo (RHMC) algorithm for the fermion kernel[p] $(\mathcal{M}^\dagger\mathcal{M})^{1/2n}$. In §2.9 we explained how the HMC algorithm alternates two Markov steps — momentum refreshment and MDMC — both of which have the desired fixed point and together are ergodic. When we include pseudofermion fields we need to add a third Markov step to ensure ergodicity, which is sampling the pseudofermion fields from a heatbath. This is easy to do because the (pseudo)fermions only appear quadratically in the action:[q] we sample the pseudofermions from a Gaussian heatbath by applying the square root of the kernel to Gaussian-distributed random noise,

$$\chi_j = (\mathcal{M}^\dagger\mathcal{M})^{1/4n}\xi_j \quad \text{with} \quad P(\xi_j) \propto e^{-\xi_j^\dagger\xi_j},$$

as then

$$\begin{aligned} P(\chi_j) &\propto \int_{-\infty}^{\infty} d\xi_j\, e^{-\xi_j^\dagger\xi_j}\,\delta\Big(\chi_j - (\mathcal{M}^\dagger\mathcal{M})^{\frac{1}{4n}}\xi_j\Big) \\ &\propto \exp\Big[-\chi_j^\dagger(\mathcal{M}^\dagger\mathcal{M})^{-\frac{1}{2n}}\chi_j\Big]. \end{aligned}$$

[p]This corresponds to $1/n^{\text{th}}$ of a fermion 'flavour'; we take advantage of the fact that it is no more work to take the $2n^{\text{th}}$ root of $\mathcal{M}^\dagger\mathcal{M}$ than to take the nth root of $\mathcal{M}$, so we can ensure that the fermion kernel is manifestly positive for no extra cost.

[q]We could not get away with this simplification if we wanted to improve the fermion action by including four-fermion operators or higher.

We then refresh the momenta[r], and integrate Hamilton's equations for the gauge fields using the force (c.f., Eq. (4))

$$\begin{aligned}\dot{\pi} &= -\frac{\partial S_{\mathrm{B}}}{\partial \phi} - \sum_{j=1}^{n} \chi_j^\dagger \frac{\partial (\mathcal{M}^\dagger \mathcal{M})^{-1/2n}}{\partial \phi} \chi_j \\ &= -\frac{\partial S_{\mathrm{B}}}{\partial \phi} + \sum_{\chi=1}^{n} \left[(\mathcal{M}^\dagger \mathcal{M})^{-1/2n} \chi_j\right]^\dagger \frac{\partial (\mathcal{M}^\dagger \mathcal{M})^{1/2n}}{\partial \phi} \left[(\mathcal{M}^\dagger \mathcal{M})^{-1/2n} \chi_j\right].\end{aligned}$$

Finally we complete the MDMC step by applying a Metropolis test.

Of course we apply all the fractional powers of the matrices that occur using optimal rational approximations expressed in partial fraction form. Note that

- we use an accurate rational approximation $r(x) \approx \sqrt[4n]{x}$ for the pseudofermion heatbath;
- we use a less accurate approximation for the MD evolution, $\tilde{r}(x) \approx \sqrt[2n]{x}$;
- we use an accurate approximation for Metropolis acceptance step.

This is because any errors in the pseudofermion heatbath or Metropolis test would effect the fixed point distribution, whereas errors in the MD only effect the acceptance rate. If the rational approximations used in the pseudofermion heatbath and the Metropolis test are accurate to machine precision then the algorithm is still as exact as HMC: that is, the only systematic errors are the rounding errors from using finite precision floating point arithmetic, and from the fact that the Markov process may not have converged well enough to its fixed point. It is not hard to generate optimal rational approximations that are good to machine precision: this should not be surprising as this is how most scalar functions (exponential, logarithm, fractional powers, etc.) are evaluated anyhow.

Let us summarize the key properties of the RHMC algorithm.

- We apply all rational approximations using their partial fraction expansions:
  - the denominators occuring in this are all just shifts of the original fermion kernel;
  - all poles of optimal rational approximations are real and positive for cases of interest (Miracle #1);

[r]There is no *a priori* reason why we have to refresh the pseudofermions and the momenta with equal frequency, but there is no obvious benefit from not doing so.

  - only simple poles appear (by construction, a double pole can only occur if we try to approximate something that was itself a square, which would not be a sensible thing to do).
- We a use multishift solver to invert all the partial fractions using a single Krylov space: the cost is dominated by the construction of the Krylov space; updating the n solution vectors is comparatively cheap, at least for $O(20)$ shifts.
- The result is numerically stable, even in 32-bit precision. This is because all the partial fractions have positive coefficients (Miracle #2).
- The MD force term is of the same form as for conventional HMC except for a constant shift for each partial fraction; therefore the method is immediately applicable to any fermion kernel for which we can implement conventional HMC.

3.11. *Comparison with R Algorithm*

Let us now look at some empirical (numerical) studies[10] that compare the performance of the RHMC and R algorithms for non-zero temperature QCD with staggered quarks (§3.11.1) and for chirally symmetric domain wall fermions (§3.11.2), where the R algorithm has been the method of choice in the past.

3.11.1. *Finite Temperature QCD*

First we compare the of the algorithms' performance near the chiral transition point. The aim of this study was to see how the RHMC algorithm behaves in this regime, and to compare its cost with that of the R algorithm.

Table 1. Values of the Binder cumulant for $\langle\bar{\psi}\psi\rangle$, B_4, and the RHMC acceptance rate P_{acc}.

Algorithm	$\delta\tau$	P_{acc}	B_4
R	0.0019		1.56(5)
R	0.0038		1.73(4)
RHMC	0.055	84%	1.57(2)

For these tests we used $2+1$ flavours of naïve staggered fermions with $m_{\text{ud}} = 0.0076$ and $m_s = 0.25$, and the Wilson gauge action on an $8^3 \times 4$ lattice. The MD trajectory length was $\tau = 1$. The results are given in Table 1, where the Binder cumulant of the chiral condensate, B_4, is shown as a function of step size for the two algorithms. The RHMC acceptance rate, P_{acc}, is also given. The correct value of the Binder cumulant is obtained with RHMC using a step size $\approx$ 29 times larger than is necessary for the R algorithm.

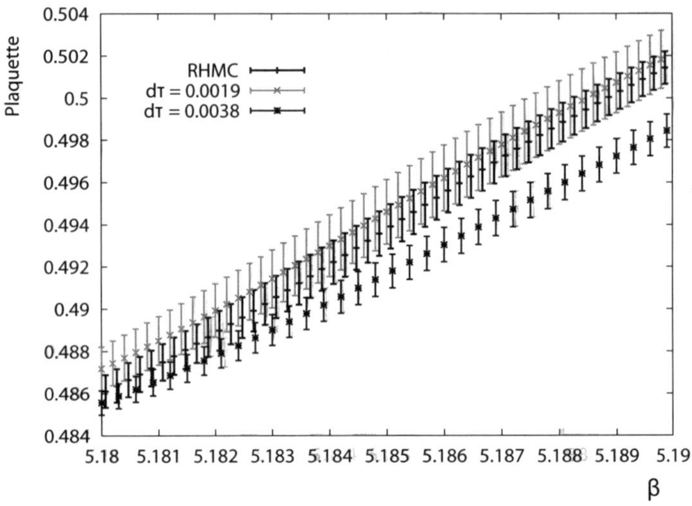

Figure 5. The mean plaquette is shown as function of β for the RHMC and R algorithms at the parameters given in the text (near the chiral transition point). The RHMC points have been displaced right so that they can be seen more clearly. For the R algorithm only the data at the smaller step size is within one standard deviation of the RHMC results.

Figures 5 and 6 are plots of the plaquette and the chiral condensate $\langle\bar{\psi}\psi\rangle$ as a function of $\delta\tau$: only for the smaller step size is the R algorithm result consistent with that from RHMC.

It is clear that RHMC is superior to the R algorithm in this case, principally because it is vital to keep the systematic step size errors under control when studying the properties of phase transitions. Even though the errors in the equilibrium distribution of the R algorithm Markov chain are of $O(\delta\tau^2)$, even the introduction of a small admixture of unwanted operators in the action can significantly change the behaviour of a system near criticality; indeed, they can even change the order of the phase transition if we are not extremely careful when we take the zero step size limit.

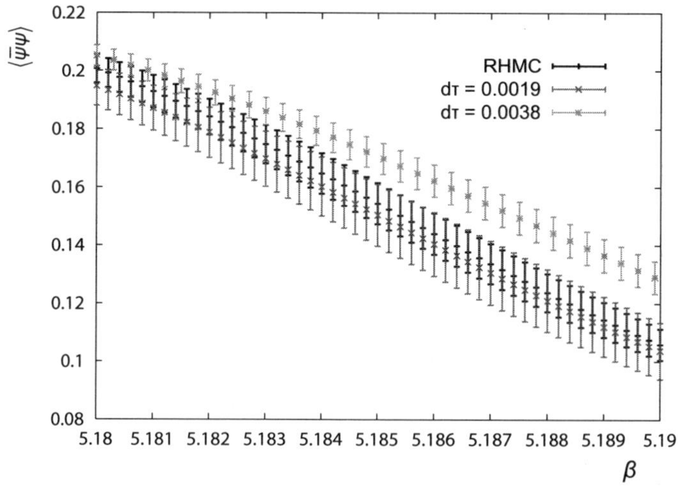

Figure 6. The chiral condensate $\langle\bar{\psi}\psi\rangle$ is shown as a function of β for the RHMC and R algorithms at the parameters given in the text (near the chiral transition point). For the R algorithm only the data at the smaller step size is within one standard deviation of the RHMC results.

3.11.2. *2+1 Domain Wall Fermions*

Lattice calculations with on-shell chiral dynamical fermions are still in their infancy, but it seems clear that they will become increasingly important as sufficient computing power becomes available. There are several methods to carry out such calculations, as we shall discuss in the next lecture (§4), but at present the only such computations carried out on large lattices with fairly light quarks have used the domain wall formulation. The results shown here were obtained as part of the UKQCD Columbia RBRC collaboration, and the parameters used are summarized in Table 2.

Table 2. The different masses at which the domain wall results were gathered, together with the step sizes $\delta\tau$, acceptance rates P_{acc}, and mean plaquette values $\langle P\rangle$. The runs were all performed on $16^3 \times 32$ lattices with $n_5 = 8$ with the DBW2 gauge action at $\beta = 0.72$

Algorithm	m_{ud}	m_s	$\delta\tau$	P_{acc}	$\langle P\rangle$
R	0.04	0.04	0.01	-	0.60812(2)
R	0.02	0.04	0.01	-	0.60829(1)
R	0.02	0.04	0.005	-	0.60817(3)
RHMC	0.04	0.04	0.02	65.5%	0.60779(1)
RHMC	0.02	0.04	0.0185	69.3%	0.60809(1)

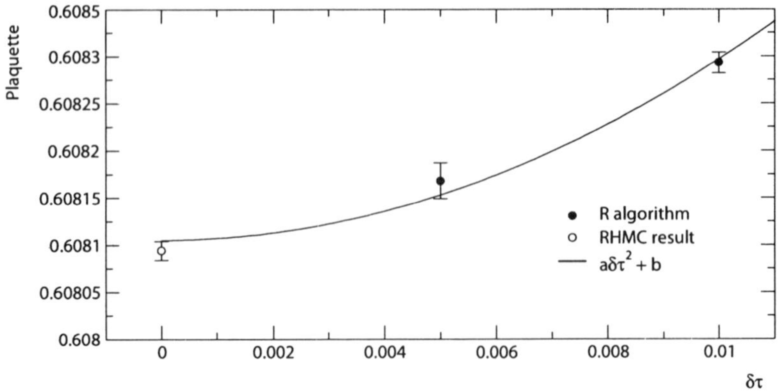

Figure 7. The step size dependence of the plaquette for domain wall fermions with the parameter values given in Table 2. Note that the RHMC data point shown at $\delta\tau = 0$ (because it has no step size errors) is really at $\delta\tau = 0.04$, which is far to the right of the graph. The quadratic extrapolation seems to work reasonably for the R algorithm data points, but the need to carry out this extrapolation significantly increases the estimated error.

The step size dependence of the plaquette is shown in Figure 7. From this we can see that in order to obtain a result consistent within error bars the R algorithm needs an integration step size more than 10 times smaller than that used for RHMC, although the zero step size extrapolation using a quadratic fit to the R algorithm data gives a reasonable value.

We expect that the autocorrelations for RHMC and R algorithms should be similar if carried out with the same trajectory lengths, except of course that the RHMC autocorrelations should be longer by a factor of the inverse acceptance rate. The integrated autocorrelation time for the π propagator from the domain wall test with $m_{ud} = 0.04$ is shown in Figure 8, and the data seems to bear this out.

The conclusion from this study too is that the RHMC algorithm leads to a significant cost reduction in this case too.

3.12. *Multiple Pseudofermions with Multiple Timescales*

We now turn to a way that the RHMC algorithm allows us to implement *UV filtering*, that is to separate the short-distance ultraviolet contributions to the pseudofermion force from the long-distance infrared ones. The semi-empirical observation is that the largest contributions to the gauge field force from the pseudofermions do not come from the partial fractions with

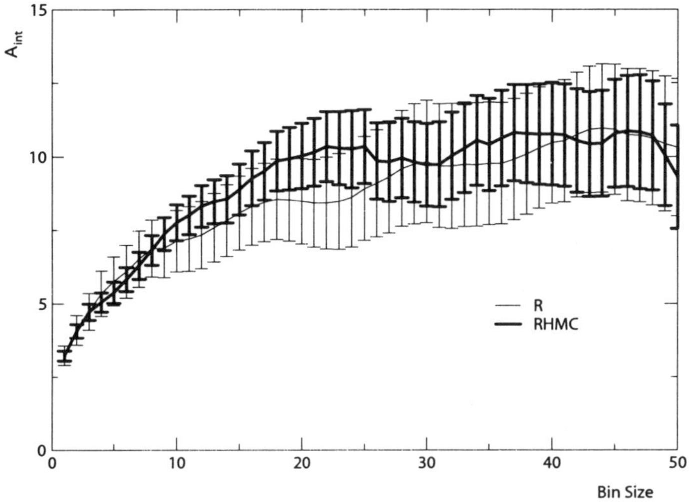

Figure 8. The integrated autocorrelation of time slice 13 of the π propagator for domain wall fermions.

small shifts, but from those with large ones. To see why this may be so, look at the numerators in the partial fraction expansion that we exhibited earlier

$$\frac{1}{\sqrt{x}} \approx 0.3904603901 + \frac{0.0511093775}{x+0.0012779193} + \frac{0.1408286237}{x+0.0286165446} + \frac{0.5964845033}{x+0.4105999719}.$$

Even for this low-order approximation the coefficient of the largest shift is more than an order of magnitude larger than that of the smallest shift. A large shift corresponds (in some sense) to a larger fermion mass, which is why the effects of the large shift partial fractions may be thought of as short distance or ultraviolet modes. Figure 9 shows the situation for a more realistic situation.

There are a couple of ways we can make use of this observation: we can use a coarser timescale for the more expensive smaller shifts, as explained in §2.14, or we can be less sophisticated (but sometimes more cost effective) and significantly increase the CG residual used in computing the force from the small shift partial fractions. We stress that this does not lead to any systematic errors provided we use a time-symmetric initial guess in constructing the Krylov space.

It is worth noting that we are putting the force contribution from different partial fractions for each pseudofermion field on different timescales: the different pseudofermions themselves are treated entirely equally.

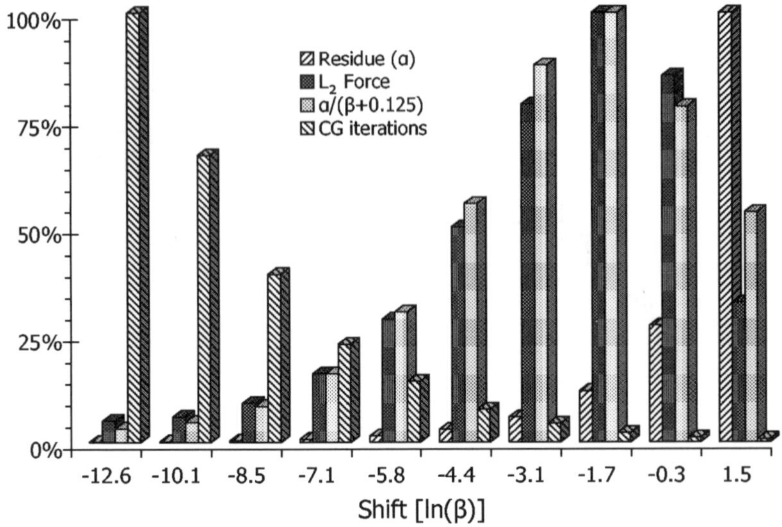

Figure 9. In this graph we plot the L^2 force as a function of the partial fraction $\alpha/(x+\beta)$. The value of the logarithm of the shift, $\ln\beta$, is plotted on the horizontal axis, and it is clear that the values of $\ln\beta$ are fairly unformly distributed. The coefficients of each pole, the residues α, are also plotted, and it can be seen that they increase rapidly with increasing values of the shift β. We also show the measured L_2 norm of the force contribution from each pole, and these also increase with β for small β, but reach a maximum at $\ln\beta = -1.7$. The L_∞ norm of the force contributions behaves similarly. We also show the values of $\alpha/(\beta+0.125)$, which can be taken as a simple model of the behaviour of the force contributions. Finally we plot the number of CG iterations required to reach a fixed residual for each pole: this is proportional to the cost of computing the force, and decreases rapidly with β. It is clear that the largest force contributions come from the poles that require only a cheap CG computation, and the expensive poles only contribute a small amount to the total force acting on the gauge field. We can taken advantage of this either by putting the small shift poles onto a coarser timescale, or by evaluating them much less accurately.

3.13. L_2 *versus* L_∞ *Force Norms*

Figure. 10 is Wilson fermion forces taken from the paper by Urbach *et al.*[11] Here the authors have used Hasenbusch's method (q.v., §3.8.1), and as can be seen there is a clear difference between the L_2 norms of the force contributions and the L_∞ norms. From our discussion of instabilities in §2.13.1 we may expect that the L_∞ is the more appropriate one to detect when a particular force contribution causes the integrator to become unstable. However, while the values of the two norms are very different the relative magnitudes of the force contributions are still fairly similar, so the difference may be moot.

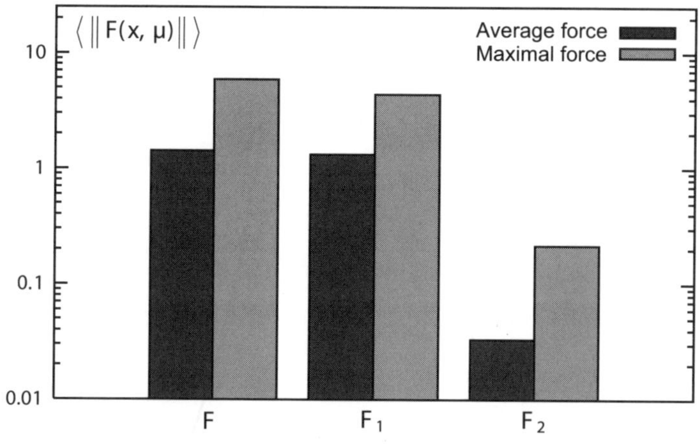

Figure 10. L_2 versus L_∞ force norms.

3.14. *Berlin Wall for Wilson fermions*

Figure 11 shows the 'Berlin wall' for Wilson fermions, and is also taken from the paper by Urbach *et al.*,[11]. The estimated cost of generating 1,000 independent configurations on a $24^3 \times 40$ lattice is plotted in units of Tflop years (3×10^{19} floating point operations) as a function of m_π/m_ρ. The solid line is the result for a conventional two-flavour HMC simulation with Wilson fermions performed by CP-PACS and JLQCD; the squares are results of work of Urbach *et al.*, where they introduced two or three pseudofermion fields using Hasenbusch's method and used different integrator step sizes for the various force contributions such that the force times step size $\delta\tau$ is roughly constant. The dashed and dotted lines are the extrapolations of these latter results using ansatze of $(m_\pi/m_\rho)^{-4}$ and $(m_\pi/m_\rho)^{-6}$ respectively. The triangles are the corresponding results for Staggered fermions, and the arrow indicates the physical pion to rho meson mass ratio.

They show that the performance of Hasenbusch trick and multiple timescale is comparable to Lüscher's SAP algorithm, which from our present point of view probably speeds up dynamical Wilson fermion computations because it too splits the pseudofermion force into three components.

3.15. *Conclusions (RHMC)*

In summary, the advantages of RHMC are that it

- is exact, there are no step size errors: hence no step size extrapola-

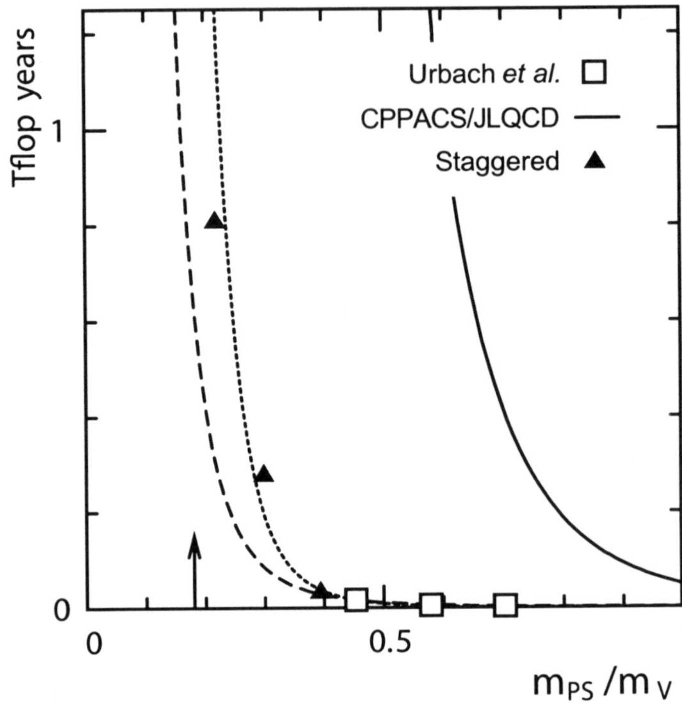

Figure 11. The Berlin wall for Wilson fermions moves to much lighter quarks when multiple pseudofermions are introduced.

tions are required;

- is significantly cheaper than the R algorithm;
- allows easy implementation of Hasenbusch-like tricks using multiple pseudofermion fields;
- is amenable to futher improvements, such as using multiple timescales for different terms in the partial fraction expansion.

So far RHMC has no obvious disadvantages.

4. 5D Algorithms for Chiral Lattice Fermions

This final lecture section is devoted to five-dimensional formulations of chiral fermions. We begin by introducing on-shell lattice chiral fermions, following a logical rather than historical approach.

4.1. *On-shell chiral symmetry*

We shall write the various forms of the Dirac operator as $\not{D} = D_\mu \cdot \gamma_\mu$ where the Dirac γ matrices in Euclidean space are necessarily Hermitian. We shall also assume all Dirac operators are γ_5 Hermitian, $\not{D}^\dagger = \gamma_5 \not{D} \gamma_5$. Note that while this is an assumption it does not seem to be very strong one, as almost all lattice Dirac operators satisfy it (except for the twisted-mass formulation).

The key idea is that it is possible to have chiral symmetry on the lattice without doublers if we only insist that the symmetry holds on-shell. Lüscher observed that on-shell chiral symmetry transformations should be of the form[12]

$$\psi \to e^{i\alpha\gamma_5[1-2\zeta a\not{D}]}\psi \quad \text{and} \quad \bar{\psi} \to \bar{\psi} e^{i\alpha[1-2(1-\zeta)a\not{D}]\gamma_5},$$

where we have introduced a free parameter ζ, for which you may choose your favourite value. As always, the choice of chiral transformation properties of a field is independent of the choice of Lorentz transformation properties, and we can assign ψ and $\bar{\psi}$ independent choices of transformation as they are independent fields, despite the notation: the dagger just indicates that $\bar{\psi} = \psi^\dagger$ has the same Lorentz transformation as the Hermitian conjugate of ψ. In order that this be a symmetry the Dirac operator occuring in it must be invariant under the transformation

$$\not{D} \to e^{i\alpha[1-2(1-\zeta)a\not{D}]\gamma_5} \not{D} e^{i\alpha\gamma_5[1-2\zeta a\not{D}]} = \not{D}.$$

As this must hold identically for all values of α the coefficient of α in its Taylor expansion must vanish, namely

$$[1 - 2(1 - \zeta)a\not{D}]\gamma_5\not{D} + \not{D}\gamma_5[1 - 2\zeta a\not{D}] = 0,$$

which is nothing but the Ginsparg–Wilson relation

$$\gamma_5\not{D} + \not{D}\gamma_5 = 2a\not{D}\gamma_5\not{D}. \tag{7}$$

4.2. *Neuberger's Operator*

We can find the unique[s] solution of the Ginsparg–Wilson relation Eq. (7) in the following way. We choose to parametrize the lattice Dirac operator by writing

$$a\not{D} = \tfrac{1}{2}(1 + \gamma_5\hat{\gamma}_5);$$

---

[s] Up to the choice of kernel, and assuming γ_5 hermiticity.

this is no more that the definition of $\hat{\gamma}_5 \equiv \gamma_5(2a\not{D} - 1)$. The γ_5 hermiticity of the Dirac operator immediately implies that $\hat{\gamma}_5$ is a Hermitian operator $\hat{\gamma}_5^\dagger = \hat{\gamma}_5$. If we also require the Dirac operator to satisfy the Ginsparg–Wilson relation, we immediately find that $\hat{\gamma}_5^2 = 1$, so $\hat{\gamma}_5$, like γ_5, is both Hermitian and unitary. $\not{D}$ must also have the correct continuum limit $\not{D} \to Z_N \not{\partial} + O(a)$, so

$$\hat{\gamma}_5 = \gamma_5(2aZ_N\not{\partial} - 1) + O(a^2) = \gamma_5\left(\frac{\not{D}_W}{M} - 1\right) + O(a^2),$$

where we have defined $M \equiv Z_W/2aZ_N$ and $\not{D}_W \to Z_W\not{\partial} + O(a)$. Here $\not{D}_W$ is some lattice Dirac operator, such as the Wilson operator, and Z_N and Z_W are wavefunction renormalizations. The large negative mass M is called the domain wall height in the language of domain wall fermions, and is an unphysical cut-off-scale quantity. Its value is to be chosen to distinguish cleanly between the physical fermions and the doublers, the two getting sent to opposite ends of the Neuberger operator spectrum. Both of these conditions are satisfied if we define

$$\hat{\gamma}_5 = \gamma_5 \frac{\not{D}_W - M}{\sqrt{(\not{D}_W - M)^\dagger(\not{D}_W - M)}} = \operatorname{sgn}\big[\gamma_5(\not{D}_W - M)\big].$$

This is the form given by Neuberger.[13,14]

It is often convenient to break chiral symmetry explicitly by adding a physical fermion mass. The massive Neuberger operator is

$$\not{D}_N(\mu, H) = \tfrac{1}{2}\big[1 + \mu + (1 - \mu)\gamma_5 \operatorname{sgn}(H)\big], \tag{8}$$

where μ is the physical mass and H is the Hermitian Dirac operator kernel including the negative mass, $H = H(-M) = \gamma_5(\not{D} - M)$. As shown in Figure 12, all the eigenvalue of the Neuberger operator lie on the unit circle.

The immediate question is whether $\not{D}_N$ is a local operator. Although it is manifestly not ultralocal, it can be shown to be a local operator if the kernel $\not{D}_W$ has a gap in its spectrum.[15] The Wilson kernel $\not{D}_W$ certainly has a gap if the gauge fields are smooth enough to satisfy the admissibility condition that $\langle P \rangle < 1/30$: this is a sufficient but not a necessary condition. Moreover, it seems reasonable that good approximations to the Neuberger operator will be local if $\not{D}_N$ is and vice versa.

4.3. *5D Chiral Fermions*

In order to measure propagators, or to compute the pseudofermionic force in HMC, we have to find the inverse of the Neuberger operator $\not{D}_N$. If we

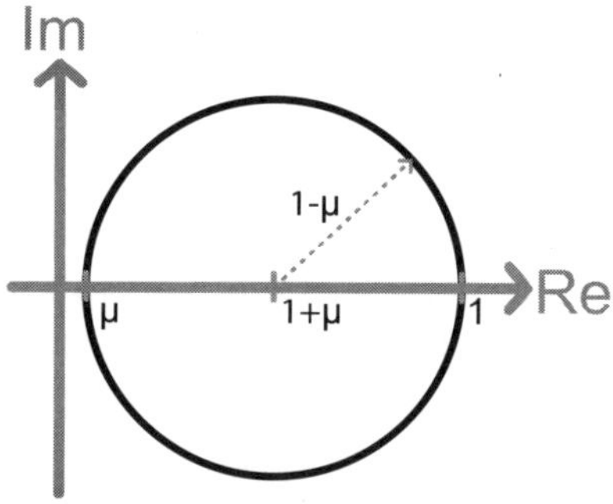

Figure 12. The operator $\gamma_5\hat{\gamma}_5$ is unitary and thus its spectrum lies on the unit circle. This means that the spectrum of the massive Neuberger operator defined by Eq. (8) lies on a circle of radius $\frac{1}{2}(1-\mu)$ whose centre is at $\frac{1}{2}(1+\mu)$, i.e., a circle whose minimum real part is $\mu = O(a)$ and whose maximum real part is one. The physical states all lie near μ and the doublers are near one.

approximate the sgn function by a rational function of some sort then just applying it using our usual partial fraction technique requires the construction of a Krylov space. When we were considering RHMC before we noted (§3.7) that it is just as cheap to compute $A^{-1/n}$ as it is to compute $A^{1/n}$, since the inverse of a rational function is another rational function. Sadly, this does not help us in the present situation, as the Neuberger operator is constructed from two non-commuting operators, $\not{D}$ and γ_5.

It transpires that one can trade-off this non-commuting property with a kind of dimensional reduction from five dimensions to four (this is of course a totally different fifth dimension from the fictitious time we introduced in §2.10). We are thus led to study five-dimensional formulations of chiral fermions where in exchange for having an extra dimension we get away with inverting the Neuberger operator within a single Krylov space.

We shall introduce a class of five-dimensional algorithms that are described by four independent choices:

- *Kernel* — what Dirac kernel is used as the argument of the sgn function in Neuberger's operator (§4.4).
- *Constraint* — whether we use the five-dimensional system solely as a means of computing propagators, or use it to construct a five-dimensional functional integral formulation (§4.6).
- *Approximation* — what rational approximation of the sgn function is used (§4.7).
- *Representation* — how we map the rational approximation into a

five-dimensional system of linear equations (§4.8).

4.4. *Choice of Kernel*

The simplest kernel is the Wilson kernel with a large negative mass, $H_W \equiv \gamma_5(\not{D}_W - M)$. Another popular kernel which arises naturally in the domain wall formalism is the Shamir kernel $H_T \equiv \gamma_5 \not{D}_T$ where

$$a\not{D}_T = \frac{a_5(\not{D}_W - M)}{2 + a_5(\not{D}_W - M)}.$$

A third kernel which interpolates between the preceding kernels is the *Möbius kernel* $H_M \equiv \gamma_5 \not{D}_M$, where

$$a\not{D}_M \equiv \frac{(b_+ + b_-)(\not{D}_W - M)}{2 + (b_+ - b_-)(\not{D}_W - M)}.$$

Special cases corresponding to some choice of the two parameters b_+ and b_- are shown in Table 3.

Table 3. Special cases of the Möbius kernel.

Kernel	b_+	b_-
Boriçi/Wilson	1	1
Shamir/DWF	a_5	0

4.5. *Schur Complement*

The key to understanding the connection between four and five-dimensional formulations is the *Schur complement* of a matrix. Let us consider the five-dimensional block matrix

$$M_5 = \begin{pmatrix} A & B \\ C & D \end{pmatrix},$$

where the blocks A, B, C and D are themselves matrices. M_5 may be block diagonalized by an LDU factorisation (also known as Gaussian elimination)

$$\begin{aligned}\begin{pmatrix} A & B \\ C & D \end{pmatrix} &= \begin{pmatrix} 1 & 0 \\ CA^{-1} & 1 \end{pmatrix} \begin{pmatrix} A & B \\ 0 & D - CA^{-1}B \end{pmatrix} \\ &= \begin{pmatrix} 1 & 0 \\ CA^{-1} & 1 \end{pmatrix} \begin{pmatrix} A & 0 \\ 0 & D - CA^{-1}B \end{pmatrix} \begin{pmatrix} 1 & A^{-1}B \\ 0 & 1 \end{pmatrix}.\end{aligned}$$

This decomposition can be done by inspection, and is unique. The bottom right block is the Schur complement. In particular, the determinant of M_5 is readily calculated by

$$\det\begin{pmatrix} A & B \\ C & D \end{pmatrix} = \det\big(AD - ACA^{-1}B\big).$$

4.6. *Constraint*

How do we make use of a representation of the Neuberger operator as the Schur complement of some five-dimensional matrix? The first way is to use a five dimensional linear system to compute the required four-dimensional propagator. Consider the five-dimensional system of linear equations

$$D_5\tilde{\Phi} \equiv D_5 \begin{pmatrix} \tilde{\phi}_1 \\ \tilde{\phi}_2 \\ \vdots \\ \tilde{\phi}_{n-2} \\ \tilde{\phi}_{n-1} \\ \psi \end{pmatrix} = \begin{pmatrix} 0 \\ 0 \\ \vdots \\ 0 \\ 0 \\ \chi \end{pmatrix} \equiv \tilde{X},$$

where D_5 is an $n \times n$ five-dimensional matrix, and $\tilde{\Phi}$ and $\tilde{X}$ are five-dimensional vectors constructed from n four-dimensional vectors as indicated. Using the block LDU factorisation (§4.5) our linear system reduces to

$$L^{-1}D_5\tilde{\Phi} = L^{-1}LDU\tilde{\Phi} = D\Phi = L^{-1}\tilde{X} = \tilde{X},$$

where

$$\Phi \equiv U\tilde{\Phi} = \begin{pmatrix} \phi_1 \\ \phi_2 \\ \vdots \\ \phi_{n-2} \\ \phi_{n-1} \\ \psi \end{pmatrix}$$

is the five-dimensional column matrix analogous to $\tilde{\Phi}$ but with the unphysical constraint fields $\tilde{\phi}$ replaced with the equally unphysical ϕ, and with the same bottom component ψ. Note that the inverse of a triangular matrix is of the same shape as the original matrix, so L^{-1} is a lower triangular matrix As D is block diagonal we have thus reduced the five-dimensional system to a family of uncoupled four-dimensional ones, the bottommost of which involves the Schur complement, which is just the Neuberger operator, $D_{n,n}\psi = \not{D}_N\psi = \chi$. Therefore by solving the five-dimensional system $D_5\tilde{\Phi} = \tilde{X}$ the bottom component ψ of $\tilde{\Phi}$ is the solution of the four-dimensional system we are interested in. Moreover, we shall see that we can construct D_5 such that it is linear in the kernel $\not{D}_W$ and has as its Schur complement any rational approximation to the Neuberger operator we desire.

Alternatively, we may introduce a five-dimensional pseudofermion field $\Phi = \left(\phi_1\ \phi_2\ \cdots\ \phi_{n-1}\ \psi\right)^T$ and write the five-dimensional pseudofermion functional integral as

$$\int d\Phi^\dagger d\Phi\, e^{-\Phi^\dagger D_5^{-1}\Phi} \propto \det D_5 = \det LDU = \det D = \prod_{j=1}^{n} \det D_{j,j}.$$

The fermion determinant thus obtained is the product of the determinants of all the diagonal blocks $D_{j,j}$, wherease we only want the determinant of the Schur complement, $D_{n,n} \approx \not{D}_N$. We therefore also introduce $n-1$ Pauli–Villars fields[t] to cancel them,

$$\prod_{j=1}^{n-1} \int d\xi_j^\dagger d\xi_j\, e^{-\xi_j^\dagger D_{j,j}\xi_j} \propto \prod_{j=1}^{n-1} \det D_{j,j}^{-1}.$$

4.7. *Approximation*

4.7.1. *Hyperbolic Tangent*

The simplest approximation to the sgn function is $\tanh\bigl(n \tanh^{-1}(x)\bigr)$. In some ways this is quite similar to the situation for Chebyshev polynomials (§3.3): there we had an apparently transcendental function that was in fact a polynomial, here we have an apparently transcendental function that is in fact a rational function. We will write the relevant formulæ for even values of n; there is no problem with using odd values of n, but the formulæ are just a little messier if we try to be too general. We first write note that

$$x \equiv \tanh z = \frac{\sinh z}{\cosh z} = \frac{e^z - e^{-z}}{e^z + e^{-z}} = \frac{1 - e^{-2z}}{1 + e^{-2z}},$$

[t]Perhaps they would be better called pseudo-pseudofermion fields, as they are boson fields with a local kernel introduced to cancel the unwanted extra pseudofermion determinants.

solving this equation gives

$$e^{-2z} = \frac{1-x}{1+x},$$

so we may write

$$\varepsilon_{n-1,n}(x) = \tanh(n \tanh^{-1} x) = \frac{1-e^{-2nz}}{1+e^{-2nz}} = \frac{1-\left(\frac{1-x}{1+x}\right)^n}{1+\left(\frac{1-x}{1+x}\right)^n}. \tag{9}$$

It should be noted that the leading term in the numerator is x^{n-1} and not x^n, as that term's coefficient vanishes for n even; this should not be surprising, as ε is an odd function of x. The denominator vanishes when

$$\left(\frac{1-x}{1+x}\right)^n = -1 = e^{(2k+1)i\pi} \quad \text{with } k \in \mathbb{Z}_n \quad \Rightarrow \quad \frac{1-x}{1+x} = e^{(2k+1)i\pi/n},$$

or

$$x^2 + \left(\tan\frac{(k+\frac{1}{2})\pi}{n}\right)^2 = 0.$$

Using a similar argument to find the roots of the numerator we obtain the factored form

$$\varepsilon_{n-1,n}(x) = xn\frac{\prod_{k=1}^{\frac{n}{2}-1}\left[x^2+\left(\tan\frac{k\pi}{n}\right)^2\right]}{\prod_{k=0}^{\frac{n}{2}-1}\left[x^2+\left(\tan\frac{(k+\frac{1}{2})\pi}{n}\right)^2\right]}.$$

If we define $\xi \equiv e^{-2z}$ then

$$\varepsilon_{n-1,n}(x) + 1 = \frac{1-\xi^n}{1+\xi^n} + 1 = \frac{2}{1+\xi^n} = \frac{2}{\prod_{k=0}^{n-1}(\xi - \xi_k)} = \sum_{k=0}^{n-1}\frac{\mu_k}{\xi - \xi_k},$$

where $\xi_k \equiv \zeta\omega^k$ satisfies $\xi_k^n = -1$ with $\zeta \equiv e^{\pi i/n}$ and $\omega \equiv \zeta^2$ being a primitive n^{th} root of unity, and

$$\mu_k \equiv \frac{2}{\prod_{\substack{j=0\\ j\neq k}}^{n-1}(\xi_k - \xi_j)} = \frac{2}{\xi_k^{n-1}\prod_{\ell=1}^{n-1}(1-\omega^\ell)} = -\frac{2\xi_k}{n}.$$

The last equality follows from the fact that $\prod_{\ell=1}^{n-1}(x-\omega_\ell)$ is a cyclotomic polynomial,

$$\frac{x^n-1}{x-1}=\sum_{k=0}^{n-1}x^k=\prod_{\ell=1}^{n-1}(x-\omega^\ell)\Rightarrow\prod_{\ell=1}^{n-1}(1-\omega^\ell)=n.$$

We thus have

$$\varepsilon_{n-1,n}(x)+1=-\frac{2}{n}\sum_{k=0}^{n-1}\frac{\xi_k}{\xi-\xi_k}=-\frac{2}{n}\sum_{k=0}^{\frac{n}{2}-1}\left\{\frac{\xi_k}{\xi-\xi_k}+\frac{\xi_{n-1-k}}{\xi-\xi_{n-1-k}}\right\}$$

$$=-\frac{2}{n}\sum_{k=0}^{\frac{n}{2}-1}\left\{\frac{\xi_k}{\xi-\xi_k}+\frac{1/\xi_k}{\xi-1/\xi_k}\right\}=-\frac{2}{n}\sum_{k=0}^{\frac{n}{2}-1}\frac{(\xi_k+1/\xi_k)\xi-2}{\xi^2-(\xi_k+1/\xi_k)\xi+1}$$

$$=\frac{2}{n}\sum_{k=0}^{\frac{n}{2}-1}\left[1+\frac{x\left(\sec\frac{(k+\frac{1}{2})\pi}{n}\right)^2}{x^2+\left(\tan\frac{(k+\frac{1}{2})\pi}{n}\right)^2}\right]=1+\frac{2x}{n}\sum_{k=0}^{n-1}\frac{1+\left(\tan\frac{(k+\frac{1}{2})\pi}{n}\right)^2}{x^2+\left(\tan\frac{(k+\frac{1}{2})\pi}{n}\right)^2},$$

which is the partial fraction expansion of $\varepsilon_{n-1,n}(x)$.

4.7.2. *Zolotarev's Formula*

The hyperbolic tangent is simple, but it is by no means the optimal approximation of a given order. The analytic form of the optimal approximation that satisfies Chebyshev's criterion (§3.4) was found by Zolotarev, who was a student of Chebyshev. Again, Zolotarev's formula is most simply written in terms of transcendental functions, this time *elliptic functions*

$$\mathrm{sn}(z/M;\lambda)=\frac{\mathrm{sn}(z;k)}{M}\prod_{m=1}^{\lfloor n/2\rfloor}\frac{1-\dfrac{\mathrm{sn}(z;k)^2}{\mathrm{sn}\left(2iK'm/n;k\right)^2}}{1-\dfrac{\mathrm{sn}(z;k)^2}{\mathrm{sn}\left(2iK'(m-\frac{1}{2})/n;k\right)^2}}.$$

The way to understand this formula is that it expresses the elliptic function $\mathrm{sn}(z/M;\lambda)$ as a function of $\mathrm{sn}(z;k)$, in just the same way that we expressed $\tanh nx$ as a function of $\tanh x$, or $\cos nx$ as a function of $\cos x$ previously. Elliptic functions are doubly periodic analytic functions, and the Jacobi elliptic function $\mathrm{sn}(z;k)$ is defined by[u]

$$z\equiv\int_0^{\mathrm{sn}(z;k)}\frac{dt}{\sqrt{(1-t^2)(1-k^2t^2)}}$$

[u]This is the analogue of defining sin as the inverse of $\sin^{-1}(x)\equiv\int_0^x dt/\sqrt{1-t^2}$. The inverse of sn is called the elliptic integral of the first kind, $E(z;k)$.

so as to have a real period of $4K(k)$ and imaginary period of $2iK(k')$, where $k^2 + k'^2 = 1$ and K is the complete elliptic integral

$$K(k) \equiv \int_0^1 \frac{dt}{\sqrt{(1-t^2)(1-k^2t^2)}}.$$

M and λ are chosen such that the imaginary period of $\text{sn}(z,k)$ is n times that of $\text{sn}(z/M;\lambda)$, $2iMK(\lambda') = 2iK(k')/n$, while the real periods are the same, $4MK(\lambda) = 4K(k)$.

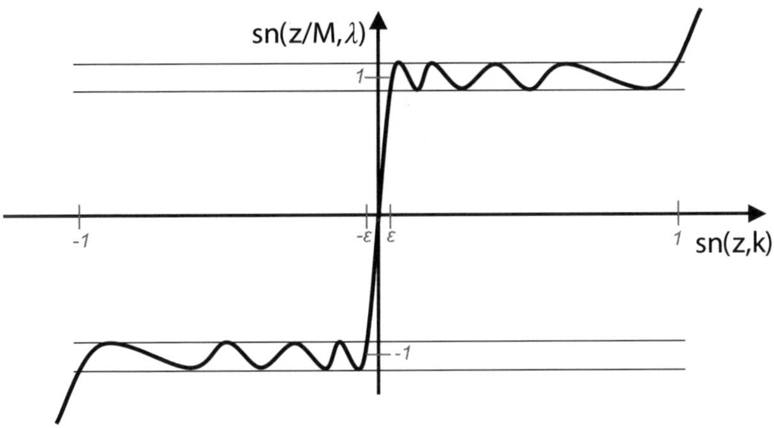

Figure 13. The degree Zolotarev approximation to $\text{sgn}\, x$ over the interval $\varepsilon \le |x| \le 1$. The real period of the elliptic function corresponds to the step in the sgn function, and the imaginary period corresponds to the equal height oscillations about ± 1.

Figure 13 shows $\text{sn}(z/M;\lambda)$ as a function of $\text{sn}(z;k)$. The rôle of the double periodicity is now apparent: the real period corresponds to the step in the sgn function at the origin, and the imaginary period corresponds to the equal height oscillations about ± 1 required by Chebyshev's criterion.

Figure 13 shows a surface plot of the real part of $\text{sn}(z/M;\lambda)$ for $n = 6$; the shading of the surface indicates the imaginary part which is zero along the indicated contour. The entire region of the complex plane shown is the fundamental region for $\text{sn}(z;k)$, and corresponds to six fundamental periods stacked in the imaginary direction for $\text{sn}(z/M;\lambda)$.

4.7.3. *Errors*

We next compare these approximations to the sign function over the interval $10^{-2} \le |x| \le 1$. For both Zolotarev and tanh approximations we take

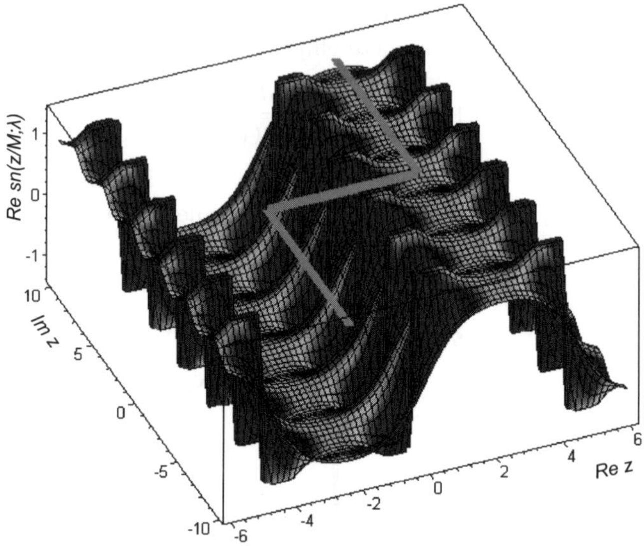

Figure 14. Surface plot showing the real part of $\mathrm{sn}(z/M;\lambda)$ over the fundamental region of $\mathrm{sn}(z;k)$. The imaginary part vanishes along the contour indicated.

rational functions of degree $(7,8)$ and compare them in Figure 15 on a log–linear plot. We can see that tanh is a better approximation near $|x| = 1$, but becomes a much worse approximation when x is small. Indeed, the errors in the tanh approximation are symmetric about $|x| = 1$: this was what motivated the introduction of the Möbius kernel, as a suitable choice of the parameters shift the spectrum of the kernel to better overlap with the region where the error for the tanh approximation is small.

4.8. *Representation*

The formalism developed in §4.6 requires that we have a (five-dimensional) matrix which has the desired approximation (§4.7) as its Schur complement. Constructing a variety of such matrices is the goal of this section.

4.8.1. *Continued Fraction*

We begin with the continued fraction representation.[16,17] Consider the block (five-dimensional) matrix with four-dimensional blocks $A_0, \ldots, A_n$ on the principal diagonal and unit matrices on the adjacent block diagonals,

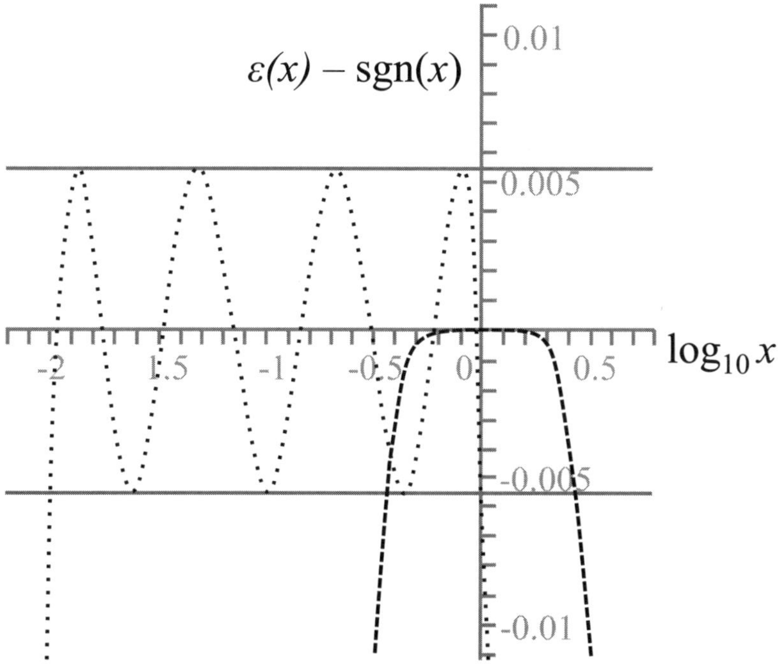

Figure 15. The errors in approximations to the sign function, $\varepsilon(x) - \mathrm{sgn}(x)$, as a function of x. The dashed curve is $\tanh\bigl(8\tanh^{-1} x\bigr) - \mathrm{sgn}\, x$ and the dotted curve is the corresponding error for the optimal (Zolotarev) approximation of the same degree over the interval $10^{-2} \le |x| \le 1$. The tanh approximation does not depend on the choice of interval.

for example with $n = 4$ this is

$$\begin{pmatrix} A_0 & 1 & 0 & 0 \\ 1 & A_1 & 1 & 0 \\ 0 & 1 & A_2 & 1 \\ 0 & 0 & 1 & A_3 \end{pmatrix};$$

it is straightforward to write its block LDU decomposition,

$$\begin{pmatrix} 1 & 0 & 0 & 0 \\ S_0^{-1} & 1 & 0 & 0 \\ 0 & S_1^{-1} & 1 & 0 \\ 0 & 0 & S_2^{-1} & 1 \end{pmatrix} \begin{pmatrix} S_0 & 0 & 0 & 0 \\ 0 & S_1 & 0 & 0 \\ 0 & 0 & S_2 & 0 \\ 0 & 0 & 0 & S_3 \end{pmatrix} \begin{pmatrix} 1 & S_0^{-1} & 0 & 0 \\ 0 & 1 & S_1^{-1} & 0 \\ 0 & 0 & 1 & S_2^{-1} \\ 0 & 0 & 0 & 1 \end{pmatrix},$$

where $S_0 = A_0$ and $S_n + 1/S_{n-1} = A_n$. The Schur complement of the matrix is thus the *continued fraction*

$$S_3 = A_3 - \frac{1}{S_2} = A_3 - \cfrac{1}{A_2 - \cfrac{1}{S_1}} = A_3 - \cfrac{1}{A_2 - \cfrac{1}{A_1 - \cfrac{1}{A_0}}}.$$

We may use this representation to linearize our rational approximations to the sgn function expressed as a continued fraction,

$$\varepsilon_{n-1,n}(H) = \beta_0 H + \cfrac{1}{\beta_1 H + \cfrac{1}{\beta_2 H + \ddots + \cfrac{1}{\beta_n H}}},$$

by setting $A_{n-j} = (-1)^j \beta_j H$ for $j = 0, \ldots, n$. We can improve the condition number of our five-dimensional matrix by making use of the freedom to introduce arbitrary parameters into the continued fraction expansion by multiplying the numerator and denominator at level j by a constant c_{j+1},

$$\varepsilon_{n-1,n}(H) = \beta_0 H + \cfrac{c_1}{c_1\beta_1 H + \cfrac{c_1 c_2}{c_2\beta_2 H + \ddots + \cfrac{c_{n-1}c_n}{c_n\beta_n H}}},$$

which is the Schur complement of the block tridiagonal five-dimensional matrix

$$\begin{pmatrix} c_n\beta_n H & \sqrt{c_{n-1}c_n} & 0 & & & \\ \sqrt{c_{n-1}c_n} & -c_{n-1}\beta_{n-1}H & \ddots & 0 & & \\ 0 & \ddots & \ddots & \ddots & 0 & \\ & 0 & \ddots & c_2\beta_2 H & \sqrt{c_1c_2} & 0 \\ & & 0 & \sqrt{c_1c_2} & -c_1\beta_1 H & \sqrt{c_1} \\ & & & 0 & \sqrt{c_1} & \beta_0 H \end{pmatrix}.$$

4.8.2. *Partial Fraction*

Next we consider the partial fraction representation.[18,19] To this end we consider a matrix of the form

$$\left(\begin{array}{cc|cc|c} \frac{x}{p_1} & 1 & 0 & 0 & 1 \\ 1 & \frac{p_1 x}{q_1^2} & 0 & 0 & 0 \\ \hline 0 & 0 & \frac{x}{p_2} & 1 & 1 \\ 0 & 0 & 1 & \frac{p_2 x}{q_2^2} & 0 \\ \hline -1 & 0 & -1 & 0 & R \end{array}\right),$$

which in general has n 2×2 blocks along the principal diagonal with the corresponding ∓ 1 in the bottom row and rightmost column, and compute its block LDU decomposition:

$$L = \left(\begin{array}{cc|cc|c} 1 & 0 & 0 & 0 & 0 \\ \frac{p_1}{x} & 1 & 0 & 0 & 0 \\ \hline 0 & 0 & 1 & 0 & 0 \\ 0 & 0 & \frac{p_2}{x} & 1 & 0 \\ \hline -\frac{p_1}{x} \quad \frac{q_1 x}{x^2 - q_1^2} & \frac{q_1}{x + q_1} & \frac{p_2}{x} \quad \frac{q_1 x}{x^2 - q_2^2} & -\frac{q_2}{x + q_2} & 1 \end{array}\right),$$

$$D = \left(\begin{array}{cc|cc|c} \frac{x}{p_1} & 0 & 0 & 0 & 0 \\ 0 & \frac{p_1(x^2 - q_1^2)}{x q_1^2} & 0 & 0 & 0 \\ \hline 0 & 0 & \frac{x}{p_2} & 0 & 0 \\ 0 & 0 & 0 & \frac{p_2(x^2 - q_2^2)}{x q_2^2} & 0 \\ \hline 0 & 0 & 0 & 0 & R + \frac{p_1 x}{x^2 - q_1^2} + \frac{p_2 x}{x^2 - q_2^2} \end{array}\right),$$

and

$$U = \left(\begin{array}{cc|cc|c} 1 & \frac{p_1}{x} & 0 & 0 & \frac{p_1}{x} \\ 0 & 1 & 0 & 0 & -\frac{q_1 x}{x^2 - q_1^2} + \frac{q_1}{x + q_1} \\ \hline 0 & 0 & 1 & \frac{p_2}{x} & \frac{p_2}{x} \\ 0 & 0 & 0 & 1 & -\frac{q_2 x}{x^2 - q_2^2} + \frac{q_2}{x + q_2} \\ \hline 0 & 0 & 0 & 0 & 1 \end{array}\right);$$

so its Schur complement is

$$R + \frac{p_1 x}{x^2 - q_1^2} + \frac{p_2 x}{x^2 - q_2^2}.$$

This allows us to represent the partial fraction expansion of our rational function as the Schur complement of a five-dimensional linear system

$$\varepsilon_{n-1,n}(H) = \sum_{j=1}^{n} \frac{p_j H}{H^2 - q_j^2}.$$

4.8.3. *Euclidean Cayley Transform*

Finally we consider a five-dimensional matrix of the following form[20,21,22,23]

$$\begin{pmatrix} 1 & 0 & 0 & -T_1^{-1}C_+ \\ -T_2^{-1} & 1 & 0 & 0 \\ 0 & -T_3^{-1} & 1 & 0 \\ 0 & 0 & -T_4^{-1} & C_- \end{pmatrix},$$

and compute its block LDU decomposition

$$L = \begin{pmatrix} 1 & 0 & 0 & 0 \\ -T_2^{-1} & 1 & 0 & 0 \\ 0 & -T_3^{-1} & 1 & 0 \\ 0 & 0 & -T_4^{-1} & 1 \end{pmatrix},$$

$$D = \begin{pmatrix} 1 & 0 & 0 & 0 \\ 0 & 1 & 0 & 0 \\ 0 & 0 & 1 & 0 \\ 0 & 0 & 0 & C_- - T_4^{-1}T_3^{-1}T_2^{-1}T_1^{-1}C_+ \end{pmatrix},$$

and

$$U = \begin{pmatrix} 1 & 0 & 0 & -T_1^{-1}C_+ \\ 0 & 1 & 0 & -T_2^{-1}T_1^{-1}C_+ \\ 0 & 0 & 1 & -T_3^{-1}T_2^{-1}T_1^{-1}C_+ \\ 0 & 0 & 0 & 1 \end{pmatrix},$$

so its Schur complement, in the general $n \times n$ case, is

$$D_{n,n} = C_- - T_n^{-1}T_{n-1}^{-1} \cdots T_2^{-1}T_1^{-1}C_+.$$

Note that L does not depend on $C_\pm$. If we define $T = T_1 \cdots T_n$ and $C_\pm = \frac{1}{2}(1-\mu) \pm \frac{1}{2}(1+\mu)\gamma_5$, then this Schur complement becomes

$$D_{n,n}(\mu) = -(1+T^{-1})\gamma_5 \times \tfrac{1}{2}\left[(1+\mu) + (1-\mu)\gamma_5 \frac{1-T}{1+T}\right].$$

The rational approximation to the Neuberger operator can thus be written as

$$\begin{aligned} \not{D}_N(H) &\approx \tfrac{1}{2}\left[(1+\mu) + (1-\mu)\gamma_5 \varepsilon(H)\right] \\ &= D_{n,n}(1)^{-1} D_{n,n}(\mu) = \tfrac{1}{2}\left[(1+\mu) + (1-\mu)\gamma_5 \frac{1-T}{1+T}\right], \end{aligned}$$

if we define $T(x)$ to be the *Euclidean Cayley transform*[v] of $\varepsilon(x) \approx \operatorname{sgn}(x)$

$$\varepsilon(x) = \frac{1-T(x)}{1+T(x)};$$

it is trivial to find $T(x)$ because this relation is its own inverse, that is

$$T(x) = \frac{1-\varepsilon(x)}{1+\varepsilon(x)}.$$

For an odd function such as $\varepsilon(x)$ we have

$$\varepsilon(-x) = -\varepsilon(x) \quad \Leftrightarrow \quad T(-x) = \frac{1}{T(x)};$$

this means that if ω_j is a zero of $T(x)$ then $-\omega_j$ must be a pole, so $T(x)$ can be written as a product of factors of the form

$$T_j(x) = \frac{\omega_j - x}{\omega_j + x}.$$

[v]A Cayley transform is the mapping

$$U = \frac{1-iH}{1+iH}, \qquad H = i\frac{U-1}{U+1},$$

between a Hermitian matrix H and a unitary matrix U.

Moreover, if $\varepsilon(0) = 0$ then $T(0) = 1$, so $T = T_1 T_2 \cdots T_n$.

The Euclidean Cayley transform of the hyperbolic tangent approximation (§4.7.1) is immediately apparent from Eq. (9),

$$T(x) = \left(\frac{1-x}{1+x}\right)^n,$$

so all the roots are degenerate, $\omega_j = 1$ $(\forall j)$. For the Zolotarev approximation (§4.7.2) the roots of its Euclidean Cayley transform are the values of x for which the $\varepsilon_{n-1,n}(x) = \pm 1$, which are readily found in terms of elliptic integrals.

We want to solve the four-dimensional linear system

$$\not{D}_N(\mu) \approx \tfrac{1}{2}\left[(1+\mu) + (1-\mu)\gamma_5 \frac{1-T}{1+T}\right]\phi = D_{n,n}(1)^{-1} D_{n,n}(\mu)\phi = \chi,$$

and we can do this by solving the five-dimensional sytem $D_5(\mu)\tilde{\Phi} = D_5(1)\tilde{X}$ with the vectors $\tilde{\Phi}$ and $\tilde{X}$ as in §4.6. The block LDU decomposition of this gives $LD(\mu)U(\mu)\tilde{\Phi} = LD(1)U(1)\tilde{X}$ where L is independent of μ (recall that we noted above that it did not depend upon $C_\pm$, which is only source of μ-dependence). Multiplying on the left by L^{-1}, and substituting $U(\mu)\tilde{\Phi} = \Phi$ and $U(1)\tilde{X} = X$ (with $X_n = \chi$), we obtain the block diagonal equation $D(\mu)\Phi = D(1)X$, whose bottom block is $D_{n,n}(\mu)\phi = D_{n,n}(1)\chi$. Since $D_{n,n}$ is the Schur complement this gives $D_{n,n}(1)^{-1}D_{n,n}(\mu)\phi = \chi$, so as before (q.v., §4.6) we find the bottom four-dimensional block ϕ of the five-dimensional solution vector $\tilde{\Phi}$ is the solution of our four-dimensional problem.

4.8.4. *Relation to Domain Wall Fermions*

The Euclidean Cayley transform representation is a generalization of the usual domain wall fermion formulation. We consider the generalised domain wall with the Möbius kernel, for which one solves the five-dimensional linear system $D_5(\mu)\tilde{\Phi} = D_5(1)\tilde{X}$ where

$$
\begin{aligned}
D_5(\mu) &= \begin{pmatrix}
D_-^{(1)} & D_+^{(1)}P_+ & 0 & 0 & -\mu D_+^{(1)}P_- \\
D_+^{(2)}P_- & D_-^{(2)} & D_+^{(2)}P_+ & 0 & 0 \\
0 & D_+^{(3)}P_- & D_-^{(3)} & D_+^{(3)}P_+ & 0 \\
0 & 0 & D_+^{(4)}P_- & D_-^{(4)} & D_+^{(4)}P_+ \\
-\mu D_+^{(5)}P_+ & 0 & 0 & D_+^{(5)}P_- & D_-^{(5)}
\end{pmatrix} \\
&= \begin{pmatrix}
D_-^{(1)} & 0 & 0 & 0 & -\mu D_+^{(1)} \\
D_+^{(2)} & D_-^{(2)} & 0 & 0 & 0 \\
0 & D_+^{(3)} & D_-^{(3)} & 0 & 0 \\
0 & 0 & D_+^{(4)} & D_-^{(4)} & 0 \\
0 & 0 & 0 & D_+^{(5)} & D_-^{(5)}
\end{pmatrix} P_- \\
&\quad + \begin{pmatrix}
D_-^{(1)} & D_+^{(1)} & 0 & 0 & 0 \\
0 & D_-^{(2)} & D_+^{(2)} & 0 & 0 \\
0 & 0 & D_-^{(3)} & D_+^{(3)} & 0 \\
0 & 0 & 0 & D_-^{(4)} & D_+^{(4)} \\
-\mu D_+^{(5)} & 0 & 0 & 0 & D_-^{(5)}
\end{pmatrix} P_+ . \qquad (10)
\end{aligned}
$$

Here $P_\pm = \frac{1}{2}(1 \pm \gamma_5)$ are the chiral projectors, and

$$D_\pm^{(s)} = \alpha_s \left[b_\pm^{(s)} (\not{D}_W - M) \pm 1 \right],$$

with

$$b_+^{(s)} + b_-^{(s)} = \frac{b_+ + b_-}{\omega_s}, \qquad b_+^{(s)} - b_-^{(s)} = b_+ - b_-,$$

and α_s is a new free parameter that we may use to minimize the condition number of D_5 as it does not effect the Schur complement. Shamir's domain wall is the special case with $b_+ = a_5$, $b_- = 0$, $\omega_s = 1$, and $\alpha_s = 1$. For Chiu's Zolotarev variant we set ω_s to be the roots of the Euclidean Cayley transform of the Zolotarev approximation to the sgn function.

We now continue our derivation from Eq. (10) by cyclically shifting the columns of the right-handed part left, $D_5 \to D_5\mathcal{P}$, where

$$\mathcal{P} \equiv \begin{pmatrix}
1&0&0&0&0 \\
0&1&0&0&0 \\
0&0&1&0&0 \\
0&0&0&1&0 \\
0&0&0&0&1
\end{pmatrix} P_- + \begin{pmatrix}
0&0&0&0&1 \\
1&0&0&0&0 \\
0&1&0&0&0 \\
0&0&1&0&0 \\
0&0&0&1&0
\end{pmatrix} P_+ ,$$

which is a unitary transformation, $\mathcal{P}^\dagger = \mathcal{P}^{-1}$. This gives

$$D_5(\mu)\mathcal{P} = \begin{pmatrix} D_-^{(1)} & 0 & 0 & 0 & -\mu D_+^{(1)} \\ D_+^{(2)} & D_-^{(2)} & 0 & 0 & 0 \\ 0 & D_+^{(3)} & D_-^{(3)} & 0 & 0 \\ 0 & 0 & D_+^{(4)} & D_-^{(4)} & 0 \\ 0 & 0 & 0 & D_+^{(5)} & D_-^{(5)} \end{pmatrix} P_-$$

$$+ \begin{pmatrix} D_+^{(1)} & 0 & 0 & 0 & D_-^{(1)} \\ D_-^{(2)} & D_+^{(2)} & 0 & 0 & 0 \\ 0 & D_-^{(3)} & D_+^{(3)} & 0 & 0 \\ 0 & 0 & D_-^{(4)} & D_+^{(4)} & 0 \\ 0 & 0 & 0 & D_-^{(5)} & -\mu D_+^{(5)} \end{pmatrix} P_+. \qquad (11)$$

If we define $Q_\pm^{(s)} \equiv D_+^{(s)} P_\pm + D_-^{(s)} P_\mp$, then

$$\begin{aligned} D_+^{(s)} P_+ &= Q_+^{(s)} P_+, \qquad & D_+^{(s)} P_- &= Q_-^{(s)} P_-, \\ D_-^{(s)} P_+ &= Q_-^{(s)} P_+, \qquad & D_-^{(s)} P_- &= Q_+^{(s)} P_-; \end{aligned}$$

so Eq. (11) becomes

$$D_5(\mu)\mathcal{P} = \begin{pmatrix} Q_+^{(1)} & 0 & 0 & 0 & Q_-^{(1)} C_+ \\ Q_-^{(2)} & Q_+^{(2)} & 0 & 0 & 0 \\ 0 & Q_-^{(3)} & Q_+^{(3)} & 0 & 0 \\ 0 & 0 & Q_-^{(4)} & Q_+^{(4)} & 0 \\ 0 & 0 & 0 & Q_-^{(5)} & Q_+^{(5)} C_- \end{pmatrix},$$

where $C_\pm \equiv \frac{1}{2}(1-\mu) \pm \frac{1}{2}(1+\mu)\gamma_5 = P_\pm - \mu P_\mp$ as before. Scaling out the matrix

$$\mathcal{Q} \equiv \begin{pmatrix} Q_+^{(1)} & 0 & 0 & 0 & 0 \\ 0 & Q_+^{(2)} & 0 & 0 & 0 \\ 0 & 0 & Q_+^{(3)} & 0 & 0 \\ 0 & 0 & 0 & Q_+^{(4)} & 0 \\ 0 & 0 & 0 & 0 & Q_+^{(5)} \end{pmatrix},$$

the domain wall operator reduces to the form introduced before

$$\mathcal{Q}^{-1} D_5(\mu)\mathcal{P} = \begin{pmatrix} 1 & 0 & 0 & 0 & -T_1^{-1} C_+ \\ -T_2^{-1} & 1 & 0 & 0 & 0 \\ 0 & -T_3^{-1} & 1 & 0 & 0 \\ 0 & 0 & -T_3^{-1} & 1 & 0 \\ 0 & 0 & 0 & -T_4^{-1} & C_- \end{pmatrix},$$

where

$$
\begin{aligned}
T_s &\equiv -Q_+^{(s)-1} Q_-^{(s)} \\
&= -\left[D_+^{(s)} P_+ + D_-^{(s)} P_-\right]^{-1} \left[D_+^{(s)} P_- + D_-^{(s)} P_+\right] \\
&= -\left[\alpha_s\big(b_+^{(s)}(\not{D}_W - M) + 1\big)P_+ + \alpha_s\big(b_-^{(s)}(\not{D}_W - M) - 1\big)P_-\right]^{-1} \\
&\quad \times \left[\alpha_s\big(b_+^{(s)}(\not{D}_W - M) + 1\big)P_- + \alpha_s\big(b_-^{(s)}(\not{D}_W - M) - 1\big)P_+\right] \\
&= -\left[(b_+^{(s)} + b_-^{(s)})(\not{D}_W - M) + \big((b_+^{(s)} - b_-^{(s)})(\not{D}_W - M) + 2\big)\gamma_5\right]^{-1} \\
&\quad \times \left[(b_+^{(s)} + b_-^{(s)})(\not{D}_W - M) - \big((b_+^{(s)} - b_-^{(s)})(\not{D}_W - M) + 2\big)\gamma_5\right] \\
&= -\left[\frac{1}{\omega_s}(b_+ + b_-)(\not{D}_W - M) + \big((b_+ - b_-)(\not{D}_W - M) + 2\big)\gamma_5\right]^{-1} \\
&\quad \times \left[\frac{1}{\omega_s}(b_+ + b_-)(\not{D}_W - M) - \big((b_+ - b_-)(\not{D}_W - M) + 2\big)\gamma_5\right] \\
&= -\gamma_5 \left[\frac{(b_+ + b_-)(\not{D}_W - M)}{2 + (b_+ - b_-)(\not{D}_W - M)}\gamma_5 + \omega_s\right]^{-1} \gamma_5 \\
&\quad \times\gamma_5 \left[\frac{(b_+ + b_-)(\not{D}_W - M)}{2 + (b_+ - b_-)(\not{D}_W - M)}\gamma_5 - \omega_s\right] \gamma_5 \\
&= \frac{\omega_s - H_M}{\omega_s + H_M},
\end{aligned}
$$

H_M being the Möbius kernel introduced in §4.4.

Note that $\not{D}_{\rm DW}^{-1} \equiv \not{D}_N^{-1} - a$ satisfies

$$\{\not{D}_{\rm DW}^{-1}, \gamma_5\} = \{\not{D}_N^{-1}, \gamma_5\} - 2a\gamma_5 = \not{D}_N^{-1}\{\not{D}_N, \gamma_5\}\not{D}_N^{-1} - 2a\gamma_5 = 0,$$

so for valence measurements the 'external' propagator $\not{D}_{\rm DW}^{-1}$ is very convenient, as it satisfies the off-shell anticommutation relation. However, for dynamical calculations this would violate the Nielsen–Ninomaya theorem, and indeed as shown by Pelisetto renormalisation induces unwanted ghost doublers that prevent the off-shell Ward identity from being preserved, so we cannot use $\not{D}_{\rm DW}$ for dynamical ('internal') propagators: we must use D_N in the quantum action instead.

4.9. *Chiral Symmetry Breaking*

In order to measure chiral symmetry breaking, we define Ginsparg–Wilson *defect*

$$\gamma_5\not{D} + \not{D}\gamma_5 - 2a\not{D}\gamma_5\not{D} \equiv \gamma_5\Delta_L.$$

For the approximate Neuberger operator $a\not{D} = \frac{1}{2}[1 + \gamma_5 \varepsilon(H)]$ the defect is $a\Delta_L = \frac{1}{2}[1 - \varepsilon^2(H)]$. If we use a Zolotarev approximation for ε that covers the spectrum of the kernel H with a minimax error of Δ this shows that[w] $\|\Delta_L\| \leq \Delta$, which enables us to bound the matrix elements of Δ_L that must be present in all (on-shell) chiral symmetry violating corrections. There is no such simple bound if the tanh approximation is used.

The *residual mass* may be defined by

$$m_{\text{res}} = \frac{\langle \operatorname{tr} G^\dagger \Delta_L G \rangle}{\langle \operatorname{tr} G^\dagger G \rangle},$$

where G is the quark propagator. For DWF this has been shown by Brower *et al.* to be exactly the usual domain wall residual mass. The residual mass m_{res} is just one moment of Δ_L.

4.10. *Numerical Studies*

Preliminary numerical studies have been carried out comparing the various five-dimensional on-shell chiral fermion formulations introduced in this lecture.[24] These were carried out on an ensemble of 15 gauge configurations from the RBRC two-flavour domain wall fermion dataset. The lattice size was $V = 16^3 \times 32$, and the parameters used to generate the ensemble were $n_s = 12$ (the size of fifth dimension), the DBW2 action with $\beta = 0.8$, a domain wall height of $M = 1.8$, and a fermion mass of $\mu = 0.02$.

The costs of inverting the various five-dimensional operators were compared as a function of the residual mass m_{res}, and in Figure 16 we plot the results for the most promising actions. When comparing different valence kernels the π mass was matched. In order to compare like with like all the Dirac operators were even-odd preconditioned, and no projection of the subspace corresponding to small eigenvalues of the kernel was attempted.

4.11. *Future Work*

There are still a lot of things to investigate about algorithms for dynamical on-shell chiral fermions:

- *Five-dimensional or four-dimensional dynamics.* Should we introduce five-dimensional pseudofermions as in §4.6, or only have four-dimensional pseudofermion fields with a five-dimensional linear equation solver? Five-dimensional pseudofermions would seem

[w] Ignoring terms of $O(\Delta^2)$.

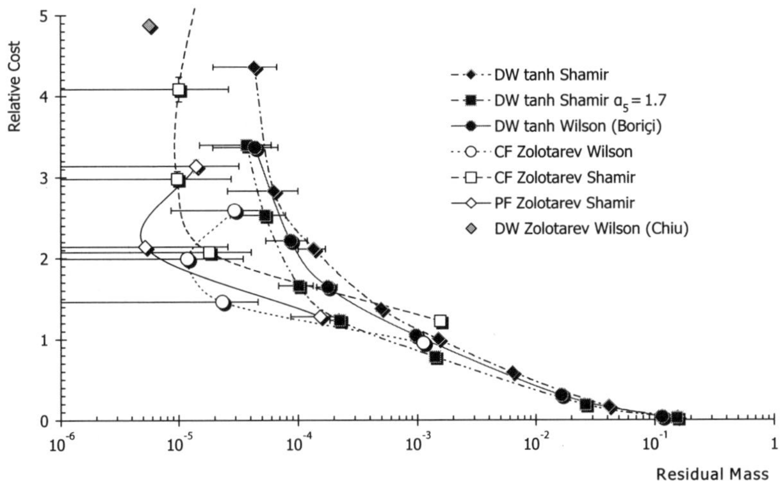

Figure 16. Relative cost in units of Wilson–Dirac matrix applications as a function of the residual mass $m_{\rm res}$ for some five-dimensional actions. Note the Zolotarev domain wall data point was only measured on a single configuration, and therefore is plotted without any error bars. The Zolotarev approximations were over an interval of $0.01 \leq |x| \leq 1$, which does not completely cover the spectrum of the kernel: if the approximation was chosen to cover the spectrum, or the eigenvectors corresponding to eigenvalues $|\lambda| < 0.01$ were projected out and handled exactly, then we would expect $m_{\rm res}$ to be much smaller, probably for little extra cost.

to introduce extra noise into the system for no obvious benefit, so the latter seems to be a more promising choice. The Pauli–Villars fields that need to be introduced to cancel the effects of the unwanted five-dimensional pseudofermions also introduce extra noise, unless they are generated from the same noise fields.

- *Multiple pseudofermion acceleration.* Do multiple pseudofermion fields allow for a significant increase in step size, as discussed in the second lecture? This has been studied for DWF, where the answer is affirmative, but it would be interesting to investigate it for four-dimensional pseudofermion dynamics.
- *Five-dimensional multishift solver.* If we want to apply the RHMC multiple pseudofermion method discussed in the second lecture to four-dimensional pseudofermion dynamics then we would like to be able to use a multishift solver to evaluate all the partial fractions

in the same Krylov space. Unfortunately we have no idea how do this, as a constant shift of the four-dimensional system, which is the Schur complement of our five-dimensional systems, is not a constant shift of the five-dimensional matrix. This makes either a Hasenbusch-like scheme (§3.8.1) or a nested four-dimensional solver look more promising at the moment.

- *Nested Krylov solvers.* Instead of using a five-dimensional Schur complement approach as discuss in this lecture we could use a nested 'inner–outer' solver. For the inner solve we use a multishift technique to apply the partial fractions of a rational representation of the sgn function, for the outer we may use it to implement multiple pseudfermion acceleration (as above), or multiple valence masses if we are carrying out valence measurements. The disadvatage is that we keep building four-dimensional inner Krylov spaces and then discarding them, which seems wasteful. At present, for a single solve (as opposed to a multishift outer solve), the five-dimensional solvers may be 5–10 times faster. There have been some studies of how the residuals for the inner solver should be chosen to ensure the convergence of the outer solver.
- *Tunneling between different topological sectors.* Whatever algorithm we use for dynamical GW fermions it will suffer from severe critical slowing down as $\mu \to 0$. The potential barriers between different topological charge sectors grow as μ gets small, and the sectors decouple for $\mu = 0$. It may be thought that for $\mu = 0$ we would never want to tunnel from the zero topological charge sector to one of a different topology, but we must be careful as it is not manifestly obvious that the topological charge zero sector is connected.
- *Reflection/refraction* algorithms have been introduced to alleviate the problems with topology change. As the problems have so far only been seen on tiny lattices, where the integration step size $\delta\tau$ may be large compared to the width of the barriers, it remains to be seen how necessary or effective these approaches are.

Acknowledgments

I would like to thank Dr. Naoto Tsutsui who prepared the excellent first draft of the write-up of these lectures, and to the organizers of the ILFTN workshop in Nara for arranging such a stimulating and interesting meeting.

References

1. A. D. Kennedy, *Parallel Computing* **25**, 1311 (1999).
2. Arthur Erdélyi, *Asymptotic Expansions*, Dover (1956).
3. J. G. Propp and D. B. Wilson, *Random Structures and Algorithms* **9**, 223 (1996).
4. S. Duane, A. D. Kennedy, B. J. Pendleton and D. Roweth, *Phys. Lett.* **B195**, 216 (1987).
5. R. G. Edwards, I. Horváth and A. D. Kennedy, *Nucl. Phys.* **B484**, 375 (1997).
6. J. C. Sexton and D. H. Weingarten, *Nucl. Phys.* **B380**, 665–678 (1992).
7. A. D. Kennedy, I. Horváth and S. Sint, *Nucl. Phys. Proc. Suppl.* **73**, 834 (1999).
8. A. D. Kennedy, *Nucl. Phys. Proc. Suppl.* **128C**, 107 (2004).
9. M. Hasenbusch and K. Jansen, *Nucl. Phys.* **B659**, 299 (2003).
10. M. A. Clark, P. de Forcrand and A. D. Kennedy, PoS LAT2005, 115 (2005).
11. C. Urbach, K. Jansen, A. Shindler and U. Wenger, *Comput. Phys. Commun.* **174**, 87 (2006).
12. M. Lüscher, *Phys. Lett.* **B428**, 342 (1998).
13. H. Neuberger, *Phys. Lett.* **B417**, 141 (1998).
14. H. Neuberger, *Phys. Lett.* **B427**, 353 (1998).
15. P. Hernandez, K. Jansen and M. Lüscher, *Nucl. Phys.* **B552**, 363 (1999).
16. H. Neuberger, *Phys. Rev.* **D60**, 065006 (1999).
17. A. Boriçi, A. D. Kennedy, B. J. Pendleton and U. Wenger, *Nucl. Phys. Proc. Suppl.* **106**, 757 (2002).
18. H. Neuberger, *Phys. Rev. Lett.* **81**, 4060 (1998).
19. R. G. Edwards, U. M. Heller and R. Narayanan, *Nucl. Phys.* **B540**, 457 (1999).
20. A. Boriçi, *Nucl. Phys. Proc. Suppl.* **83**, 771 (2000).
21. R. G. Edwards and U. M. Heller, *Phys. Rev.* **D63**, 094505 (2001).
22. T. W. Chiu, *Phys. Rev. Lett.* **90**, 071601 (2003).
23. R. C. Brower, H. Neff and K. Orginos, *Nucl. Phys. Proc. Suppl.* **140**, 686 (2005).
24. R. G. Edwards, B. Jóo, A. D. Kennedy, K. Orginos and U. Wenger, PoS LAT2005, 146 (2005).

APPLICATIONS OF CHIRAL PERTURBATION THEORY TO LATTICE QCD

STEPHEN. R. SHARPE
Physics Department,
University of Washington,
Seattle, WA 98195-1560, USA
E-mail: sharpe@phys.washington.edu

These lectures describe the use of effective field theories to extrapolate results from the parameter region where numerical simulations of lattice QCD are possible to the physical parameters (physical quark masses, infinite volume, vanishing lattice spacing, etc.). After a brief introduction and overview, I discuss three topics: 1) Chiral perturbation theory (χPT) in the continuum; 2) The inclusion of discretization effects into χPT, focusing on the application to Wilson and twisted-mass lattice fermions; 3) Extending χPT to describe partially quenched QCD.

1. Overview and Aims

More than 30 years after Wilson introduced lattice QCD[1], and more than 25 years after Creutz's pioneering numerical studies of non-abelian gauge theories[2], we can now simulate lattice QCD, including quarks, with parameters that approach their physical values. This is the result not only of advances in computer power but also of improvements in algorithms and actions.[a] In particular, we can simulate QCD with pion masses of 250 MeV or lower, with the minimum value depending on the choice of fermion action. Such masses should allow a controlled extrapolation to the physical pion masses, one that can give errors at the few percent level[3]. One of the aims of the field is to provide results with this accuracy for many hadronic quantities, allowing both tests of the method and predictions for unmeasured quantities. This goal has begun to be attained[4].

Despite the successes just outlined, it is important to keep in mind the limitations of LQCD (lattice QCD). Simulations are, and will remain for the foreseeable future, limited in scope—one or two particle states in a box

[a]Some of these improvements are reviewed in Tony Kennedy's lectures.

unlikely to exceed $L = 3 - 5$ fm, with lattice spacings unlikely to be smaller than $a \approx 0.05$ fm, and pion masses unlikely to drop below 200 MeV.[b] In order to connect these results to those for physical quark masses in the continuum and infinite volume limits one needs a quantitative theoretical understanding of how to extrapolate.[c] Such an understanding can be provided by chiral perturbation theory generalized to include discretization errors, and is the topic of these lectures.

One way to think of this situation is that LQCD is a powerful tool with several adjustable parameters ("knobs"). While we are able to turn these knobs independently (unlike in the physical world where they are fixed), we cannot turn them to their physical values. Thus we are stuck simulating theories with unphysical values of the parameters, and we need additional theoretical input.

In fact, there are several other knobs (beyond quark masses, a, L and $\alpha_{\rm EM}$) that we can adjust independently. We can use different sea and valence quark masses—giving partially quenched (PQ) theories—or we can go further and use different actions for valence and sea quarks—"mixed action" simulations. An interesting example of the latter is to use valence fermions with good chiral symmetry (Domain-Wall or Overlap) and cheaper sea quarks (staggered or Wilson-like). Both PQ and mixed action theories are "really" unphysical: they not only have unphysical values of the parameters but they are also not unitary. Nevertheless, they are well-defined Euclidean statistical systems, with long-distance correlations, and it is plausible that they can be described by an effective chiral theory. Furthermore, in both cases there are points in parameter space for which the theories are physical, which "anchor" the effective theories. I will discuss this in detail for PQQCD in Sec. 4, and for now only illustrate the situation with Fig. 1. The aim is to use the freedom provided by having extra knobs, which are relatively cheap to turn, in order to improve the accuracy of the extrapolation to the physical point: "physical results from unphysical simulations"[6]. This is an essential feature of the MILC collaboration's work on decay constants and quark masses[3].

A different example of unphysical theories is the use of "rooted" staggered fermions. Each staggered flavor leads to four degenerate "tastes" in the continuum limit, and the standard approach to obtain a single contin-

[b]There are important exceptions, such as the very small lattice spacings used to match QCD with a b quark onto heavy quark effective theory[5], which are possible because L can also be reduced.

[c]For percent accuracy one must also account for the effects of electromagnetism.

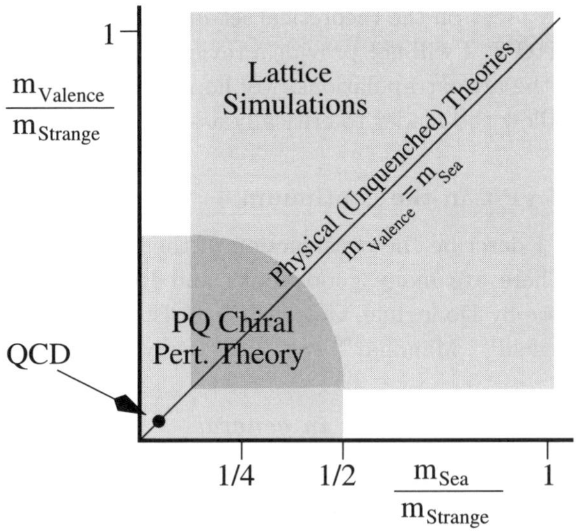

Figure 1. PQQCD: using $m_{\rm val} \neq m_{\rm sea}$ and PQχPT to better extrapolate (from the overlap of the shaded regions) to the physical theory.

uum fermion per flavor is to take the fourth root of the fermion determinant. As the taste symmetry is broken for $a \neq 0$, this rooting leads, however, to a *non-local* single-flavor fermion action on the lattice[7]. The implications of a non-locality that formally vanishes as $a \to 0$ are controversial—does one remain in the universality class of QCD? I will not discuss this issue here, but only note that the theory at $a \neq 0$ is undoubtedly unphysical, and the chiral-continuum extrapolation can only be done if one has an effective theory which describes the unphysical features. This is provided by "(rooted) staggered χPT"[8,9,10]. Whatever the outcome of the rooting controversy, this is another example of using effective field theory to obtain physical results from unphysical simulations.

Due to limitations of time and space, I will discuss only a subset of applications of χPT to LQCD in these lectures. I begin with a brief introduction to χPT in the continuum, with an emphasis on lessons for the lattice. I follow that with the example of incorporating discretization errors into χPT for twisted-mass fermions, which includes Wilson and improved Wilson fermions as a subset. In this case the theory is physical, but gives a nice example of the power of adding an extra knob (the twist angle) and of the utility of χPT. Finally, I discuss PQχPT, i.e. chiral perturbation theory for PQQCD.

I will mainly focus on the theoretical set-up and on general issues of the applicability of χPT. I will not provide a review of the status and accuracy of the state-of-the-art extrapolations. My hope is that this introduction to the tools will allow the reader to critically assess current work.

2. Review of χPT in the continuum

In this section I describe the construction of the chiral Lagrangian in the continuum. There are many good books and lectures on this topic. I have found those by Donoghue, Golowich & Holstein[11], Ecker[12], Georgi[13], Kaplan[14], Kronfeld[15], Manohar[16] and Pich[17] very useful.

2.1. *Effective Field Theories in general*

In these lectures I consider two examples of effective field theories (EFTs): χPT as an EFT for QCD, and Symanzik's effective continuum theory for lattice QCD (the latter to be discussed in Sec. 3.4). Thus it is useful to begin with a discussion of EFTs in general. If you are unfamiliar with the subject then some of this section may be hard to follow in detail, but my aim is to begin with a broad-brush sketch, which will be filled in later.

The generic situation is that we have an underlying theory in which there is a separation of scales. In the theories of interest we have:

$$\chi\text{PT:}\ \ p_\pi \sim m_\pi \ll m_\rho, m_N\,; \tag{1}$$

$$\text{Symanzik:}\ \ p_{\text{quark}}, p_{\text{gluon}} \sim \Lambda_{\text{QCD}} \ll \pi/a\,. \tag{2}$$

Note that in the former case there is a separation of masses, with the "pions" (by which I mean the light pseudoscalars: π, K and η) being lighter than all other hadrons, while in Symanzik's theory we choose to consider momenta much smaller than the lattice cut-off. In both cases there is a good reason to split off the low-scale physics. For χPT it is because the pion sector changes most rapidly as we approach the chiral limit (as we will see in detail). For Symanzik's theory we want to understand the impact of lattice spacing errors on the quarks and gluons which dominate the non-perturbative contributions to hadronic quantities.

Crudely speaking, we now introduce a momentum cut-off Λ lying between the two scales, and "integrate out" the high-momentum degrees of freedom. This process is illustrated schematically in Fig. 2. It leaves only pions with low momenta in χPT, and continuum-like quarks and gluons in the Symanzik theory. These degrees of freedom interact via vertices that are quasi-local, with a physical size Λ^{-1}. This quasi-locality follows because we consider only external momenta satisfying $p \ll \Lambda$, so the high-momentum degrees of freedom are always highly virtual. The vertices are then expanded in powers of p/Λ, yielding local operators with increasing numbers of derivatives. The low-momentum modes themselves can become nearly on shell (or exactly on shell if we continue to Minkowski space), but the resulting analytic structure of correlation functions (leading to cuts in Minkowski space) is maintained in the effective theory.

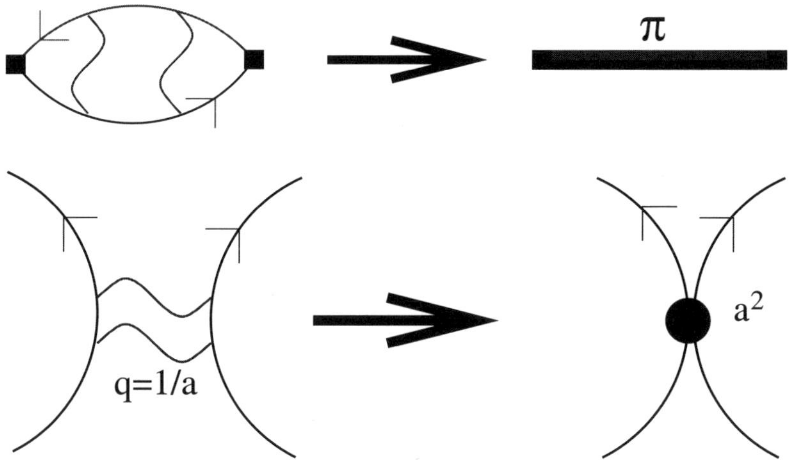

Figure 2. Generation of effective field theories.

This description is impractical to implement in most cases. In particular, we do not know how to integrate out quarks and gluons from QCD analytically to yield a theory of pions, since confinement is a non-perturbative phenomenon. Even in the Symanzik theory, where one might have expected that quarks and gluons with $p \sim \pi/a$ would have been perturbative since $1/a \gtrsim 2$ GeV ($a \lesssim 0.1$ fm), it turns out that accurate results mostly require non-perturbative calculations.[d] The beauty of the EFT method, however, is that we do not actually need to do the integrations. Instead, following Weinberg[18] we can rely on the general properties of EFTs. If the underlying theory is physical, its S-matrix will be unitary, Lorentz-invariant, satisfy cluster decomposition, and transform appropriately under the internal symmetries. These properties must be maintained by the EFT, which is, after

[d] This is found when implementing the improvement program for Wilson fermions [19].

all, designed to reproduce the S-matrix of the low-momentum degrees of freedom. The only known way to do this is with a local, Lorentz-invariant Lagrangian. It should be constructed solely from the low-energy degrees of freedom, and satisfy the same internal symmetries as the underlying theory. All possible terms consistent with these symmetries must be included—this precludes the need to explicitly integrate-out degrees of freedom, at the price of introducing unknown constants.

One notable feature of the resulting $\mathcal{L}_{\rm eff}$ is that it is not renormalizable, and thus valid only over a limited energy range. This is an intrinsic part of the construction: we know that the EFT breaks down when $p \gtrsim \Lambda$. Non-renormalizability does not, however, imply a lack of calculability. As we will see, one can expand quantities in powers of p/Λ, with a finite number of unknown coefficients at each order. The limitations of the method are then (a) the need to introduce unknown coefficients and (b) an unavoidable truncation error. This error, however, decreases as the separation in scales increases (i.e. as $m_\pi \to 0$ or $a \to 0$).

As just described, the justification of EFT is based on properties of the S-matrix, and thus rooted in Minkowski space. While this is fine for the development of continuum χPT (my first topic), the natural objects in lattice simulations (my second topic) are Euclidean (finite-volume) correlation functions of local operators. In particular, the discretization errors are constrained by the symmetries of a Euclidean lattice. Thus an alternative approach to developing and justifying an EFT is needed. This has been provided for the case at hand by Symanik[20], using an extension of renormalization theory. The result (established to all orders in perturbation theory) is that the recipe given above still applies: keep all local terms consistent with the symmetries of the underlying theory (in this case the discrete symmetries of a Euclidean lattice). My third topic, PQQCD, is also strictly limited to Euclidean space, but here neither of the previous justifications apply, and one must make further assumptions.

Because my second and third topics involve Euclidean theories, I have chosen to couch the discussion of the first (χPT) also in Euclidean space. This allows later sections to build on the earlier notation. In fact, it is perfectly legitimate, having determined the Minkowski-invariant local effective Lagrangian, to rotate this to Euclidean space. The result will be the most general Euclidean-invariant local Lagrangian (consistent with the other symmetries, which are unaffected by the rotation). This Lagrangian will reside in the functional integral which generates the Euclidean correlation functions of the theory.

2.2. *Chiral symmetry in QCD and its breaking*

Without further ado, let me turn to the first concrete example, χPT. The fermionic part of Euclidean Lagrangian for QCD is given by

$$\mathcal{L}_{QCD} = \overline{Q}_L \not{D} Q_L + \overline{Q}_R \not{D} Q_R + \overline{Q}_L M Q_R + \overline{Q}_R M^\dagger Q_L \,. \tag{3}$$

where I have included only the $N = 3$ light quarks, $Q^{tr} = (u, d, s)$, $\overline{Q} = (\overline{u}, \overline{d}, \overline{s})$. I will also consider the $N = 2$ theory without the strange quark. Left- and right-handed fields are defined with projectors $P_\pm = (1 \pm \gamma_5)/2$, and are $Q_{L,R} = P_\mp Q_{L,R}$ and $\overline{Q}_{L,R} = \overline{Q}_{L,R} P_\pm$. There is no problem with $\overline{Q}_L$ and Q_L being defined with the different projectors, since $\overline{Q}$ and Q are independent fields. The key fact is that, in the massless limit, left- and right-handed quarks can be rotated independently, so the Lagrangian has a $G = SU(3)_L \times SU(3)_R$ chiral symmetry under which

$$Q_{L,R} \to U_{L,R} Q_{L,R} \,, \quad \overline{Q}_{L,R} \to \overline{Q}_{L,R} U_{L,R}^\dagger : \quad U_{L,R} \in SU(3)_{L,R} \,. \tag{4}$$

There is also the overall vector $U(1)$ symmetry, which counts quark number, while the apparent axial $U(1)$ symmetry is broken by the anomaly.

Quark masses enter through the mass matrix M, which is conventionally taken to be $M = \text{diag}(m_u, m_d, m_s) = M^\dagger$. These masses break the chiral symmetry: any non-zero values violate the axial symmetries (those with $U_L = U_R^\dagger$), while the vector symmetries ($U_L = U_R$) are broken unless the masses are degenerate. We can, however, formally retain the chiral symmetry of $\mathcal{L}_{QCD}$ by treating M as a complex "spurion field" transforming as $M \to U_L M U_R^\dagger$ and $M^\dagger \to U_R M^\dagger U_L$. This is a convenient trick for keeping track of the symmetry-breaking caused by the mass term.

Since chiral symmetry is key to all that follows, and quark masses break this symmetry, we must require that M be small. What does small mean? One criterion is that M should be small compared to the QCD scale, $m_q \ll \Lambda_{QCD} \sim 300$ MeV. A more precise criterion will arise from χPT: $m_{\pi,K,\eta} \ll \Lambda_\chi \equiv 4\pi f_\pi \approx 1200$ MeV. It follows that in physical QCD, with $(m_u + m_d)/2 \approx 4$ MeV, $SU(2)_L \times SU(2)_R$ is a very good approximate symmetry, while $SU(3)_L \times SU(3)_R$ is more badly broken since $m_s \approx 100$ MeV and $m_{K,\eta} \approx \Lambda_\chi/2$. This brings up an important question for lattice applications of χPT: can approximate chiral symmetry can be used to determine the strange quark mass dependence when $m_s^{\text{lat}} \approx m_s$? If not, then we can only use chiral symmetry to guide extrapolations in m_u and m_d.

Chiral perturbation theory is an expansion about the chiral limit, $M = 0$, so I first discuss massless QCD. It is expected in this theory that the exact chiral symmetry is spontaneously broken by the vacuum. This is based on an accumulation of evidence about QCD itself: the lightness of the π, K and η are consistent with their being *pseudo*-Nambu-Goldstone bosons (PGBs) of the broken *approximate* chiral symmetry of real QCD; the absence of (approximate) parity-doubling in the hadron spectrum, e.g. $m_N(P=+) \neq m_N(P=-)$, as would be required by an unbroken (approximate) chiral symmetry; and the accumulated successes of χPT, especially in the $SU(2)$ sector. It is also expected that the order parameter for chiral symmetry breaking is the condensate:[e]

$$\langle \overline{q} q\rangle = \langle (\overline{q}_L q_R + \overline{q}_R q_L)\rangle \sim \Lambda^3_{\rm QCD} \neq 0\,, \quad q = u, d, s\,. \tag{5}$$

The vector symmetry is not expected to be spontaneously broken, based on the presence of approximate $SU(3)_V$ multiplets in the hadron spectrum, and on theoretical considerations[23]. This implies that the condensates are equal in the massless theory, $\langle \overline{u}u\rangle = \langle \overline{d}d\rangle = \langle \overline{s}s\rangle$.

The form of the condensate in Eq. (5) is a convention. There is in fact a manifold of equivalent vacua related by chiral transformations, and parameterized by the orientation of the condensate in flavor space:

$$\Omega_{ij} = \langle Q_{L,i,\alpha,c}\overline{Q}_{R,j,\alpha,c}\rangle \xrightarrow{G} U_L\,\Omega\,U_R^\dagger\,. \tag{6}$$

Here I have shown the summed Dirac and color indices explicitly (α and c, respectively), since Q and $\overline{Q}$ are in the opposite order from usual. This order makes the flavor matrix structure (indices i and j) transparent, and is particularly useful when discussing PQχPT below. The assumption of unbroken vector symmetry implies that $\Omega_{ij} = \omega\,\delta_{ij}$ must lie in the manifold, and so the general point is $\omega U_L U_R^\dagger$. In this language, chiral symmetry breaking is equivalent to ω being non-zero, for then the vacuum is left invariant only by a subgroup H of G:

$$\underbrace{SU(3)_L \times SU(3)_R}_{G} \longrightarrow \underbrace{SU(3)}_{H}\,, \tag{7}$$

The nature of H is simplest using the conventional vacuum orientation, $\Omega_{ij} = \omega\,\delta_{ij}$, with $\omega = -\langle \overline{q}q\rangle$. Then $H = SU(3)_V$ (with $U_L = U_R$), while the axial transformations with $U_L = U_R^\dagger$ are broken ($\Omega = \omega \to \omega U_L^2$).

Goldstone's theorem then implies that there are 8 massless Nambu-Goldstone bosons (NGB—labeled π^b, and corresponding to the π, K, and

---

[e]There has been some controversy over whether the condensate is the dominant order parameter[21], but this standard picture is now strongly favored[22], and is supported by lattice calculations of $\langle \overline{q}q\rangle$, so I will accept it here.

η), each coupled to one of the eight broken axial generators. For the conventional vacuum orientation, one has

$$\langle \pi^b(p)|\overline{Q}\gamma_\mu\gamma_5 T^a Q(0)|0\rangle = -if_\pi p_\mu \delta^{ba}\,, \tag{8}$$

with T^a being $SU(3)$ generators. The masslessness of the NGB follows from the fact that rotations of the condensate with four-momentum p cost zero energy as $p \to 0$, in which limit they become global rotations.

We now have the ingredients with which to construct an EFT: a separation of scales ($m_{\rm GB} = 0$ compared to the scale of other hadron masses—m_ρ, m'_η, $m_{\rm proton}$, etc., with $m_{\rm had} \approx 1$ GeV), and a knowledge of the symmetries of the underlying theory. The EFT will contain only NGB as dynamical degrees of freedom, and should be valid as long as $p_{\rm GB} \ll m_{\rm had}$. The spurion field M should also be added to include the effects of quark masses.[f]

Representing NGB fields is probably the most conceptually non-trivial step of the construction, because (a) the underlying theory is written in terms of quarks rather than mesons (one is not just "thinning" degrees of freedom, one is also changing basis) and (b) the choice is not unique[24,25]. Because of point (b), the strategy is simply to find any representation which works—it turns out that by field redefinitions one can then switch to other choices as desired.

Since there are no precise rules to follow, it is useful to proceed by analogy. To this end, I recall the canonical example of spontaneous symmetry breaking: a complex scalar field with a "Mexican hat" potential

$$V = -\mu^2\phi^\dagger\phi + \lambda(\phi^\dagger\phi)^2/2\,. \tag{9}$$

There is a $G = U(1)$ phase symmetry, $\phi \to e^{i\alpha}\phi$, which is spontaneously broken if $\mu^2 > 0$, for then there is a non-zero vacuum expectation value (VEV) $|\langle\phi\rangle| = v = \sqrt{\mu^2/\lambda}$. The vacuum manifold consists of the phase of $\langle\phi\rangle$, and is thus $U(1)$. The symmetry breaking is $G \longrightarrow H = 1$, implying a single NGB corresponding to phase rotations. The EFT for this massless mode is analogous to that we wish to write down for the pions in QCD.

The advantage of this theory, compared to QCD, is that we can directly construct the EFT by integrating out heavy fields, as long as λ is small enough that we can use perturbation theory. Having done so, we can see how the EFT could be obtained using the symmetries alone, and use this to guide the construction for QCD.

---

[f] The EFT can also contain *static* sources, representing heavy particles with $m \gtrsim m_{\rm had}$, off which the NGBs can scatter. These can represent the interesting cases of vector mesons, baryons, or heavy-light hadrons.

I decompose the field in a way which differs from that used, say, in studying the Higgs: $\phi(x) = ve^{\rho(x)}e^{i\theta(x)}$ rather than $\phi(x) = v + h(x)$. This choice picks out the NGB degree of freedom, θ, explicitly, and allows the phase symmetry to act linearly: $e^{i\theta[x]} \to e^{i\alpha}e^{i\theta[x]}$. If we integrate out the heavy radial degree of freedom, ρ, we obtain an effective Lagrangian in terms of $e^{i\theta}$. But we do not need to do any work to determine the general form of $\mathcal{L}_{\rm eff}$—we need only require locality, reality, Euclidean and $U(1)$ invariance. The result is

$$\mathcal{L}_{\rm eff} = c_2\partial_\mu(e^{i\theta})\partial_\mu(e^{-i\theta}) + c_4\partial_\mu(e^{i\theta})\partial_\mu(e^{-i\theta})\partial_\nu(e^{i\theta})\partial_\nu(e^{-i\theta}) + \dots, \quad (10)$$

where c_i are unknown constants.[g] Terms without derivatives on every factor of $e^{\pm i\theta}$ can be brought into the form shown (up to total derivatives) using $e^{i\theta}e^{-i\theta} = 1$ and the abelian nature of the group. The result is a massless NGB having interactions proportional to p^4. It is an interesting exercise to check the latter result in perturbation theory using the conventional expansion in terms of $h(x)$—the p^4 arises from cancellations between non-derivative interactions.

We learn two things from the $U(1)$ example. First, to use the exponential of the "pion" fields, since it transforms linearly under G, and simplifies the implementation of the symmetries. Second, that a "fixed radius" form ($|e^{i\theta}| = 1$ above) automatically includes only the NGB, excludes heavy degrees of freedom, and enforces the spontaneous breakdown of the symmetry (i.e. $U(1)$ is broken for any value of $\langle\theta\rangle$).

The QCD analog of $\langle\phi\rangle$ is the condensate Ω_{ij} of Eq. (6)—both map out the corresponding vacuum manifolds. The analog of fixed length angular fluctuations is obtained by promoting Ω/ω to a dynamical field $\Sigma_{ij}(x)$, corresponding roughly to fluctuations in the condensate. Just as Ω/ω is in $SU(3)$, so is Σ, and the chiral transformation properties are the same:

$$\Sigma(x) \in SU(3): \qquad \Sigma(x) \xrightarrow{G} U_L\Sigma(x)U_R^\dagger\,. \quad (11)$$

Any VEV of Σ breaks G to $H = SU(3)$, leading to the desired number of NGBs. The fixed radius of Σ (i.e. $\Sigma\Sigma^\dagger = \Sigma^\dagger\Sigma = 1$) implies that the *only* degrees of freedom in Σ are the NGBs. For example, if $\langle\Sigma\rangle = 1$ we can expand as

$$\Sigma(x) = \exp\left(2i\Pi(x)/f\right) = \exp\left(2i\pi^a(x)T^a/f\right), \qquad a = 1, 8, \quad (12)$$

[g]Since $U(1)$ is abelian, $\mathcal{L}_{\rm eff}$ can be simplified to $c_2(\partial_\mu\theta)^2 + c_4[(\partial_\mu\theta)^2]^2$. I do not pursue this as similar manipulations fail for the non-abelian chiral groups relevant for QCD.

in terms of the eight "pion" fields and a constant f to balance dimensions. Note that although Σ transforms linearly, this is not the case for the pion fields (e.g. $\delta\pi_a$ contains terms with any odd powers of π_b). Thus constructing $\mathcal{L}_{\rm eff}$ in terms of the pion field directly would be very difficult.

2.3. *Constructing the pionic effective Lagrangian*[26]

2.3.1. *Building blocks for $\mathcal{L}_{\rm eff}$*

We are now in business. The ingredients are Σ and $\Sigma^\dagger$, as well as the spurions M and $M^\dagger$. I recall their transformation properties under the chiral symmetry group $G = SU(3)_L \times SU(3)_R$:

$$\Sigma \to U_L \Sigma U_R^\dagger , \quad \Sigma^\dagger \to U_R \Sigma^\dagger U_L^\dagger , \quad M \to U_L M U_R^\dagger , \quad M^\dagger \to U_R M^\dagger U_R^\dagger . \tag{13}$$

It is useful to construct objects which transform solely under the left-handed (LH) or right-handed (RH) sub-groups (and which I call respectively LH and RH building blocks), since they simplify enumeration of operators:[h]

$$\begin{aligned}
&\text{LH: } L_\mu = \Sigma\partial_\mu\Sigma^\dagger = -\partial_\mu\Sigma\Sigma^\dagger = -L_\mu^\dagger \longrightarrow U_L L_\mu U_L^\dagger \\
&\text{LH: } M\Sigma^\dagger \longrightarrow U_L(M\Sigma^\dagger)U_L^\dagger , \qquad \Sigma M^\dagger \longrightarrow U_L(\Sigma M^\dagger)U_L^\dagger \\
&\text{RH: } R_\mu = \Sigma^\dagger\partial_\mu\Sigma = -\partial_\mu\Sigma^\dagger\Sigma = -R_\mu^\dagger \longrightarrow U_R R_\mu U_R^\dagger \\
&\text{RH: } M^\dagger\Sigma \longrightarrow U_R(M^\dagger\Sigma)U_R^\dagger , \qquad \Sigma^\dagger M \longrightarrow U_R(\Sigma^\dagger M)U_R^\dagger .
\end{aligned}$$

where I have repeatedly used the unitarity of Σ. From the fact that $\det(\Sigma) = 1$ one learns that L_μ and R_μ are traceless, e.g.

$$0 = \partial_\mu(\det\Sigma) = \partial_\mu(\exp \operatorname{tr}\ln\Sigma) = \det\Sigma \ \operatorname{tr}(\partial_\mu\Sigma\Sigma^{-1}) = -\operatorname{tr}(L_\mu) . \tag{14}$$

Thus L_μ and R_μ ("Weyl derivatives") are elements of the Lie algebra, $su(3)$.

Another symmetry of QCD is parity. Since $\Sigma \sim q_L\overline{q}_R$ and $\Sigma^\dagger \sim q_R\overline{q}_L$, the transformations in the EFT are [with $x_P = (-\vec{x}, x_4)$]

$$\Sigma(x) \leftrightarrow \Sigma^\dagger(x_P), \quad M \leftrightarrow M^\dagger, \quad L_i(x) \leftrightarrow -R_i(x_P), \quad L_4(x) \leftrightarrow R_4(x_P). \tag{15}$$

If one expands about $\langle\Sigma\rangle = 1$, as in Eq. (12), then the pion field transforms as $\Pi(x) \to -\Pi(x_P)$. There are also C and T symmetries, which I do not show explicitly.

I now enumerate terms which are local, real and satisfy the symmetries of QCD. These are just products of the building blocks above, with LH and RH blocks combined separately into traces to make them invariant under

---

[h]I use the convention throughout that derivatives only act on the objects immediately to their right. The arrows implicitly denote transformation under G.

G. In fact, for the terms I display, one need only use LH building blocks as the results equal their "parity conjugates" (p.c.). Since in the end we expand in powers of momenta, it is useful to classify terms according to the number of derivatives. Similarly, as M is treated as small, one should classify according to the number of spurion insertions. We will see that one should usually count two derivatives for each spurion.

There are no non-trivial terms without derivatives or spurions: these would be constructed from $\mathrm{tr}[(\Sigma\Sigma^\dagger)^n]$ or powers or $\det\Sigma$, but both are constants. Euclidean invariance rules out a single derivative. The only independent term with two derivatives is

1. $\mathrm{tr}(L_\mu L_\mu) = -\mathrm{tr}(\partial_\mu\Sigma\partial_\mu\Sigma^\dagger) = \mathrm{tr}(R_\mu R_\mu)$,

while the only term with no derivatives and one mass insertion is

2. $\mathrm{tr}(M\Sigma^\dagger) + \mathrm{tr}(\Sigma M^\dagger)$.

There are five terms with four derivatives:

3. $[\mathrm{tr}(L_\mu L_\mu)]^2$
4. $\mathrm{tr}(L_\mu L_\nu)\mathrm{tr}(L_\mu L_\nu)$
5. $\mathrm{tr}(L_\mu L_\mu L_\nu L_\nu)$ [not independent for two light flavors]]
6. $\mathrm{tr}(L_\mu L_\nu L_\mu L_\nu)$ [not independent for 2 or 3 light flavors]]
7. The Wess-Zumino-Witten (WZW) term involving $\epsilon_{\mu\nu\rho\sigma}$[27];

two terms with two derivatives and one mass insertion:

8. $\mathrm{tr}(L_\mu L_\mu)\ \mathrm{tr}(M\Sigma^\dagger + \Sigma M^\dagger)$
9. $\mathrm{tr}(L_\mu L_\mu[M\Sigma^\dagger + \Sigma M^\dagger])$;

and three terms with two mass insertions:

10. $[\mathrm{tr}(M\Sigma^\dagger + \Sigma M^\dagger)]^2$
11. $[\mathrm{tr}(M\Sigma^\dagger - \Sigma M^\dagger)]^2$
12. $\mathrm{tr}(M\Sigma^\dagger M\Sigma^\dagger + M^\dagger\Sigma M^\dagger\Sigma)$.

Each of these terms appears in L_{eff} with an independent unknown coefficient, except for terms 5 and 6 which, as noted, are not independent for certain chiral groups. I will not discuss the interesting structure of the WZW term, as it is complicated, and does not contribute to the simple processes considered here. I will also not continue the enumeration beyond this point. This has been done as part of next-to-next-to-leading order (NNLO) calculations[28], but is beyond the scope of this introduction.

With the enumeration of operators in hand, I turn to the predictions. I will assume a power counting with $\partial^2 \sim M$ and justify this *a posteriori*. At leading order (LO) we have (the superscript on $\mathcal{L}$ counting derivatives):

$$\mathcal{L}_\chi^{(2)} = \frac{f^2}{4}\mathrm{tr}\left(\partial_\mu \Sigma \partial_\mu \Sigma^\dagger\right) - \frac{f^2 B_0}{2}\mathrm{tr}(M\Sigma^\dagger + \Sigma M^\dagger)\,, \tag{16}$$

where the unknown "low energy constants" (LECs) have been given their standard names f and B_0. Since we are expanding about massless QCD, the only scale that can appear is $\Lambda_{\rm QCD}$, so we expect $f \sim B_0 \sim \Lambda_{\rm QCD}$. Up to this stage, M is a complex spurion field. To include the effects of quark masses we set it to its physical value: $M \to M_0 = \mathrm{diag}(m_u, m_d, m_s) = M_0^\dagger$. This makes the potential—the second term in Eq. (16)—depend on the direction of Σ, and we must determine its VEV by minimizing this potential. In other words, the quark masses, which break the chiral symmetry, pick out a preferred direction in the vacuum manifold. If all quark masses can be chosen positive (as is apparently the case in reality), then one finds that the VEV is $\langle\Sigma\rangle = 1$. I stress that this result is convention dependent: we could choose $M = U_L M_0 U_R^\dagger$, in which case $\langle\Sigma\rangle = U_L U_R^\dagger$.

2.3.2. *Brief aside on vacuum structure*

It is instructive to consider the vacuum structure of *two flavor* theory in a little more detail. Then we can write $\Sigma = \exp(i\theta\vec{n}\cdot\vec{\tau})$, implying $\Sigma + \Sigma^\dagger = 2\cos\theta \times 1$, and thus $V^{(2)} \propto -\mathrm{tr}(M)\cos\theta$. This is minimized by $\langle\Sigma\rangle = 1$ if $\mathrm{tr}M > 0$, and by $\langle\Sigma\rangle = -1$ if $\mathrm{tr}M < 0$. Note that at LO all that matters is the average quark mass $\mathrm{tr}M/2$; the difference $m_u - m_d$ does not enter. There is a first order phase transition when $\mathrm{tr}M$ changes sign at which the condensate flips sign but maintains its magnitude. Note, however, that the physical theory is the same for both signs of $\mathrm{tr}M$: one can go between them with a chiral rotation satisfying $U_L U_R^\dagger = -1$. I will discuss how discretization errors effect this transition in Sec. 3.9 below.

With three flavors (or any odd number), the situation is more complicated because $\Sigma = -1$ is not an element of $SU(3)$. Changing the sign of M leads to a different theory (one with the original M plus a θ−term with $\theta = \pi$). Without going into details, I show in Fig. 3 the phase structure if m_s is fixed and positive while the other two masses change. The shaded region is where CP is spontaneously broken. In the real world we are very likely in the right-hand upper quadrant, but it is striking that such interesting physics lurks not far away and is contained in the LO potential.

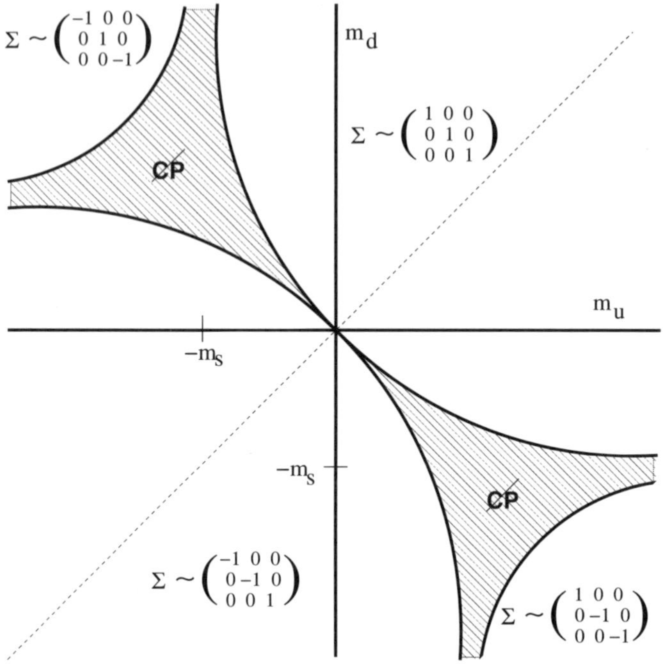

Figure 3. Phase structure at LO with 3 flavors and fixed m_s (from Creutz[29]).

2.3.3. *Properties of pseudo-Nambu-Goldstone bosons at leading order*

Assuming positive quark masses so that $\langle\Sigma\rangle = 1$, we can study pion properties by inserting Eq. (12) into $\mathcal{L}^{(2)}$ and expanding:

$$\begin{aligned}\mathcal{L}_\chi^{(2)} &= \mathrm{tr}(\partial_\mu\Pi\partial_\mu\Pi) + \frac{1}{3f^2}\mathrm{tr}([\Pi,\partial_\mu\Pi][\Pi,\partial_\mu\Pi]) + O([\partial\Pi]^2\Pi^4)\\ &\quad +2B_0\,\mathrm{tr}(M\Pi^2) - \frac{2B_0}{3f^2}\mathrm{tr}(M\Pi^4) + O(M\Pi^6). \qquad (17)\end{aligned}$$

We can now understand the choice of factors multiplying the kinetic term in (16): they are chosen so that the pion kinetic term is correctly normalized [using $\mathrm{tr}(T^aT^b) = \delta^{ab}/2$]. If $M = 0$ [the first line of (17)], the pions are massless as required by Goldstone's theorem, and their interactions all involve derivatives, as exemplified by $(\partial\Pi)^2\Pi^2$ term shown. Note that the non-abelian nature of the group allows there to be interaction terms with two (as opposed to four) derivatives, in contrast to the $U(1)$ example described above. This interaction term is non-renormalizable, as are subsequent terms involving more pion fields, with the dimensions balanced by factors of f.

Including M, the pions become massive pseudo-Goldstone bosons (PGB). χPT predicts at LO that the pion mass *squared* is proportional to the quark mass. The explicit form is simplest for degenerate quarks: $m_\pi^2 = 2B_0 m_q$. This answers the following potential puzzle: how can physical quantities like m_π^2 be related to scheme- and scale-dependent quantities like m_q? The answer is that B_0 cancels the scheme dependence in m_q. This works for all the terms in the second line of (17), as they contain the common factor $B_0 M$. For this reason it is useful to give the combination a name, specifically $\chi_q = 2B_0 m_q$.

The spurion terms also give rise to higher-order, non-renormalizable interactions among pions. Note that all vertices in $\mathcal{L}^{(2)}$ contain an even number of pions, because $\mathcal{L}^{(2)}$ is invariant under $\Pi(x) \leftrightarrow -\Pi(x)$. This is an accidental symmetry, which does not correspond to a symmetry of QCD (note that it differs from parity), and is broken by NLO terms in χPT. More precisely, it is broken by the WZW term, which allows, for example, interactions between five PGB.

LO χPT makes a number of predictions. First, it is clear that once f and χ have been determined (e.g. from PGB masses and scattering amplitudes), all higher order vertices are predicted. But there are also predictions from the structure of the quadratic and quartic interactions alone. The former give relations between PGB masses, that latter between pion scattering in different channels (e.g. $I = 0, 1, 2$ in the two-flavor theory). I will discuss the mass relations here.

First we need to place the physical particles in the pion matrix Π. This can be done using the vector symmetry corresponding to diagonal phase rotations, $U(1)_{V,u} \times U(1)_{V,d} \times U(1)_{V,s} \in U(3)_V$, which is unbroken by quark masses, and under which the pion field transforms linearly: $\Pi \to U_V \Pi U_V^\dagger$. The $\pi^\pm$, $K^\pm$, K^0 and $\overline{K^0}$ are charged under these symmetries[i] (e.g. the π^+ has u-ness $+1$ and d-ness -1), and so live in definite off-diagonal positions in Π. Isospin then determines the position of the π_0, and orthogonality that of the η. Including normalizations, the result is

$$\Pi = \begin{pmatrix} \frac{\pi^0}{2} + \frac{\eta}{\sqrt{12}} & \frac{\pi^+}{\sqrt{2}} & \frac{K^+}{\sqrt{2}} \\ \frac{\pi^-}{\sqrt{2}} & -\frac{\pi^0}{\sqrt{2}} + \frac{\eta}{\sqrt{12}} & \frac{K^0}{\sqrt{2}} \\ \frac{K^-}{\sqrt{2}} & \frac{\overline{K^0}}{\sqrt{2}} & -\frac{2\eta}{\sqrt{12}} \end{pmatrix} . \tag{18}$$

[i] In the following, I refer to all such particles as "charged", having in mind this generalized definition rather than electric charge.

Inserting this into $-2B_0\mathrm{tr}(M\pi^2)$, we find that charged particle masses are proportional to the average mass of the quarks they contain: $m^2_{q_iq_j} = B_0(m_i + m_j)$, $i \neq j$. While there are no predictions (masses of three pairs of CPT conjugate mesons are given in terms of three quark masses), one can determine quark mass ratios from the experimental PGB masses, e.g.

$$\frac{m^2_{K^+} + m^2_{K^0}}{2m^2_{\pi^+}} = \frac{m_\ell + m_s}{2m_\ell} \approx 13 \qquad \left(m_\ell = \frac{m_u + m_d}{2}\right). \tag{19}$$

This implies $m_s/m_\ell \approx 25$, up to NLO χPT and electromagnetic (EM) corrections. This is how we know that the strange quark is so much heavier than the up and down quarks. The corresponding determination of $(m_u - m_d)/m_\ell$ (or equivalently m_u/m_d) from $(m^2_{K^+} - m^2_{K^0})/m^2_{\pi^+}$ fails, however, since the NLO corrections ($\Delta(m_u/m_d) \propto m_s/\Lambda_{\mathrm{QCD}}$[30]) and EM contributions are potentially as large as the LO term in χPT. A determination of m_u/m_d can be achieved by a direct lattice calculation of the meson masses using non-degenerate quarks together with an estimate of the EM contributions. The most accurate results at present[3] are $m_u/m_d = 0.43(8)$ and $m_s/m_\ell = 27.4(4)$.

The first predictions of χPT occur in the neutral sector. The π^0 and η mix, but with an angle $\theta \sim (m_u - m_d)/m_s$ that (despite the uncertainty in m_u/m_d) we know to be very small. Thus

$$m^2_{\pi^0} = m^2_{\pi^+} + O(\theta^2 m^2_K) + \dots, \tag{20}$$

$$\underbrace{m^2_\eta}_{(548\ \mathrm{MeV})^2} = \underbrace{(2[m^2_{K^+} + m^2_{K^0}] - m^2_{\pi^+})/3}_{(566\ \mathrm{MeV})^2} + O(\theta^2 m^2_K) + \dots. \tag{21}$$

Both predictions are well satisfied. The first, that of approximate isospin symmetry for the pions, holds not because $m_u/m_d \approx 1$ (which is not true), but because $(m_u - m_d)/\Lambda_{\mathrm{QCD}} \ll 1$. The second is the famous Gell-Mann–Okubo (GMO) relation.

2.3.4. *Lessons for lattice simulations*

(I) Leading order χPT works to $\sim 10\%$ in GMO relation, despite the fact that this is a three-flavor relation involving the strange quark. This gives hope that three-flavor χPT can be used to extrapolate from the lattice simulations that are presently being undertaken, which have $m_s^{\mathrm{phys}}/2 \gtrsim m_\ell^{\mathrm{lat}} \gtrsim 2m_\ell^{\mathrm{phys}}$ and $m_s^{\mathrm{lat}} \approx m_s^{\mathrm{phys}}$. The issue is whether the NLO corrections are generically this small, and I return to this below.

(II) Assuming the validity of χPT, the ratio $m_{\pi^+}^2/m_q$ determines the *physical* B_0 (in whatever scheme the quark mass is defined in) *even when the quarks are degenerate and have masses differing from (usually larger than) their physical values.* Such a determination is an example of obtaining physical results from simulations with unphysical parameters. It works as long as there are $N = 3$ dynamical quarks: B_0 (like all LECs) depends on N, and so one must simulate with the same number as in QCD. Of course, one also needs NLO corrections in χPT to be small.
(III) The isospin limit, $m_u = m_d$, is close to physical QCD. Working in this limit simplifies simulations, e.g. by reducing the number of adjustable parameters, and by canceling disconnected contributions to neutral correlators such as in the π^0 propagator:

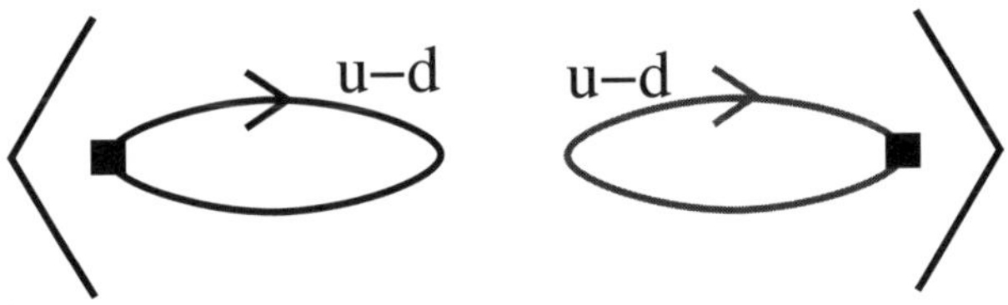

The error one makes in hadron masses by setting $m_u = m_d$ is generically $\sim (m_u - m_d)/\Lambda_{\rm QCD} \sim 1\%$, comparable to those from EM contributions. Of course, once one can attain 1% accuracy in the isospin limit, further improvement requires the calculation of disconnected and EM contributions

2.3.5. *Power counting in χPT ($M = 0$)*

I now turn to the questions of power counting and predictivity of non-renormalizable theories: how are contributions ordered, and by what factor are higher order terms suppressed? As already noted, the ordering turns out to be in powers of momenta-squared and mass insertions, so the NLO effective Lagrangian, $\mathcal{L}^{(4)}$, contains the terms proportional to ∂^4, $\partial^2 M$ and M^2 enumerated above. Setting $M = 0$ to simplify discussion, and expanding, one finds, schematically:

$$\mathcal{L}^{(2)} \sim f^2 \mathrm{tr}(L_\mu L_\mu) \sim (\partial\Pi)^2 + \frac{\Pi^2(\partial\Pi)^2}{f^2} + \dots \tag{22}$$

$$\mathcal{L}^{(4)} \sim L_{GL}\mathrm{tr}(L_\mu L_\mu)^2 + \dots \sim L_{GL}\left[\frac{(\partial\Pi)^4}{f^4} + \frac{\Pi^2(\partial\Pi)^4}{f^6}\right] + \dots, \tag{23}$$

where L_{GL} are unknown dimensionless LECs, first enumerated by Gasser and Leutwyler[26]. Taking $\pi\pi$ scattering as an example (with, say, dimensional regularization to avoid power divergences), the contributions up to quartic in momenta are:

$\mathcal{L}^{(2)}_{\text{tree}}$: 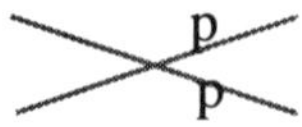 $\sim \frac{p^2}{f^2}$ $\quad$ $\mathcal{L}^{(4)}_{\text{tree}}$: 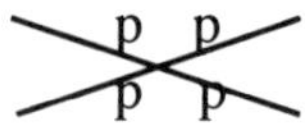 $\sim L_{GL} \left(\frac{p^2}{f^2}\right)^2$

$\mathcal{L}^{(2)}_{1-\text{loop}}$:  $\sim \left(\frac{p^2}{f^2}\right)^2 \frac{\ln(p^2/\mu^2)}{(4\pi)^2}$

where I have shown only a representative loop diagram, and the positions of the "p's" in the diagrams indicate whether they refer to external or loop momenta. Contributions from $\mathcal{L}^{(4)}$ at tree-level, and $\mathcal{L}^{(2)}$ at one-loop, are proportional to p^4 (up to logs), where p is a generic external momentum. These terms are suppressed relative to the tree-level contribution from $\mathcal{L}^{(2)}$ by p^2/f^2 (up to logs).

It is straightforward to generalize from this example to a power-counting scheme, using the fact that each pion field brings with it a factor of $1/f$. One finds that, for all processes involving PGB, the expansion parameters are (up to logs) p^2/f^2 and (if we reintroduce M) $\chi/f^2 \sim m_\pi^2/f^2$. At LO contributions come from $\mathcal{L}^{(2)}_{\text{tree}}$, at NLO from $\mathcal{L}^{(4)}_{\text{tree}}$ and $\mathcal{L}^{(2)}_{1-\text{loop}}$, and at NNLO from $\mathcal{L}^{(6)}_{\text{tree}}$, $(\mathcal{L}^{(2)} + L^{(4)})_{1-\text{loop}}$ and $\mathcal{L}^{(2)}_{2-\text{loop}}$, etc.. These are, respectively, "trivial", "easy" and "hard" to calculate (though the latter is done), with NNNLO being "very hard".

We can now understand the nature of predictions from χPT. At each order there are a finite number of LECs (2 at LO, 10 at NLO, 90 at NNLO). We pick an order to work at, say NLO. We determine the LECs from the appropriate number of physical quantities, and then make predictions for all other quantities. These predictions will be accurate up to truncation errors (of NNLO size in our example). In the continuum, these errors can only be estimated. On the lattice, we can attempt to fit them as part of the extrapolations.

The example discussed above also illustrates the general relationship between terms analytic and non-analytic in p^2 and M. Non-analytic terms (often called "chiral logs") come from pion loops and involve lower order vertices (e.g. LO vertices in the NLO calculation of the example). In particular, they do not involve the LECs of the order being worked at, and in this sense are predictions. They do, however, depend on the renormalization scale μ, as exemplified by the $p^4 \ln \mu$ term in the example. A physical result cannot depend on μ, and indeed the dependence can be canceled by renormalizing the LECs: $L_{GL} \to L_{GL}(\mu)$. Renormalization theory shows that all divergences can be removed in this way as long as all terms consistent with the symmetries are included in $\mathcal{L}_{\text{eff}}$, which was, of course, our starting point. Thus the LECs can be thought of as counterterms which contain our ignorance about the short distance parts of loops, crudely speaking the parts with $|p| > \mu$. It is then natural to set $\mu \approx m_{\text{had}}$, since $m_{\text{had}} \approx 1$ GeV is the scale above which we know the EFT to be ineffective.

One can argue for a natural value for this "cut-off" scale from within χPT itself, and at the same time estimate the LECs at this scale. The above example should make plausible the result that the LECs satisfy renormalization group equations of the form $dL_{GL}/d\ln(\mu) = O(1) \times (4\pi)^{-2}$. The $(4\pi)^{-2}$ comes from the four-dimensional loop integral. Now we do not know the value of the L_{GL} at the natural matching scale $\mu \sim m_{\text{had}}$, nor do we know what this matching scale is within a factor of two or so. But we do know how the LECs change with μ, $|L_{GL}(2\mu) - L_{GL}(\mu)| \approx 1/(4\pi)^2$, so that even if $L_{GL} \approx 0$ at a possible matching scale, it would be of order $1/(4\pi^2)$ at another. This motivates assuming that the natural size is $|L_{GL}(\mu)| \approx 1/(4\pi)^2$. This can be generalized into a self-consistent scheme to all orders. Of course, this argument does not rule out LECs with larger magnitudes, but it turns out that those LECs that have been determined from experiment have the expected magnitude.

If we make this assumption, then the NLO contributions are suppressed relative to the LO by $L_i p^2/f^2 \sim p^2/\Lambda_\chi^2$ (and m_π^2/Λ_χ^2 when we include M), with $\Lambda_\chi = 4\pi f$. We will shortly see that $f \approx f_\pi$, in which case $\Lambda_\chi = 1.2$ GeV. This value is completely consistent with our original expectation that χPT would break down for $|p| \gtrsim m_{\text{had}}$.

2.3.6. *Lessons for lattice simulations (continued)*

(IV) We can use χPT to extend the reach of lattice calculations to multi-particle processes. Simulations can only access properties of single or two particle states, with the latter being a challenge away from threshold. However we can use lattice calculations to determine the LECs, and then use χPT to calculate inaccessible processes. For example, one can determine $\mathcal{A}(K \to \pi\pi)$ using unphysical, but more accessible, matrix elements.

(V) The downside of relying on χPT is the inevitable presence of a truncation error. If we want to estimate this error, we need fits including the next higher order terms. For example, to reliably determine the L_{GL} discussed above, one must do fits including, at least approximately, NNLO terms. In fact, the full NNLO expression in PQχPT is available for PGB masses and decay constants[31].

2.3.7. *Technical aside: adding sources*

Electroweak currents can be used to probe PGBs. For example, the weak leptonic decay $\pi \to \ell\overline{\nu}$ is proportional to the matrix element of the axial current, $\langle 0|A_\mu|\pi\rangle$, itself proportional to the pion decay constant f_π. It is thus important to incorporate vector and axial currents into the χPT framework. Other operators are of interest too, and I discuss here the scalar and pseudoscalar densities. The aim is to map such operators at the quark level into operators in the EFT in such away that matrix elements agree, up to truncation errors in χPT. This mapping should be based only on the symmetries of the operators and the action.

A convenient method for doing the mapping is to start with the QCD action including position-dependent sources for left and right-handed currents (l_μ and r_μ, respectively), and for the densities (s and p):

$$\mathcal{L}_{\rm QCD} = \overline{Q}_L(\not{D}-i\gamma_\mu l_\mu)Q_L+\overline{Q}_R(\not{D}-i\gamma_\mu r_\mu)Q_R+\overline{Q}_L(s+ip)Q_R+\overline{Q}_R(s-ip)Q_L\,. \tag{24}$$

Functional derivatives of the partition function $Z_{\rm QCD}(l_\mu, r_\mu, s, p)$ with respect to the sources (which are hermitian matrices in flavor space) bring down the corresponding currents and densities. Note that s and p are just a (position-dependent) rewriting of the mass spurion, $M \to s + ip$. After derivatives with respect to the sources have been taken, they are to be set to zero, except for $s + ip$, which is set to M.[j]

One nice feature of $Z_{\rm QCD}(l_\mu, r_\mu, s, p)$ is that it represents what one can actually calculate in QCD—correlation functions of gauge-invariant operators. It allows one to access PGB physics, because the axial currents (and pseudoscalar densities) couple to pions. For example, a 4-point correlator of axial currents, analytically continued to Minkowski momenta, can be amputated and then evaluated at the pole positions so as to obtain the pion scattering amplitude using the LSZ formula. Similarly one can obtain matrix elements of currents and densities between PGBs (or other states).

[j]For the standard diagonal mass term, s is set to M and p to zero.

Thus it is natural to phrase the matching of QCD with the EFT in terms of this partition function[26]

$$Z_\chi(l_\mu, r_\mu, s, p) = Z_{\rm QCD}(l_\mu, r_\mu, s, p) + \text{truncation errors} \quad (p_\pi, m_\pi \ll \Lambda_\chi)\,. \tag{25}$$

This simply says that the correlation functions of the two theories must match for momenta such that PGBs give the dominant contributions, up to errors due to the truncation of χPT. The matching ensures that masses, scattering amplitudes and matrix elements agree in the two theories.

This is nice packaging, but to give it content we need to learn how to add sources to Z_χ. This can be done by generalizing the spurion method to enforce a *local* $SU(3)_L \times SU(3)_R$ symmetry. $\mathcal{L}_{\rm QCD}$ in (24) is invariant if l, r_μ transform as gauge fields,

$$l_\mu \to U_L l_\mu U_L^\dagger + iU_L\partial_\mu U_L^\dagger\,, \quad r_\mu \to U_R r_\mu U_R^\dagger + iU_R\partial_\mu U_R^\dagger \tag{26}$$

(with $U_{L,R} = U_{L,R}(x)$), while s and p transform as before, $\chi = 2B_0(s + ip) \to U_L \chi U_R^\dagger$. Here I have introduced the convenient matrix χ. Note that this invariance only works because the left and right-handed currents are Noether currents for the corresponding symmetries. One would like to argue that, in order to satisfy Eq. (25), $\mathcal{L}_\chi$ must also be invariant under local chiral transformations. This turns out to be correct, but the argument, given by Leutwyler[32], is subtle. First one must deal with anomalies: $Z_{\rm QCD}$ is not invariant under local chiral transformations, but has a variation which is a known functional of the sources alone. This must be matched by the variation in Z_χ, which in turn requires the presence of the WZW term in $\mathcal{L}_\chi$. This allows the remainder of $\mathcal{L}_\chi$ to be invariant, but this is not automatic. Leutwyler shows that it is possible to bring it to an invariant form if one uses the freedom of adding total derivatives (which do not change the action) and changing variables.

The invariance of $\mathcal{L}_\chi$ under local transformations can be accomplished in the usual way: by replacing normal derivatives with covariant derivatives which transform homogeneously, e.g.

$$D_\mu\Sigma = \partial_\mu\Sigma - il_\mu\Sigma + i\Sigma r_\mu \to U_L(D_\mu\Sigma)U_L^\dagger\,. \tag{27}$$

Note that this fixes the normalization of the l, r_μ terms, implying that the normalization of the currents in the EFT is known.[k] This is possible because the currents generate a non-abelian group, whose algebra is

[k] One could also obtain the correct normalization in the absence of symmetry breaking by matching the Noether currents in the two theories.

schematically $[L, L] \sim L$, in which normalizations are fixed. Terms without derivatives, e.g. mass terms, are automatically invariant under the local symmetry, so do not need to be changed. The inclusion of sources also allows new types of terms in $\mathcal{L}_\chi$, as will be seen shortly.

2.3.8. *Final form of chiral Lagrangian*

We now have the ingredients to construct the EFT through NLO including sources. The LO result differs from the earlier form (16) only by $\partial_\mu \to D_\mu$ and $2B_0 M \to \chi$:

$$\mathcal{L}_\chi^{(2)} = \frac{f^2}{4} \mathrm{tr}\left(D_\mu \Sigma D_\mu \Sigma^\dagger\right) - \frac{f^2}{4} \mathrm{tr}(\chi\Sigma^\dagger + \Sigma\chi^\dagger)\,, \tag{28}$$

The addition of the sources allows us to match currents with QCD, e.g. equating $\delta Z/\delta l_\mu(x)|_{l=r=p=0,s=M}$ in QCD and χPT implies

$$\overline{Q}_L \gamma_\mu T^a Q_L \simeq (if^2/2)\mathrm{tr}(T^a \Sigma \partial_\mu \Sigma^\dagger) = -(f/2)\partial_\mu \pi^a + \ldots\,. \tag{29}$$

This result can also be obtained using the Noether procedure. It allows us to determine f: from the vacuum to pion matrix element one finds $f = f_\pi \approx 93$ MeV. Similarly, we can relate B_0 to quark-level quantities. Taking $\delta Z/\delta s(x)|_{l=r=p=0,s=M}$ gives

$$\overline{Q}Q \simeq -(f^2 B_0/2)\mathrm{tr}(\Sigma + \Sigma^\dagger) = -N f^2 B_0 + O(\pi^2)\,, \tag{30}$$

the VEV of which gives $\langle \overline{q}q \rangle = -f^2 B_0$ (with $q = u$, d, or s). This is a place where the lattice can contribute, since the condensate is not experimentally measurable but can be directly calculated on the lattice. Alternatively, since the PGB masses give the combination $B_0 m$, as seen above, a lattice determination of the quark masses gives another result for B_0. Comparing the two determinations tests the accuracy of χPT at LO.

At NLO the addition of sources leads to new terms. For three flavors, the enumeration in Sec. 2.3.1 gave 8 independent terms plus the WZW term. Sources allow two additional terms involving the field-strength tensors $L, R_{\mu\nu}$ (constructed from the source gauge fields l, r_μ in the usual way), as well two terms involving sources alone. The latter are multiplied by so-called high-energy coefficients (HECs). In sum, one has

$$\begin{aligned}\mathcal{L}^{(4)} = &-L_1\left[\mathrm{tr}(D_\mu\Sigma D_\mu\Sigma^\dagger)\right]^2 - L_2\,\mathrm{tr}(D_\mu\Sigma D_\nu\Sigma^\dagger)\mathrm{tr}(D_\mu\Sigma D_\nu\Sigma^\dagger)\\ &+L_3\,\mathrm{tr}(D_\mu\Sigma D_\mu\Sigma^\dagger D_\nu\Sigma D_\nu\Sigma^\dagger)\\ &+L_4\,\mathrm{tr}(D_\mu\Sigma^\dagger D_\mu\Sigma)\mathrm{tr}(\chi^\dagger\Sigma+\Sigma^\dagger\chi) + L_5\,\mathrm{tr}(D_\mu\Sigma^\dagger D_\mu\Sigma[\chi^\dagger\Sigma+\Sigma^\dagger\chi])\\ &-L_6\left[\mathrm{tr}(\chi^\dagger\Sigma+\Sigma^\dagger\chi)\right]^2 - L_7\left[\mathrm{tr}(\chi^\dagger\Sigma-\Sigma^\dagger\chi)\right]^2 - L_8\,\mathrm{tr}(\chi^\dagger\Sigma\chi^\dagger\Sigma+p.c.)\\ &+L_9\,i\mathrm{tr}(L_{\mu\nu}D_\mu\Sigma D_\nu\Sigma^\dagger+p.c.) + L_{10}\,\mathrm{tr}(L_{\mu\nu}\Sigma R_{\mu\nu}\Sigma^\dagger)\\ &+H_1\,\mathrm{tr}(L_{\mu\nu}L_{\mu\nu}+p.c.) - H_2\,\mathrm{tr}(\chi^\dagger\chi) + \mathcal{L}_{WZW}\,,\end{aligned} \tag{31}$$

There is also a term $\propto \mathrm{tr}(D_\mu\chi^\dagger D_\mu\Sigma)$, which contributes only to the mass dependence of the currents, but it can be removed by a change of variables for Σ, and is thus redundant. The dimensionless L_i are the well-known "Gasser-Leutwyler coefficients". A subset of them can be determined experimentally to good accuracy and a different subset is straightforward to determine on the lattice, as illustrated below.

What about the HECs, $H_{1,2}$? Since these multiply terms which do not involve Σ they give contact terms in correlation functions (e.g. H_2 gives a contribution to $C_P(x) = \langle\overline{Q}\gamma_5T^aQ(x)\overline{Q}\gamma_5T^aQ(0)\rangle$ proportional to $H_2\delta(x)$, and also contributes to the mass dependence of the condensate). Thus they do not contribute to physical scattering amplitudes or matrix elements. Why does one need them? They are required if one wants to describe the mass dependence of the condensate (which can be calculated on the lattice) within χPT, and in order that quark level Ward identities are satisfied (e.g. $m\int d^4x C_P(x) \propto \langle\overline{q}q\rangle$).[1] For the matching to work, the HECs must depend on the regulator used for QCD. For example, if QCD is regulated using a lattice, then $H_2 \propto 1/a^2$ because it must reproduce $\langle\overline{q}q\rangle \propto m/a^2$. This is in stark distinction to the LECs which do not depend on the underlying regulator (leading to their different name), and are physical parameters.

2.4. *Examples of NLO results*

With $\mathcal{L}^{(2,4)}$ in hand, it is straightforward to calculate NLO results for physical quantities. For example, the pole in the two-point function of

[1]One might be concerned that, because the short-distance behavior in QCD and χPT are different (operators having different dimensions in the two theories), the whole notion of matching the partition functions with sources is flawed. How can correlators be matched when the operator product expansions differ? The answer is that the matching is done only for $|p| < \Lambda_\chi$, so one avoids the short-distance regime in the underlying theory.

the left-handed current gives the PGB mass, while the residue is proportional to $f^2_{\rm PGB}$ (just as in lattice simulations). Recall that the LO result is $m^2_{\rm PGB,0} = (\chi_{q1}+\chi_{q2})/2 = 2B_0(m_{q1}+m_{q2})/2$. At NLO, there is a tree-level contribution

$$\delta m^2_{\rm PGB} \sim$$ 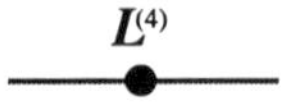 $$\sim \chi L \tfrac{\chi}{f^2} \sim \chi(16\pi^2 L)\frac{m^2_{\rm PGB,0}}{\Lambda^2_\chi},$$

and a 1-loop contribution

$$\delta m^2_{NG} \sim$$ 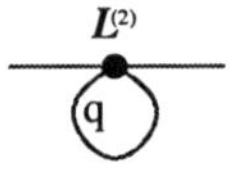 $$\sim \tfrac{\chi}{f^2}\int_q \frac{1}{q^2+m^2_{\rm PGB,0}} \sim \chi\frac{m^2_{\rm PGB,0}}{\Lambda^2_\chi}\ln\left(\frac{m^2_{\rm PGB,0}}{\mu^2}\right).$$

I quote the final result for the $\pi^\pm$ as an example[26]:

$$m^2_{\pi^\pm} = \chi_\ell\left\{1+\frac{8}{f^2}[\underbrace{(2L_8-L_5)\chi_\ell}_{\text{valence}}+\underbrace{(2L_6-L_4)(2\chi_\ell+\chi_s)}_{\text{sea}}]+\underbrace{\frac{3L_\pi-L_\eta}{6}}_{\text{logs}}\right\}, \tag{32}$$

where $\chi_l = (\chi_u+\chi_d)/2$, and the chiral logs are (using dim. reg.)

$$L_\pi = \frac{m^2_\pi}{\Lambda^2_\chi}\ln\left(\frac{m^2_\pi}{\mu^2}\right), \qquad L_\eta = \frac{m^2_\eta}{\Lambda^2_\chi}\ln\left(\frac{m^2_\eta}{\mu^2}\right). \tag{33}$$

The μ dependence is absorbed by the L_i, as discussed above. Looking ahead to the discussion of PQQCD, I have separated the analytic terms into those that arise from the masses of the quarks which compose the pion ("valence") and those of the quarks in loops ("sea"). How this is done will be explained in Sec. 4. Within QCD itself this separation is not useful, as corresponding valence and sea quarks have the same masses.

Another example is the ratio of decay constants, which is

$$\frac{f_K}{f_\pi} = 1+\frac{2}{f^2}\underbrace{(L_5)(\chi_s-\chi_\ell)}_{\text{valence}}+\underbrace{\frac{5}{8}L_\pi-\frac{1}{4}L_K-\frac{3}{8}L_\eta}_{\text{logs}}. \tag{34}$$

This result allows L_5 to be determined from experiment. The result depends on the choice of μ, a conventional value being $\mu = m_\rho$.

These results illustrate the general structure at NLO: there are corrections analytic in χ and dependent on the masses of valence and sea quarks, and chiral logs that are non-analytic in χ. The expansion parameter $m^2_{\rm PGB}/\Lambda^2_\chi$ is clear in the logs, but is obscured in the analytic terms by the convention for the L_i (which are numerically of size $\lesssim 1/(16\pi^2)$.)

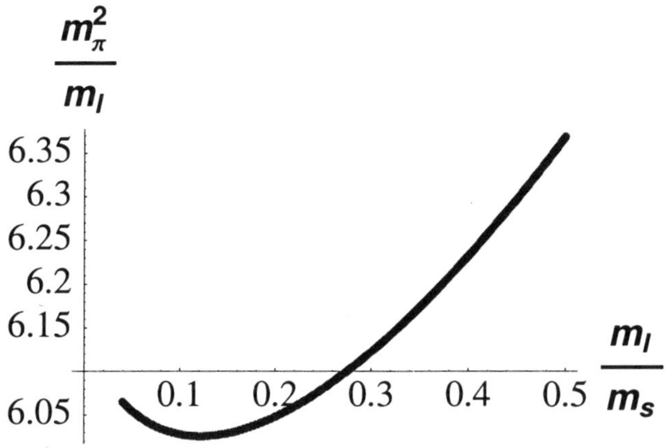

Figure 4. Typical behavior of m_π^2/m_l as a function of m_l/m_s at NLO.

2.4.1. *Lessons for lattice simulations (continued)*

(VI) Non-analytic terms become important at small masses. To illustrate this, I plot $m_{\pi^\pm}^2/m_l$ versus m_l/m_s in Fig. 4, with m_l the average light-quark mass. I hold the strange mass fixed at $m_s = 0.08$ GeV, and use values for the LECs that are representative of those from χPT analyses: $f = 0.093$ GeV, $L_5 = 1.45 \times 10^{-3}$, $L_8 = 10^{-3}$, $L_4 = L_6 = 0$ (with $\mu = m_\rho$ here and below). The curve's lower end is approximately the physical point. At LO the χPT prediction is a constant. The analytic NLO corrections lead to linear dependence on m_l, and the logs to curvature. Clearly, to obtain 1% accuracy one must simulate down to $m_l/m_s \approx 0.1$ in order to see and fit to the predicted curvature.[m] This has been achieved in the MILC simulations. In my opinion, seeing curvature consistent with χPT predictions is a necessary check on lattice techniques. Similar comments hold for f_K/f_π, plotted in Fig. 5 using the same parameters except now $f =$ 0.085 GeV (chosen so as to better match the experimental value at the lower end of the curve). Here the linear terms are larger, but an extrapolation accurate to a few percent still requires inclusion of the curvature.

(VII) The results further illustrate the utility of the lattice for obtaining LECs. In nature one can vary the valence content by considering different PGBs, while the sea content is fixed. Thus L_4 and L_6 are not accessible

[m]Extrapolation can be simplified in some cases by considering "golden (silver) ratios" in which the chiral logs completely (partially) cancel[33]. By contrast, for some quantities the chiral logs are enhanced, e.g $\langle r^2\rangle_\pi \sim \ln(m_\pi^2/\mu^2)$ rather than $m_\pi^2 \ln(m_\pi^2)$.

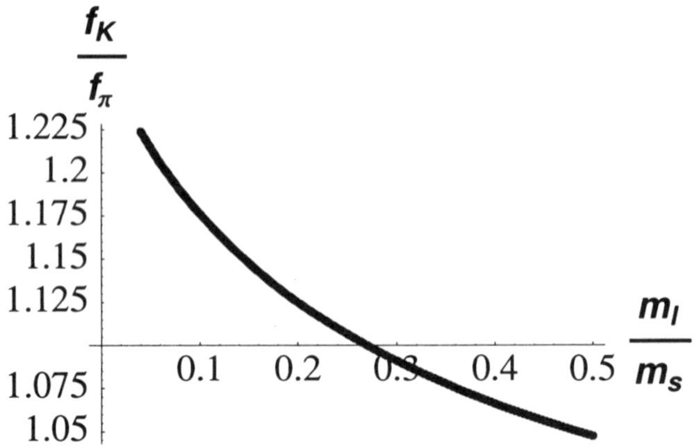

Figure 5. Typical behavior of f_K/f_π as a function of m_l/m_s at NLO.

using the results above. On the lattice, however, one can change the sea quark masses and thus determine these LECs more easily. Using PQ simulations (varying valence and sea masses independently) further simplifies the determinations, as will be discussed in Sec. 4.5.

2.4.2. *Volume dependence from* χPT

For single particle matrix elements, PGB loops loops give the leading correction due to the use of a finite spatial volume. To obtain the leading volume dependence, one simply replaces the momentum integrals with sums, e.g.

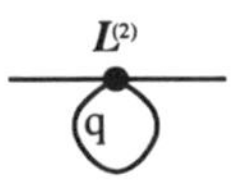

$$\rightarrow \int_q \left(\frac{1}{q^2+m_{NG}^2}\right) \rightarrow \int_{q_4} \sum_{\vec{q}=2\pi\vec{n}/L} \left(\frac{1}{q^2+m_{NG}^2}\right).$$

This is now done routinely when fitting lattice data. Figure 6 illustrates the rapid growth in finite volume shifts as the quark mass is reduced.

Unfortunately, recent work suggests that an accurate estimate of volume corrections requires the inclusion of at least the dominant part of the two-loop contributions[34]. The requisite calculation has only been done for a few quantities, so for others one may be forced to work in volumes large enough that finite volume effects are negligible. For determining such volumes one-loop results should be a reliable guide if used conservatively.

If the quark mass is reduced at fixed box size satisfying $L \gg 1/\Lambda_{QCD}$, one eventually enters the so-called "$\epsilon-$regime" where $m_{NG}L \lesssim 1$. Here the pion propagator is completely distorted by finite volume effects. It turns

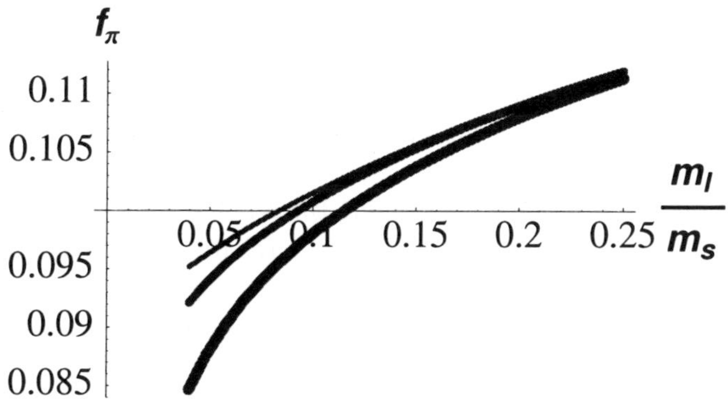

Figure 6. f_π vs. m_l/m_s at $a = 0.1$ fm, with $L = 2.4$ fm (thick), 3.2 fm (medium) and ∞ (thin), using $m_s = 0.08$ GeV, $f = 0.08$ GeV, $L_5 = 1.45 \times 10^{-3}$, $L_4 = 0$.

out, however, that one can still use χPT to predict the form of correlation functions[35]. I will not discuss this regime further, but note that there is an ongoing effort to determine the LECs of QCD (including electroweak interactions) by comparing the results of simulations in the ϵ−regime to the predictions of χPT[36].

2.4.3. *Convergence of χPT*

I have only scratched the surface of calculations in χPT, which have been extended to include the electroweak Hamiltonian in the PGB sector and to NNLO (as reviewed, e.g., by Bijnens[37]). Many quantities are relevant for lattice simulations—I give a list below in Sec. 4.5.1 when discussing PQχPT. Here I only discuss what has been learned about the important question of convergence of χPT.

I quote one example[38], obtained from a fit of the $N = 3$ NNLO formulae to a number of experimental inputs, but with the NNLO LECs estimated approximately using "single resonance saturation". One of the input quantities is f_K/f_π, and this turns out to have a chiral expansion[37]

$$f_K/f_\pi = \underbrace{1}_{\text{LO}} + \underbrace{0.169}_{\text{NLO}} + \underbrace{0.051}_{\text{NNLO}}, \tag{35}$$

showing reasonable convergence. The convergence is less good, however, for the PGB masses.

The naive conclusion from this and similar results is that NNLO terms in χPT are needed for good accuracy. This is, I think, correct if one does a global fit to several quantities using $SU(3)$ χPT. One might, in practice, be able to get away with including only the *analytic* terms at NNLO (whose form is easy to determine) rather than the full two-loop expression. This is the approach used in the MILC analysis[3]. This amounts to mocking up the two-loop contributions by changing the NLO and NNLO LECs. While this makes the results for these LECs less reliable, I expect it to impact the extrapolated results for physical quantities only at the level of NNNLO corrections.[n] Clearly, though, a full NNLO fit would be preferable.

Another approach which reduces the impact of NNLO terms is to use $SU(2)$ χPT alone, treating m_s as heavy. After all, the actual extrapolation being done in present simulations is for the light quarks alone, with m_s fixed near its physical value. In this approach the kaon and eta are treated as heavy particles, and one makes no assumption about the convergence of the expansion in m_s. The idea is that this removes the dominant contribution to the corrections in (35) and sums them to all orders. In practice, this approach has been used primarily in the baryon sector.

2.4.4. *Extension to "heavy" particles*

I will not describe χPT technology for including heavy particles here, but I do want to mention the form of the results. "Heavy" means $m_{\rm had} \gtrsim \Lambda_\chi$, and the approach is to expand in $1/m_{\rm had}$ so that at LO the hadron is a static source for PGBs. In this way one can include the dominant long-distance physics which gives rise to curvature at small light-quark mass.

The form of the resulting chiral expansion depends on the quantity considered. For heavy-light meson decay constants it is similar to that for

[n]This is based on the following argument. The dominant NNLO terms are those involving m_s^2, either explicitly or through factors of m_K or m_η. These are of size $(m_K/\Lambda_\chi)^4 \approx 0.03$ relative to LO terms, consistent with the result in Eq. (35). (This is to be compared to purely light quark NNLO contributions—$(m_\pi/\Lambda_\chi)^4 \approx 0.0002$—and mixed light-strange contributions—$m_\pi^2 m_K^2/\Lambda_\chi^4 \approx 0.003$.) The m_s^2 terms can involve logarithms of m_K or m_η, but not m_π, since they cannot be singular when $m_u = m_d \to 0$. It follows that the dominant NNLO logarithms are being evaluated far from the meson masses where they are non-analytic ($m_K, m_\eta = 0$), and thus can be well represented by analytic terms. This will be especially true if NNNLO LECs are included, as in some MILC fits. The subleading NNLO logarithms involving m_π will be much less well represented by analytic terms, but these are numerically smaller than the NNNLO contributions proportional to m_s^3.

PGB properties, e.g.

$$F_B \sim F_{B,0}(1 + \underbrace{m_\pi^2}_{\text{analytic}} + \underbrace{m_\pi^2 \ln(m_\pi)}_{\text{chiral log}} + \dots) \tag{36}$$

One new feature is that the non-analytic terms are not predicted in terms of the LO LECs, but involve an additional coefficient, $g_{\pi BB^*}$.

For baryons and vector meson masses the expansion differs further, involving odd powers of m_{PGB}.[o]

$$M_H \sim M_0 + \underbrace{m_\pi^2}_{\text{analytic}} + \underbrace{g_{\pi HH'}\, m_\pi^3}_{\text{leading loop}} + \underbrace{m_\pi^4 \ln(m_\pi)}_{\text{subleading loop}} + m_\pi^4 + \dots \tag{37}$$

This means that the expansion is in powers of m_π/Λ_χ (c.f. $(m_\pi/\Lambda_\chi)^2$ for PGBs and heavy-light mesons), so that the convergence is generically poorer. Thus it is even more important to use light quark masses when studying baryon properties.[p]

3. Incorporating discretization errors into χPT

In this lecture I describe how for Wilson and twisted-mass fermions one can incorporate discretization errors into χPT, and what one learns by doing so. The method is general, and has been applied also to staggered fermions[8,9], and to mixed-action theories[39]. See also the review by Bär[40].

3.1. *Why incorporate discretization errors?*

At first sight, it may seem strange to incorporate the effects of the *ultraviolet* (UV) cut-off of the underlying theory into the EFT describing its *infrared* (IR) behavior. The key point is that the UV effects break the chiral symmetry which determines the IR behavior. One way of saying this is that discretization errors lead to a non-trivial potential in the vacuum manifold which is otherwise flat due to the symmetry. As we will see, symmetry breaking due to quark masses and discretization errors have comparable effects on the potential if $m/\Lambda \approx (a\Lambda)^2$, with Λ a scale of $O(\Lambda_{\rm QCD})$. The appropriate value of Λ depends on the action, and it is quite possible that this condition is satisfied even for relatively fine lattices and light quarks.

[o]There is a similar "m_π^3" contribution to heavy-light meson masses but there the leading term is $M_{\rm heavy-light} \gg \Lambda_{\rm QCD}$, so the correction is less important.

[p]For vector mesons, and unstable baryons, the chiral expansion is yet more complicated because of the opening of the decay channel as the quark mass is reduced.

For example, if $\Lambda = 500$ MeV and $a^{-1} = 2$ GeV, then it is satisfied when $m = 30$ MeV. Thus it is imperative to study the impact of discretization errors.[q]

One question that often arises in the present context is whether one should first extrapolate $a \to 0$ and then use continuum χPT, or do combined extrapolation in $a \to 0$ and $m \to m_{\text{phys}}$. The possibilities are illustrated to the right.

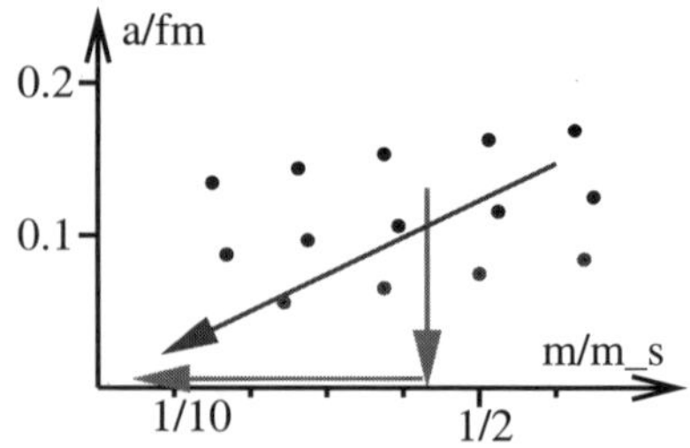

An apparent advantage of the first approach is that one does not have to rely on the validity of χPT to do the continuum extrapolation; one simply uses a standard polynomial *ansatz.* There are, however, several reasons to use the second, "combined", approach if χPT formulae are available:

- It incorporates relations between discretization errors in different quantities that follow from the specific way in which chiral symmetry is broken.
- It accounts for non-analyticities in a which arise because of pion loops (e.g. for staggered fermions one has, schematically, $m_\pi^2 \sim m_q[1 + (m_q + a^2)\ln(m_q + a^2) + \ldots])$. These might well be missed in a simple polynomial continuum extrapolation. The "a^2" in the chiral logs reduce the curvature, as clearly observed in the MILC results[3]. It should be kept in mind, however, that "a^2" always means "up to logs", so not all non-analyticities are included.
- It accounts for changes in orientation of the condensate, which can be rapid with twisted-mass fermions if $m \sim a^2$, as discussed below.

3.2. *General strategy*

One proceeds in two steps[41]. First, following Symanzik[20], determine the continuum EFT describing the interactions of quarks and gluons with $|p| \ll 1/a$. Discretization errors enter with explicit factors of a, and are

[q]As an aside, I note that if one uses lattice fermions with an exact on-shell chiral symmetry (overlap, perfect or Domain-wall fermions with $N_5 \to \infty$) then the considerations of this lecture become almost trivial. These fermions are described by *continuum* χPT, but with LECs that depend on a^2 and must be extrapolated to the continuum limit. The only exception is that there are additional terms induced by the breaking of Euclidean symmetry, but these are of very high order in the meson sector, as discussed below.

controlled by the symmetries (or lack thereof) of the underlying lattice theory. Second, use standard techniques to develop χPT for the Symanzik EFT. Since the latter is a continuum theory, this is no different conceptually from determining the effect of beyond-the-standard-model physics on the IR properties of QCD. The two steps are illustrated in Fig. 2 above.

3.3. *Application to Wilson & twisted mass fermions*

Twisted mass lattice QCD (tmLQCD)[42] has received a lot of recent attention because of its improved algorithmic properties and because, at maximal twist, physical quantities (including matrix elements) are automatically $O(a)$ improved[43]. Here it also serves as an excellent example, particularly as it contains Wilson fermions as a subset.[r]

Twisting the mass in continuum QCD simply means doing an $SU(3)_L \times SU(3)_R$ rotation. The standard diagonal mass M_0 (which is hermitian assuming real m_q) is rotated into $M = U_L M_0 U_R^\dagger$ which is not hermitian. The example I consider in detail has two degenerate flavors and

$$M = m_q e^{i\tau_3\omega} = m_q(\cos\omega + i\sin\omega\tau_3) \equiv m + i\mu\tau_3\,, \tag{38}$$

resulting in a Lagrangian mass term containing a $\gamma_5\tau_3$ part:

$$\overline{Q}_L M Q_R + \overline{Q}_R M^\dagger Q_L = \overline{Q}(m + i\mu\gamma_5\tau_3)Q\,. \tag{39}$$

Note that m_q is the physical quark mass, with m and μ respectively the untwisted and twisted components.

The "geometry" of the parameters is shown to the right. Although it naively appears that parity and flavor are broken, we know this is not the case since physics is unchanged by the chiral rotation. Thus ω is a redundant parameter. Usually, we keep the symmetries manifest by working at $\omega = 0$, but it is important to know that we do not need to do so. The continuum

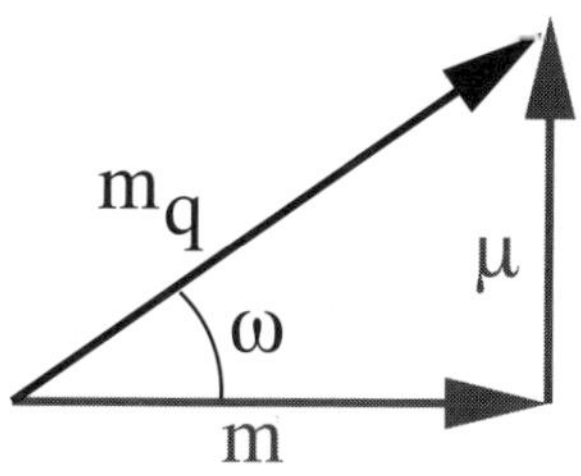

χPT analysis described above goes through for any ω, as long as we expand about the rotated vacuum.

[r]For more extensive discussion of tmLQCD and its applications see the lectures by Sint, and recent reviews[44,45].

The situation is quite different after discretization. The lattice action is

$$S_{l,\text{glue}} + a^4 \sum_x \overline{\psi}_l D_W \psi_l + \overline{\psi}_{l,L} M \psi_{l,R} + \overline{\psi}_{l,R} M^\dagger \psi_{l,L}\,, \tag{40}$$

with M the twisted mass, and the subscript "l" indicating "lattice". Here D_W is Wilson's doubler-free derivative,

$$\not{D} \longrightarrow D_W = \frac{1}{2}\sum_\mu \gamma_\mu(\nabla^*_\mu + \nabla_\mu) - \frac{1}{2}\sum_\mu (\nabla^*_\mu \nabla_\mu) \tag{41}$$

(∇ and ∇^* are forward and backward derivatives, respectively). Since the second term in D_W (the "Wilson term", in which I have set the Wilson parameter $r = 1$) breaks chiral symmetry, one cannot rotate away the twist in the mass. The theories with mass term M_0 and $U_L M_0 U_R^\dagger$ are different on the lattice. In fact, the full fermion matrix $D_W + MP_R + M^\dagger P_L$ has positive determinant (and is thus useful in practice) only for special M. One such choice is two flavors with the twisted mass (38), for which the lattice action is

$$S_{\text{tmLQCD}} = S_{l,\text{glue}} + a^4 \sum_x \overline{\psi}_l \left(D_W + m_0 + i\gamma_5\tau_3\mu_0\right)\psi_l\,. \tag{42}$$

Here m_0 and μ_0 are, respectively, the bare untwisted and twisted mass (in lattice units). For the remainder of this lecture I will focus entirely on this theory.

3.4. *Determining the local effective Lagrangian*

Symanzik[20] showed how to study the approach of the lattice theory to its continuum limit. The first step is to understand this limit itself. The lattice provides a legitimate regularization of QCD (one that is awkward from a perturbative point of view, but has the great advantage of being non-perturbative), and so one obtains continuum QCD (in this case tmQCD) as the cut-off $1/a$ is sent to infinity. This has been established to all orders in perturbation theory[46] and it assumed to hold non-perturbatively. One must appropriately tune the ("relevant") lattice bare parameters to reach the continuum limit. In particular, since the Wilson term mixes with the identity operator, m_0 is additively renormalized, and must be tuned to $m_c(a) < 0$, while the twisted mass has nothing to mix with and is multiplicatively renormalized.[s] The resulting continuum theory is

$$\mathcal{L}_{\text{tmQCD}} = \mathcal{L}_{\text{glue}} + \overline{\psi}(\not{D} + m + i\gamma_5\tau_3\mu)\psi\,, \tag{43}$$

[s]One must also tune the bare coupling $g_0 \to 0$ in the usual way.

with $\mathcal{L}_{\text{glue}}$ the usual continuum gluon action, and with the continuum field and physical masses being

$$\psi = a^{-3/2} Z\, \psi_l\,, \quad m = Z_S^{-1}(m_0 - m_c)/a\,, \quad \text{and} \quad \mu = Z_P^{-1}\mu_0/a\,. \tag{44}$$

Here Z, Z_S and Z_P are renormalization factors relating quantities in the lattice regularization to those in the chosen continuum scheme. The corresponding geometry is illustrated below.

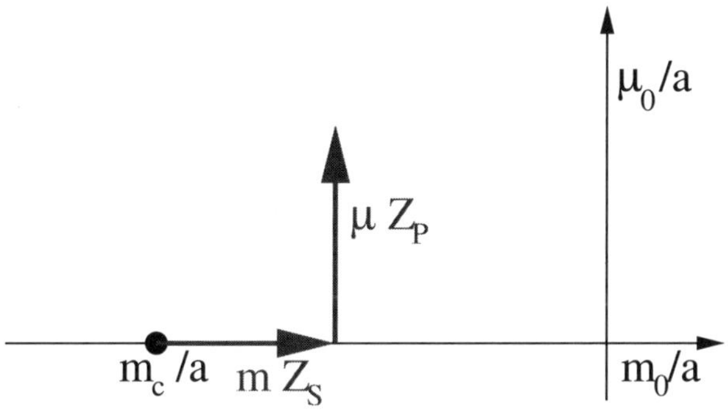

Corrections to the continuum limit are suppressed by inverse powers of the cut-off, i.e. by positive powers of a. Symanzik showed how these can collected into a local effective Lagrangian

$$\mathcal{L}_{\text{Sym}} = \mathcal{L}_{\text{tmQCD}} + a\mathcal{L}^{(5)} + a^2\mathcal{L}^{(6)} + \ldots\,, \tag{45}$$

where $\mathcal{L}_{\text{tmQCD}}$ is the desired continuum Lagrangian (note the absence of the "L" in the subscript), and $\mathcal{L}^{(5)}$ and $\mathcal{L}^{(6)}$ contain terms of dimension 5 and 6, respectively.[t] The effective theory must be regularized, either by a standard continuum regulator such as dimensional regularization, or possibly with a finer lattice having $a' \ll a$. We will not actually use $\mathcal{L}_{\text{Sym}}$ for concrete calculations so do not need to be more specific. A key feature of $\mathcal{L}_{\text{Sym}}$ is that all factors of a are explicit—the effective theory does not "know" about the lattice spacing in any other way. Note, however, that "a" includes logarithms, of the form $\sim a[1 + g(a)^2 \ln a + \ldots]$. I return to this point below.

The content of Eq. (45) is that all discretization errors in all correlation functions can be reproduced by a set of *local* insertions. This is established

[t]The number of effective Lagrangians in these lectures is approaching a confusing level. Note that I always use the subscript χ for the chiral Lagrangian, so that the six-derivative contribution $\mathcal{L}_\chi^{(6)}$ can be distinguished from $\mathcal{L}^{(6)}$ in Eq. (45).

by a procedure akin to renormalization (in which one determines divergent parts of graphs by doing a Taylor expansion in the external momenta, or a lattice variant of this procedure[46], and then subtracts them), except that one subtracts more terms in the Taylor expansion ("over-subtraction"), thus including those proportional to powers of a. As with renormalization, the consistency of this procedure requires that one include *all terms in* $\mathcal{L}^{(n)}$ *of the appropriate dimension which are invariant under the symmetries of the theory*—here, of tm**L**QCD. These terms will have coefficients such that reflection positivity is satisfied, since they arise from a theory in which it is satisfied.[u] This procedure also works if one includes sources for external operators, which should be treated using the spurion trick. The procedure has been demonstrated to all orders in perturbation theory, and is assumed to work non-perturbatively.

The preceding discussion is nothing other than (a sketch of) a derivation of an EFT. The usual EFT words—"separation of scales"—were not mentioned but were implicit. $\mathcal{L}_{\mathrm{Sym}}$ is only useful if $p \ll 1/a$, for otherwise successive terms in the expansion, which give contributions of relative size ap, are not suppressed. Thus $\mathcal{L}_{\mathrm{Sym}}$ is an EFT for quarks and gluons with energies far below the cut-off scale. The set-up is the same as when considering the impact of new short-distance physics on QCD, except that the new physics here violates rotation and translation symmetries. It indicates how one can derive an EFT in a Euclidean context, at least order-by-order in perturbation theory. One does not need to rely on the S-matrix argument of Weinberg. This is important because the underlying lattice theory is discretized in Euclidean space.

Let me illustrate these general words with a simple example. Consider the quark-gluon vertex shown in Fig. 7. In EFT language, the counterterms $\mathcal{L}^{(5,6,\dots)}$ are to be determined by matching correlation functions with those of the lattice theory. At tree-level, the $O(a)$ terms in the lattice vertex can be matched by adjusting the coefficients of operators in $\mathcal{L}^{(5)}$ such as $\overline{\psi}D^2\psi$ as long as $|p| \ll 1/a$. At one-loop, for $|q| \ll 1/a$, the integrands match by construction, giving a logarithmic divergence. For larger $|q|$, however, the matching fails, leading to a finite difference in the results (finite since both theories are regulated). To match one must add an $O(g^2)$ contribution to

---

[u] This is true for the action used in the text. Reflection positivity is violated with improved Wilson fermions or improved gauge actions, but it is expected that this does not effect the long distance physics which is being captured by $\mathcal{L}_{\mathrm{Sym}}$.

the coefficients of the operators in $\mathcal{L}^{(5)}$. One loop matching is schematically:

$$apg^2(\ln[pa] + r_{\rm lat}) = apg^2(\ln[pa'] + r_{\rm EFT}) + apg^2 c^{(2)}_{\rm EFT} \tag{46}$$

$$\Rightarrow \quad c^{(2)}_{\rm EFT} = r_{\rm lat} - r_{\rm EFT} + \ln[a/a'] \,, \tag{47}$$

where $r_{\rm lat,EFT}$ are the finite parts of the loop diagram, and I have used a lattice regularization of the EFT with spacing a'. We can now see the generic form of the a dependence of the coefficients in $\mathcal{L}^{(5)}$ (whose one-loop contribution is given here by $g^2 c^{(2)}_{\rm EFT}$). There is explicit logarithmic dependence, and an implicit logarithmic dependence through g, which is evaluated in the lattice calculation at a scale $\sim a$.

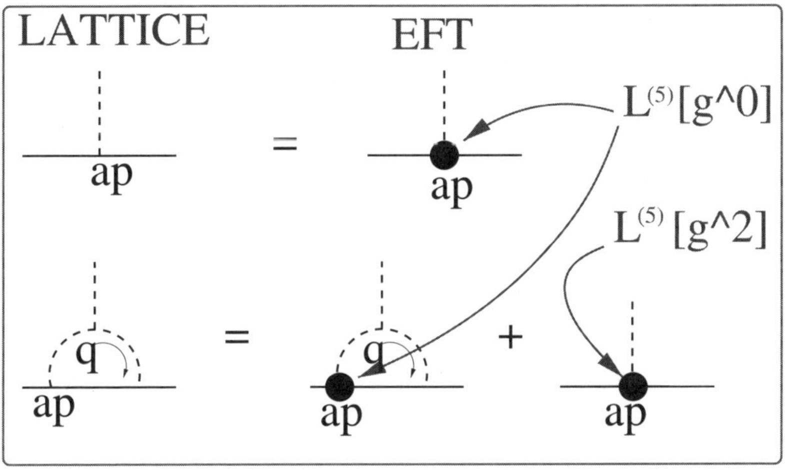

Figure 7. Matching lattice QCD with the Symanzik EFT. Dashed lines are gluons, solid lines quarks, $p \ll 1/a$ an external momentum, q a loop momentum.

As an aside, I note that, having determined the form of $\mathcal{L}^{(5)}$ in the EFT, one can add corresponding terms to the lattice theory (i.e. terms having operators in $\mathcal{L}^{(5)}$ as their classical continuum limit), and then adjust the coefficients of the lattice terms to set those in $\mathcal{L}^{(5)}$ to zero. This is the "improvement program" at $O(a)$[20], and it has been implemented non-perturbatively[19,47]. In the "new physics" context this would be called "fine tuning", with negative connotations, but here we can again turn all the knobs at our disposal to improve extrapolations. The program can be extended, in principle, to any order, but has in practice not been extended to $\mathcal{L}^{(6)}$. Since $\mathcal{L}^{(6)}$ will play a key role in the following, this means that the considerations below are essentially unaffected by improvement.

3.5. *Symanzik effective action for tmLQCD*

We are now ready to determine the operators in $\mathcal{L}^{(5)}$ and $\mathcal{L}^{(6)}$. These must satisfy the symmetries of tmLQCD, Eq. (42), and be reflection positive. The symmetries are gauge invariance, lattice rotations and translations, charge conjugation and fermion number, but *not* flavor $SU(2)$ nor parity (for generic ω). Only the $U(1)$ flavor subgroup generated by τ_3, and combinations of parity with a discrete flavor rotations survive:

$$\mathcal{P}_F^{1,2}: \quad \psi_l(x) \to \gamma_0(i\tau_{1,2})\psi_l(x_P)\,, \quad \overline{\psi_l}(x) \to \overline{\psi_l}(x_P)(-i\tau_{1,2})\gamma_0\,, \tag{48}$$

Also useful is $\widetilde{\mathcal{P}}$: parity combined with $[\mu_0 \to -\mu_0]$.

The continuum Lagrangian consistent with these symmetries is tmQCD, Eq. (43). To determine $\mathcal{L}^{(5)}$ one simply enumerates all allowed operators, generalizing the work done for Wilson fermions[19]. The result is[48]

$$\begin{aligned}\mathcal{L}^{(5)} &= b_1\overline{\psi}i\sigma_{\mu\nu}F_{\mu\nu}\psi + b_2\overline{\psi}(\not{D} + m + i\gamma_5\tau_3\mu)^2\psi \\ &\quad + b_3 m\overline{\psi}(\not{D} + m + i\gamma_5\tau_3\mu)\psi + b_4 m L_{\text{glue}} + b_5 m^2\overline{\psi}\psi \\ &\quad + b_6\mu\overline{\psi}\big[(\not{D} + m + i\gamma_5\tau_3\mu), i\gamma_5\tau_3\big]_+\psi + b_7\mu^2\overline{\psi}\psi\,,\end{aligned} \tag{49}$$

where I use continuum masses m, μ rather than bare masses. The coefficients b_i (the analog of the LECs of χPT) are real (from reflection positivity), and depend on $g^2(a)$ and $\ln a$. Among them $b_{6,7}$ are "new" compared to Wilson case. Many terms have been forbidden by lattice symmetries: $\widetilde{P}$ forbids $m\mu\overline{\psi}\psi$ and $m^2\overline{\psi}i\gamma_5\tau_3\psi$, and it requires $\overline{\psi}\tau_3\gamma_5\psi$ to come with factor $\mu_0 \propto \mu$, and the twisted Pauli term $\overline{\psi}\sigma_{\mu\nu}F_{\mu\nu}\tau_3\psi$ to have factor of μ (so that it appears in $\mathcal{L}^{(6)}$); flavor $U(1)$ forbids $\overline{\psi}\tau_{1,2}\psi$; and $\mathcal{P}_F^{1,2}$ forbids $\overline{\psi}\gamma_5\psi$, $\widetilde{F}_{\mu\nu}F_{\mu\nu}$ and $\overline{\psi}\tau_3\psi$.

$\mathcal{L}^{(5)}$ looks rather forbidding, with 7 unknown coefficients, but this proliferation is misleading, for two reasons. First, we will be doing a joint continuum-chiral expansion, and working in the "generic small mass" (GSM) regime in which $m \sim \mu \sim a\Lambda_{\text{QCD}}^2$. Thus each factor of m or μ counts as an additional power of a. We will work at NLO in this power counting. Since $\overline{\psi}\not{D}\psi$ and $\mathcal{L}_{\text{glue}}$ map into LO operators in χPT, and $\mathcal{L}^{(5)}$ comes with an overall factor of a, any further factors of m or μ make the operator of next-to-next-to-leading order (NNLO). This allows us to drop all except the b_1 term, and the $\overline{\psi}\not{D}^2\psi$ parts of the b_2 and b_3 terms.[I] Second, we will be mapping $\mathcal{L}_{\text{Sym}}$ into χPT, at which point all that matters is the

[I] In the usual discussion of on-shell Symanzik improvement, one drops terms vanishing by the LO equations of motion, which only contribute to contact terms. This is not necessary here, but explains the basis used in Eq. (49).

chiral transformation properties of the operators. Now the Pauli (b_1) term and $\overline{\psi}\not{D}^2\psi$ transform the same way, so, since the coefficients in the mapping to χPT are unknown, we can drop the latter operator. The outcome is

$$\mathcal{L}^{(5)}_{\rm NLO} = b_1 \overline{\psi} i\sigma_{\mu\nu}F_{\mu\nu}\psi\,, \tag{50}$$

which is unchanged from the result for (untwisted) Wilson fermions.

Moving onto $\mathcal{L}^{(6)}$, there are now gluonic terms[49]

$$\begin{aligned}\mathcal{L}^{(6)}_{\rm glue} \sim\ & {\rm Tr}(D_\mu F_{\rho\sigma} D_\mu F_{\rho\sigma}) + {\rm Tr}(D_\mu F_{\mu\sigma} D_\rho F_{\rho\sigma}) \\ & + \underbrace{{\rm Tr}(D_\mu F_{\mu\sigma} D_\mu F_{\mu\sigma})}_{\text{Lorentz violating}} + \underbrace{(m^2,\mu^2){\rm Tr}(F_{\mu\nu}F_{\mu\nu})}_{O(a^2m^2,\,a^2\mu^2)\text{ so NNNLO}}\,, \end{aligned} \tag{51}$$

where I use a schematic notation without coefficients, and fermionic terms (obtained by generalizing the analysis for Wilson fermions[50,51])

$$\begin{aligned}\mathcal{L}^{(6)}_q \sim\ & \underbrace{\overline{\psi} D^3_\mu \gamma_\mu \psi}_{\text{Euclidean non-invariant}} + \overline{\psi} D_\mu \not{D} D_\mu \gamma_\mu \psi + \dots \\ & + \underbrace{m\overline{\psi}\not{D}^2\psi + \mu\overline{\psi}\not{D}^2 i\gamma_5\tau_3\psi}_{O(a^2m,\,a^2\mu)\text{ so NNLO}} + \underbrace{m\overline{\psi}i\sigma_{\mu\nu}F_{\mu\nu}\psi + \mu\overline{\psi}\sigma_{\mu\nu}F_{\mu\nu}\gamma_5\tau_3\psi}_{O(a^2m,\,a^2\mu)\text{ so NNLO}} \\ & + \underbrace{(m^2,\mu^2)\overline{\psi}\not{D}\psi + m\mu\overline{\psi}\not{D}i\gamma_5\tau_3\psi}_{O(a^2m^2),\text{ etc. so NNNLO}} + \underbrace{(m^3,m\mu^2)\overline{\psi}\psi + (\mu^3,\mu m^2)i\gamma_5\tau_3\psi}_{O(a^2m^3),\text{ etc. so NNNNLO}} \\ & + (\overline{\psi}\psi)^2 + (\overline{\psi}\gamma_\mu\psi)^2 + \dots\,, \end{aligned} \tag{52}$$

where the first ellipsis indicates other Euclidean invariant terms with three derivatives and the second other four-fermion operators. As can be seen, in the GSM regime, most of the fermionic operators in $\mathcal{L}^{(6)}$ are of at least NNLO. In particular, no flavor-parity breaking terms appear, since they require a factor a μ. The net result is that the part of $\mathcal{L}^{(6)}$ of NLO in the GSM regime is the same as that for untwisted Wilson fermions (with the ellipses having the same meaning as above):

$$\begin{aligned}\mathcal{L}^{(6)}_{\rm NLO} \sim\ & {\rm Tr}(D_\mu F_{\rho\sigma} D_\mu F_{\rho\sigma}) + {\rm Tr}(D_\mu F_{\mu\sigma} D_\rho F_{\rho\sigma}) \\ & + \overline{\psi} D_\mu \not{D} D_\mu \gamma_\mu \psi + \dots + (\overline{\psi}\psi)^2 + (\overline{\psi}\gamma_\mu\psi)^2 + \dots \\ & + \underbrace{{\rm Tr}(D_\mu F_{\mu\sigma} D_\mu F_{\mu\sigma}) + \overline{\psi} D^3_\mu \gamma_\mu \psi}_{\text{Euclidean non-invariant}}\,. \end{aligned} \tag{53}$$

Note that the only symmetry broken by $\mathcal{L}^{(6)}_{\rm NLO}$ that is not already broken by $\mathcal{L}^{(5)}_{\rm NLO}$ is Euclidean invariance. In fact, we will see that in the PGB sector the Euclidean non-invariant terms pick up an additional factor of p^2 and (given the overall a^2) are of NNLO.

This takes care of the action, but what about currents and densities? The matching of these between the lattice theory and the EFT can be worked out using symmetries[20], and I quote only the relevant results[19,54]:

$$V_\mu^b = \overline{\psi}\gamma_\mu T^b\psi + a\tilde{c}_V\partial_\nu\overline{\psi}i\sigma_{\mu\nu}T^b\psi\,, \tag{54}$$
$$A_\mu^b = \overline{\psi}\gamma_\mu\gamma_5 T^b\psi + a\tilde{c}_A\partial_\mu\overline{\psi}\gamma_5 T^b\psi\,, \tag{55}$$
$$S^0 = \overline{\psi}\psi + a\tilde{g}_S\mathrm{Tr}(F_{\mu\nu}F_{\mu\nu})\,, \qquad P^b = \overline{\psi}\gamma_5 T^b\psi \tag{56}$$

Here I work at NLO in the GSM regime, so $am \sim a\mu$ terms are dropped. I also drop the mixing of S^0 with the identity operator, as it does not contribute to connected matrix elements. The coefficients $\tilde{c}_{V,A}$ and $\tilde{g}_S$ depend on $g(a)$ and on $\ln(a)$, just like the b_i above. The content of these equations is that the on-shell matrix elements of the operators shown, evaluated in the EFT, will reproduce those of the lattice currents and densities, including the leading discretization error. These forms apply for any choice of lattice currents and densities (e.g. ultra-local or smeared) as long as they have been multiplied by appropriate Z-factors so as to be correctly normalized. The numerical values of the coefficients $\tilde{c}_V$ etc. will, of course, depend on the form of the lattice operators, and on the lattice action. In particular, if the action and operators have been $O(a)$ improved, then these coefficients will vanish. Note that the density P^b is automatically improved.

When we map the operators in (54-56) into χPT we are free to do this for the $O(1)$ and $O(a)$ parts separately and then combine at the end. It is straightforward to see that the $\tilde{c}_V$ and $\tilde{g}_S$ terms map into operators which are of NNLO and can be dropped. For the former, the argument is given by Wu and I[55], and follows because of the need to have three derivatives in order to match the quark-level operator. For the $\tilde{g}_S$ term the argument is even more simple: the matching of chiral singlet $\mathrm{Tr}(F_{\mu\nu}F_{\mu\nu})$ gives the LO chiral kinetic term $\mathrm{Tr}(\partial_\mu\Sigma\partial_\mu\Sigma^\dagger)$. Thus the $\tilde{g}_S$ term is of size ap^2, and so of NNLO compared to $\overline{\psi}\psi$, which maps to $\mathrm{Tr}(\Sigma + \Sigma^\dagger) \sim O(1)$.

I conclude that, at NLO in the GSM regime, the currents and densities in the Symanzik EFT have the same form as in the continuum, aside from the $\tilde{c}_A$ term. Thus, as long as one treats the $\tilde{c}_A$ term separately, one can include the currents and densities using sources just as in continuum QCD, and the resulting theory will have a *local* chiral symmetry.[II]

[II]There is a one subtlety here. After adding in the $O(1)$ parts of the currents and densities, the local invariance is only true for the continuum part of the Symanzik action, $\mathcal{L}_{\mathrm{tmQCD}}$. It can be extended to $\mathcal{L}^{(5)}_{\mathrm{NLO}}$, however, by allowing the corresponding spurion, called $\widetilde{A}$ below, to transform like χ under the local chiral symmetry. It can also be

3.6. *Mapping the Symanzik action into χPT*

I now turn to the second step of the procedure—taking the Symanzik EFT and determining the chiral EFT which describes it at long distances. This was done for continuum Lagrangian, $\mathcal{L}_{\rm tmQCD}$, in Sec. 2—a twisted mass was already included by the generality of the formalism. The result is $\mathcal{L}_\chi^{(2)}$ of Eq. (28) at LO and $\mathcal{L}_\chi^{(4)}$ of Eq. (31) at NLO. The task here is to include the effects of $\mathcal{L}^{(5,6)}$, as well as the $\tilde{c}_A$ term in Eq. (55). To do so systematically requires a power counting scheme, to which I now turn.

3.6.1. *Power counting and terminology*

As anticipated above, discretization errors introduce a new parameter into the power counting in χPT. In addition to the usual chiral expansion in powers of $p^2/\Lambda_\chi^2 \sim m_{NG}^2/\Lambda_\chi^2 \sim m_q/\Lambda_{\rm QCD}$, we must include $a\Lambda_{\rm QCD}$. Note that $\Lambda_{\rm QCD}$ is the only scale available to balance dimensions when we map quark-level operators in $\mathcal{L}_{\rm Sym}$ into χPT. I stress that, once one has the Symanzik EFT in hand, m and a are on a similar footing—both are simply small parameters in a continuum Lagrangian.

How we should weight discretization errors relative to mass corrections? The numerical comparison is shown in Fig. 8. I have been conservative by having "present simulations" range down to $m_s/10$, which is yet to be achieved with Wilson-like fermions.[III] I conclude from the figure that the appropriate power counting for the coming decade is $a^2\Lambda_{\rm QCD}^3 \lesssim m_q \lesssim a\Lambda_{\rm QCD}^2$, and that to disentangle quark mass dependence from discretization effects we need to remove errors of $O(a)$ and understand those of $O(a^2)$.

I will consider in the following two regimes (sketched below):

extended to $\mathcal{L}_{\rm NLO}^{(6)}$, but this is not actually necessary because the resulting contributions to the currents are of NNLO. I should note that there is some disagreement on the validity of this approach for mapping currents and densities[56,57], which is why I have given here a more detailed discussion than is present in the literature[55].

[III]Note that the comparison is not precise: the relative coefficients of mass and discretization effects in the EFT are expected to be of $O(1)$, but could easily be $2-3$.

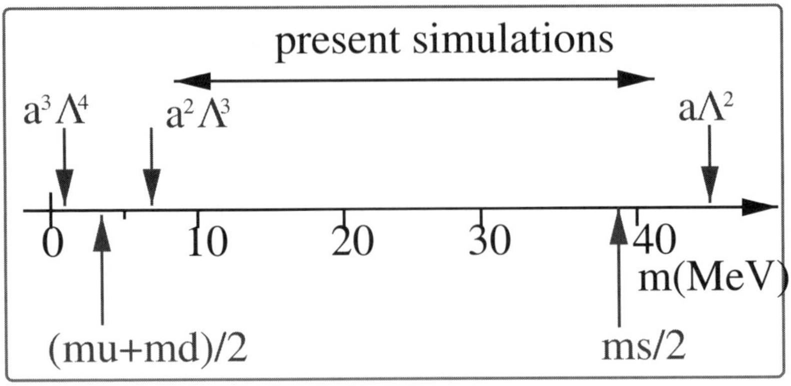

Figure 8. Comparing discretization errors to mass corrections. I use $a^{-1} = 2$ GeV and $\Lambda = 300$ MeV.

I. The GSM regime, already introduced above, which I define more precisely as $a\Lambda^2_{\rm QCD} \lesssim m_q \ll \Lambda_{\rm QCD}$, so that it includes $a\Lambda^2_{\rm QCD} \ll m_q$ (where we would like to be) as well as $a\Lambda^2_{\rm QCD} \approx m_q$ (where we actually are). This is the regime in which we want to learn how to remove $O(a)$ errors.

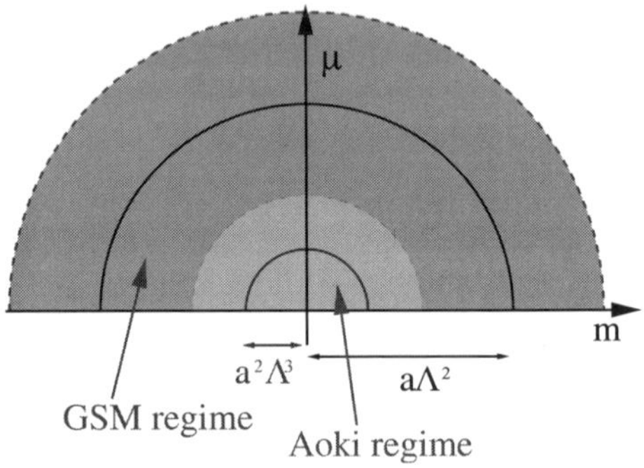

II. The "Aoki regime", in which $m_q \lesssim a^2\Lambda^3_{\rm QCD} \ll \Lambda_{\rm QCD}$ (including $m_q \ll a^2\Lambda^3_{\rm QCD}$). This is where we cannot avoid discretization errors, and where they lead to non-trivial phase structure. Entering this regime changes the relative weight given to operators in χPT, but not the operators themselves.

I begin by working in the GSM regime at NLO. This requires keeping terms of $O(a^2)$, and thus up to $\mathcal{L}^{(6)}$ in the Symanzik expansion. The enumeration will turn out to suffice also for a LO study of the Aoki regime.

3.6.2. *Mapping $\mathcal{L}^{(5)}_{\rm NLO}$ and $\mathcal{L}^{(6)}_{\rm NLO}$ into χPT*

$\mathcal{L}^{(5)}_{\rm NLO}$ transforms just like a mass term under $SU(2)_L \times SU(2)_R$ (recall that we set $N = 2$, although the considerations are easily generalized):

$$a\mathcal{L}^{(5)}_{\rm NLO} \sim a\overline{\psi} i\sigma_{\mu\nu}F_{\mu\nu}\psi = \overline{\psi}_L \widetilde{A} i\sigma_{\mu\nu}F_{\mu\nu}\psi_R + \overline{\psi}_R \widetilde{A}^\dagger i\sigma_{\mu\nu}F_{\mu\nu}\psi_A\,. \tag{57}$$

Here $\widetilde{A}$ is a spurion transforming like M, i.e. $\widetilde{A} \to U_L \widetilde{A} U_R^\dagger$, so that $\mathcal{L}^{(5)}_{\rm NLO}$

is invariant. At the end we set $\widetilde{A} = a$. The enumeration of operators in the chiral Lagrangian is like that for M, yielding the new terms[52,51,53,55]

$$\mathcal{L}^{(2)}_{\chi,A} = -\frac{f^2}{4}\mathrm{tr}(\hat{A}^\dagger\Sigma + \Sigma^\dagger\hat{A}) \tag{58}$$

$$\begin{aligned}\mathcal{L}^{(4)}_{\chi,A} &= W_{45}\mathrm{tr}(D_\mu\Sigma^\dagger D_\mu\Sigma)\mathrm{tr}(\hat{A}^\dagger\Sigma + \Sigma^\dagger\hat{A}) - W'_{68}\left[\mathrm{tr}(\hat{A}^\dagger\Sigma + \Sigma^\dagger\hat{A})\right]^2 \\ &-W_{68}\mathrm{tr}(\chi^\dagger\Sigma + \Sigma^\dagger\chi)\mathrm{tr}(\hat{A}^\dagger\Sigma + \Sigma^\dagger\hat{A}) + W_{10}\mathrm{tr}(D_\mu\hat{A}^\dagger D_\mu\Sigma + D_\mu\Sigma^\dagger D_\mu\hat{A}) \\ &-H'_2\mathrm{tr}(\hat{A}^\dagger\chi + \chi^\dagger\hat{A}) - H_3\mathrm{tr}(\hat{A}^\dagger\hat{A}),\end{aligned} \tag{59}$$

where $\hat{A} = 2W_0\widetilde{A}$ is the analog of $\chi = 2B_0(s + ip)$, with W_0 a new LEC at LO. We will not need to use $\hat{A}$ as a source (since we already have χ), so we will always be set $\hat{A} \to \hat{a} = 2W_0 a$. $SU(2)$ simplifications can then reduce the number of terms. In particular $(\hat{A}^\dagger\Sigma + \Sigma^\dagger\hat{A})$ is proportional to the identity, allowing single trace terms involving this combination to be rewritten with two traces. These simplifications have been used in (59).

Note that, following the discussion at the end of Sec. 3.5, I have included sources for currents and densities and enforced *local* chiral invariance by using covariant derivatives D_μ [defined as in Eq. (27)]. This incorporates all discretization effects in lattice matrix elements except those due to the $\tilde{c}_A$ term in Eq. (55), which I will treat separately below.

$\mathcal{L}^{(6)}_{\mathrm{NLO}}$ contains three types of terms. First, those that are invariant under Euclidean and chiral symmetries [the gluonic terms on the first line of Eq. (53) and some of the four-fermion operators]. These match into χPT as follows:[8]

$$a^2\mathrm{Tr}(D_\mu F_{\rho\sigma}D_\mu F_{\rho\sigma}) + \cdots + a^2\overline{\psi}D_\mu \not{D} D_\mu\gamma_\mu\psi + \ldots \longrightarrow a^2\mathrm{tr}(D_\mu\Sigma D_\mu\Sigma^\dagger), \tag{60}$$

i.e. one obtains the leading order continuum result multiplied by a^2. This leads to an $O(a^2)$ correction to the LEC f, and is present for any fermion discretization (including chirally invariant ones).

Second, there are four-fermion operators which violate chiral symmetry (e.g. those having LR-LR structure). Their matching can be analyzed using two $\hat{A}$ spurions, leading to[51]

$$(\overline{\psi}\psi)^2 + (\overline{\psi}\gamma_\mu\psi)^2 + \ldots \longrightarrow \mathrm{tr}\left[(\hat{A}^\dagger\Sigma + \Sigma^\dagger\hat{A})\right]^2. \tag{61}$$

This operator is already present in $\mathcal{L}^{(4)}_{\chi,A}$, having been produced by two insertions of $\mathcal{L}^{(5)}_{NLO}$. This illustrates that what is relevant for matching are the symmetries broken by the operators (here, chiral symmetry), and not their detailed form. The four-fermion operators simply change an unknown coefficient, W'_{68}, by an unknown amount. The only exception is if one uses a non-perturbatively $O(a)$ improved quark action, as discussed below, when W'_{68} would vanish were it not for the four-fermion operators.

Finally, there are the terms violating Euclidean symmetry. These can be decomposed into Euclidean singlet and non-singlet parts. The former match as in Eq. (60), while the latter give rise to Euclidean non-invariant chiral operators[8]

$$a^2 \mathrm{Tr}(D_\mu F_{\mu\sigma} D_\mu F_{\mu\sigma}) + a^2 \overline{\psi} D_\mu^3 \gamma_\mu \psi \longrightarrow a^2 \mathrm{tr}(D_\mu^2 \Sigma D_\mu^2 \Sigma^\dagger) \,. \qquad (62)$$

Since one needs four factors of D_μ to make a non-invariant operator, the result, when combined with the two powers of a, is an operator of NNNLO in χPT, two orders higher than we are working.

We thus find that $\mathcal{L}^{(6)}_{\rm NLO}$ adds no new operators, so the results (58) and (59) are complete. They are to be added to $\mathcal{L}^{(2)}_\chi$ [Eq. (28)] and $\mathcal{L}^{(4)}_\chi$ [Eq. (31)], respectively, to obtain the full LO and NLO contributions to the chiral Lagrangian in the GSM regime. Note that using a twisted mass had no impact on the analysis of this subsection, since $\mathcal{L}^{(5,6)}_{\rm NLO}$ have the same form as for untwisted Wilson fermions.

I now return to the $\tilde{c}_A$ term in the axial current, Eq. (55). To obtain the full axial current in the EFT one must separately match this term into χPT and add it to the result obtained by taking derivatives of the Lagrangian obtained above with respect to sources. It is a simple exercise to show, however, that the result is simply to change the coefficient W_{10}, since the operator it multiplies is exactly of the form $a\partial_\mu P^b$, at linear order in the sources. Thus the final form of the previous paragraph remains complete, albeit with somewhat changed (although still unknown) coefficients.

There are thus five new LECs introduced by discretization errors:[IV] W_0 at LO, W_{45}, W_{68}, W'_{68} and W_{10} at NLO. What do we know about their values? Of course this depends on the choice of fermion and gauge actions, so we can only make order of magnitude estimates. Perturbing in a and m after rotating to Minkowski space, one finds

$$\frac{W_0}{B_0} \sim \frac{\langle \pi | \overline{\psi} i \sigma_{\mu\nu} F^{\mu\nu} \psi | \pi \rangle}{\langle \pi | \overline{\psi} \psi | \pi \rangle} \sim \Lambda^2_{\rm QCD} \,. \qquad (63)$$

The first relation is not an equality because there are unknown coefficients multiplying numerator and denominator, while the second is a dimensional

[IV]This becomes ten new LECs in $SU(3)$ or PQ theories[51].

estimate. One might be tempted to include a loop factor because the matrix element in the numerator requires at least one gluon loop, but this is a non-perturbative matrix element so such counting is inappropriate.

The four NLO W_i are dimensionless, because the dimensions needed to balance powers of a are provided by the W_0 residing in $\hat{A}$. Like the Gasser-Leutwyler coefficients L_i, they depend on the renormalization scale, μ. We will find that the combinations

$$\widetilde{W} = W_{45} - L_{45}\,, \quad W = W_{68} - 2L_{68}\,, \text{ and } W' = W'_{68} - W_{68} + L_{68} \quad (64)$$

are μ independent, so the scale dependence of the W_i themselves is comparable to that of the L_i. Using the argument of Sec. 2.3.5, we then estimate $|W_i| \sim |L_i| \sim 1/(4\pi)^2$. Another line of argument gives a similar estimate. In continuum χPT, $\overline{\psi}\psi$ maps into $\mathcal{O} = f^2 2B_0 \mathrm{tr}(\Sigma + \Sigma^\dagger)/4$. Thus we expect the mapping of $a^2(\overline{\psi}\psi)^2$ to contain $O(1) \times a^2\mathcal{O}^2$ plus other operators. In fact, this four-fermion operator (along with others) maps into $a^2\mathcal{O}^2 16 W'_{68} W_0^2/(f^4 B_0^2)$. Comparing, and using Eq. (63), I find $W'_{68} \sim (1/16)(f/\Lambda_{\mathrm{QCD}})^4$. It is reassuring that this estimate, albeit crude, agrees with the order of magnitude of that above.

The analysis to this point has assumed that the fermion action is not non-perturbatively improved. Since many simulations now use such improvement, it is interesting to ask how it changes the analysis. Improvement sets $\mathcal{L}^{(5)}_{NLO}$ to zero. Since this was the source of almost all terms linear in a in $\mathcal{L}^{(2,4)}_{\chi,A}$ above, the impact is to set $\mathcal{L}^{(2)}_{\chi,A} = W_{45} = W_{68} = 0$.[V] The term quadratic in a, with coefficient W'_{68} survives, since this also comes from matching with $\mathcal{L}^{(6)}_{NLO}$. Two coefficients of terms linear in a do not vanish: W_{10} and H'_3. Why? As already noted, the former multiplies a term which only contributes if the source for the axial current is non-vanishing (the part containing the vector current cancels). It thus represents discretization errors in the matrix elements of this current. But these matrix elements require improvement *additional to that of the action*, namely the addition of the "c_A term" to the axial current[19]. If this improvement has not been implemented, then $W_{10} \neq 0$. Note that, following the discussion at the end of Sec. 3.5, the matrix elements of the vector current and the densities are automatically improved (at NLO) if the action is improved.

The non-vanishing of H'_3 even with non-perturbative improvement is a consequence of the improvement being on-shell. Contact terms in cor-

[V] Note that the vagaries of the notation do not allow one to set $W_0 = 0$, for then all discretization effects would vanish.

relation functions are not improved, and this translates into the HECs of $O(a)$ being non-zero. This is, however, an academic point, since the only quantity of interest affected by H_3' is the scalar condensate[55], and this is very difficult to calculate in practice on the lattice.

3.7. *Results for $m_q \sim a\Lambda^2_{\rm QCD}$ (GSM regime)*

We are now ready to reap the benefits of the work we have done setting up χPT for tmLQCD, which I will call tmχPT. We can already see one benefit of the χPT technology: although there are several unknown LECs describing discretization errors, no more are needed for the entire twisted mass plane than for the untwisted Wilson mass axis.

In this section I discuss the results in the GSM regime, first at LO and then at NLO.

3.7.1. *TmχPT at LO*

The complete LO Lagrangian include discretization errors is

$$\mathcal{L}^{(2)}_{\chi,\rm GSM} = \frac{f^2}{4}{\rm tr}(D_\mu\Sigma D_\mu\Sigma^\dagger) - \frac{f^2}{4}{\rm tr}(\chi^\dagger\Sigma + \Sigma^\dagger\chi) - \frac{f^2}{4}{\rm tr}(\hat{A}^\dagger\Sigma + \Sigma^\dagger\hat{A}) \quad (65)$$

It will be useful to change the notation for sources:

$$\chi = 2B_0(s+ip) = 2B_0(m+i\mu\tau_3) + \delta\chi\,, \qquad \delta\chi = 2B_0(\delta s + i\delta p)\,. \quad (66)$$

We now make the simple but important observation that the factors of $\hat{A}$ can be absorbed by using the shifted variable[41] $\chi' = \chi + \hat{A}$:

$$\mathcal{L}^{(2)}_{\chi,\rm GSM} = \frac{f^2}{4}{\rm tr}(D_\mu\Sigma D_\mu\Sigma^\dagger) - \frac{f^2}{4}{\rm tr}(\chi'^\dagger\Sigma + \Sigma^\dagger\chi')\,. \quad (67)$$

The result has exactly the same form as the LO continuum Lagrangian. The shift $\chi \to \chi'$ corresponds to an $O(a)$ shift in the untwisted mass, $m \to m' = m + aW_0/B_0$, but leaves δs (and, indeed, $\delta\chi$) unchanged.. Recalling the definition of m, Eq. (44), this is equivalent to a shift in the critical mass, $\Delta m_c = -a^2 Z_S W_0/B_0$. This shift is not measurable, however, since m_c is not known *a priori*. It must be determined non-perturbatively from the simulation itself. The traditional definition is that m_c is the bare mass at which $m_\pi^2 \to 0$ on the Wilson axis. Since at LO $\mathcal{L}^{(2)}_{\chi,\rm GSM}$ predicts $m_\pi^2 \propto m'$, we discover that this "m_π^2" definition of m_c *automatically* includes the $O(a^2)$ shift in m_c, and chooses the untwisted quark mass to be m'.

The upshot is that, with the standard numerical definition of m_c, the pion and vacuum sectors are automatically $O(a)$ improved at LO in tmχPT for any twist angle. The analysis of this theory is as in the continuum and I sketch it quickly. The twist angle ω_0 is defined by[VI]

$$(m' + i\tau_3\mu) \equiv m_q e^{i\omega_0\tau_3}\,, \quad (m_q \text{ real and positive}) \tag{68}$$

giving the geometry shown below (note the renormalized axes).

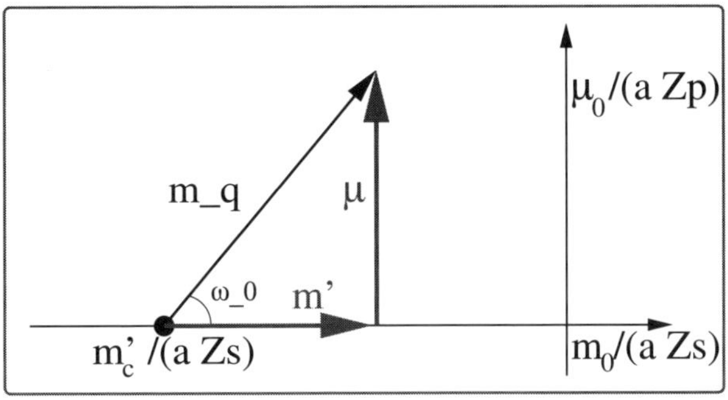

The potential is minimized when Σ is aligned with χ': $\langle\Sigma\rangle = \exp(i\omega_0\tau_3)$. It is conventional to expand Σ about the condensate in a symmetric way,

$$\Sigma = \xi_0\,\Sigma_{ph}\,\xi_0\,, \quad \xi_0 \equiv exp(i\omega_0\tau_3/2)\,, \quad \Sigma_{ph} \equiv \exp(i\vec{\pi}\cdot\vec{\tau}/f)\,, \tag{69}$$

since this corresponds diagonalizing the mass matrix with an axial transformation.[VII] The resulting theory is

$$\mathcal{L}_{\chi,\mathrm{LO}} = \frac{f^2}{4}\mathrm{tr}(D_\mu\Sigma_{ph}D_\mu\Sigma_{ph}^\dagger) - \frac{f^2}{4}\mathrm{tr}\left\{\left[\widehat{m_q} + (\delta\chi)_{ph}^\dagger\right]\Sigma_{ph} + \mathrm{h.c.}\right\} \,, \tag{70}$$

where $\widehat{m_q} \equiv 2B_0 m_q$ and "h.c." is hermitian conjugate. The masses and interactions of the pions are manifestly independent of ω_0, since the mass term has been "untwisted". At the same time, however, sources have been twisted: $(\delta\chi)_{ph}^\dagger = \xi_0(\delta\chi)^\dagger\xi_0$ and

$$D_\mu\Sigma_{ph} = \partial_\mu\Sigma_{ph} - il_\mu^{ph}\Sigma_{ph} + i\Sigma_{ph}r_\mu^{ph}\,, \text{ with } l_\mu^{ph} = \xi_0^\dagger l_\mu\xi_0\,,\ r_\mu^{ph} = \xi_0 r_\mu\xi_0^\dagger\,. \tag{71}$$

The operators used to determine physical matrix elements (obtained by functional derivatives with respect to the physical sources) are thus related

---

[VI]This is another unfortunate notation[55]. This is a renormalized twist angle with a good continuum limit, because of the Z-factors in the definitions of m' and μ. The subscript on ω_0 does not indicate a bare quantity.

[VII]One could equally well use, say, a LH transformation to diagonalize M, and a correspondingly asymmetric form for Σ. The difference is a $U(1)$ vector transformation that has no effect on the physics, although there will be extra phases associated with operators to those in the lattice theory (obtained using the original sources) by an ω_0-dependent twist. This reproduces the twisting one finds at the quark level: fields in the original (or "twisted") basis [that with $M = m_q \exp(i\omega_0\tau_3)$] are related to those in the "physical basis" [$M = m_q$] by an axial transformation [$\psi_{ph} = \exp(i\omega_0\tau_3\gamma_5/2)\psi$], so that operators are also transformed[42], e.g.

$$\overline{u}_{ph}\gamma_\mu\gamma_5 d_{ph} = \cos\omega_0(\overline{u}\gamma_\mu\gamma_5 d) - i\sin\omega_0(\overline{u}\gamma_\mu d)\,. \tag{72}$$

Thus, at maximal twist ($\omega_0 = \pi/2$) the charged pion should be created with the lattice *vector current.*

3.7.2. *tmχPT at NLO*

The NLO Lagrangian, rewritten in terms of χ', and dropping HECs, is

$$\begin{aligned}
\mathcal{L}^{(4)}_{\chi,\mathrm{GSM}} &= \mathcal{L}^{(4)}_{\chi} + \mathcal{L}^{(4)}_{\chi,A} \\
&= -L_{13}\mathrm{tr}(D_\mu\Sigma D_\mu\Sigma^\dagger)^2 - L_2\mathrm{tr}(D_\mu\Sigma D_\nu\Sigma^\dagger)\mathrm{tr}(D_\mu\Sigma D_\nu\Sigma^\dagger) \\
&\quad + L_{45}\mathrm{tr}(D_\mu\Sigma^\dagger D_\mu\Sigma)\mathrm{tr}(\chi'^\dagger\Sigma + \Sigma^\dagger\chi') - L_{68}\left[\mathrm{tr}(\chi'^\dagger\Sigma + \Sigma^\dagger\chi')\right]^2 \\
&\quad + \widetilde{W}\mathrm{tr}(D_\mu\Sigma^\dagger D_\mu\Sigma)\mathrm{tr}(\hat{A}^\dagger\Sigma + \Sigma^\dagger\hat{A}) - W\mathrm{tr}(\chi'^\dagger\Sigma + \Sigma^\dagger\chi')\mathrm{tr}(\hat{A}^\dagger\Sigma + \Sigma^\dagger\hat{A}) \\
&\quad - W'\left[\mathrm{tr}(\hat{A}^\dagger\Sigma + \Sigma^\dagger\hat{A})\right]^2 + W_{10}\mathrm{tr}(D_\mu\hat{A}^\dagger D_\mu\Sigma + D_\mu\Sigma^\dagger D_\mu\hat{A})\,.
\end{aligned} \tag{73}$$

Here I used $SU(2)$ relations to combine terms, Eq. (31), with $L_{13} = L_1 + L_3/2$, $L_{45} = L_4 + L_5/2$ and $L_{68} = L_6 + L_8/2$. I have also used the shifted W's defined in Eq. (64). The latter, as noted above, turn out to be independent of the renormalization scale. If we are using a non-perturbatively $O(a)$ improved action then $W = \widetilde{W} = 0$, and if we also improve the axial current then $W_{10} = 0$.

Subsequent results are simplified by the observation that W_{10} is redundant[55]. A change of variables, $\delta\Sigma = (2W_{10}/f^2)\left(\Sigma\hat{A}^\dagger\Sigma - \hat{A}\right)$, which keeps $\Sigma \in SU(3)$ up to NNLO corrections, cancels the W_{10} term while shifting the other LECs: $W \to W + W_{10}/4$ and $\widetilde{W} \to \widetilde{W} + W_{10}/2$. Because of this, I set $W_{10} = 0$ henceforth. Note that if one uses a non-perturbatively $O(a)$ improved action but an unimproved axial current, then after this change of variables W and $\widetilde{W}$ no longer vanish, but are related by $2W - \widetilde{W} = 0$.

I now present a sampling of NLO results[53,58,55,56]. These require as a first step the determination of the condensate about which to expand, and this is realigned by NLO terms. In the continuum there is no realignment, because the L_{68} term is already extremized by the LO condensate, $\langle\Sigma\rangle = \exp(i\omega_0\tau_3) \propto \chi'$. The "discretization terms" (those proportional to W and

W') do, however, lead to a realignment, because they "pull" the condensate either towards or away from the identity direction. The result is:

$$\langle\Sigma\rangle = e^{i(\omega_0+\epsilon)\tau_3}\,, \quad \epsilon = -\frac{16\hat{a}\sin\omega_0}{f^2}\left(W + 2W'\cos\omega_0\frac{\hat{a}}{\widehat{m_q}}\right) \tag{74}$$

Note that ϵ vanishes on the Wilson axis ($\omega_0 = 0, \pi$), and that the W' contribution is enhanced if $\widehat{m_q} \ll \hat{a}$ (in which case one enters the Aoki regime, where an $O(1)$ vacuum realignment is possible, as discussed below).

While ϵ is not measurable in simulations (as it is defined within χPT and not in terms of observables), it illustrates the typical magnitude of NLO effects. Given that we expect $|W|, |\widetilde{W}|, |W'| \sim 1/(4\pi)^2$ and $W_0 \sim \Lambda_{\rm QCD}^3$, it follows that $\epsilon \sim a\Lambda_{\rm QCD}$. To implement these expectations, it is useful to use rescaled variables[59]

$$\delta_W = \frac{16\hat{a}W}{f^2} \sim a\Lambda_{\rm QCD},\; \delta_{\widetilde{W}} = \frac{16\hat{a}\widetilde{W}}{f^2} \sim a\Lambda_{\rm QCD},\; w' = \frac{16\hat{a}^2W'}{f^2} \sim a^2\Lambda_{\rm QCD}^4. \tag{75}$$

The general form of NLO results for observables is illustrated by

$$\begin{aligned} m_{\pi^\pm}^2 &= \widehat{m_q}\left[1 + \frac{1}{2}L_\pi + \frac{16}{f^2}\widehat{m_q}(2L_{68} - L_{45})\right] \\ &\quad + \widehat{m_q}\cos\omega_0(2\delta_W - \delta_{\widetilde{W}}) + 2(\cos\omega_0)^2 w'\,. \end{aligned} \tag{76}$$

The first line is the continuum NLO result [the $SU(2)$ version of Eq. (32), with chiral logs defined in (33)], while the second shows the impact of discretization. Recall that $\widehat{m_q}$ is defined to be positive. The scale dependence of the chiral log is absorbed by $2L_{68} - L_{45}$, leaving $2\delta_W - \delta_{\widetilde{W}}$ scale invariant. Chiral logs do not contain discretization corrections at this order because the LO discretization errors can be absorbed into χ'.

The result (76) shows the different possibilities for removing $O(a)$ errors.

- Non-perturbatively $O(a)$ improve the quark action, in which case $2\delta_W - \delta_{\widetilde{W}} = 0$ and the $O(a)$ term vanishes.
- Use "mass averaging"[43] in which one averages over ω_0 and $\omega_0 + \pi$ at fixed $\widehat{m_q}$. This flips the sign of both m' and μ, and thus of $\cos\omega_0$, and cancels the $O(a)$ term. It has been shown to lead to $O(a)$ improvement for all physical quantities, including matrix elements[43].
- Work at maximal twist, $\omega = \pm\pi/2$ (both choices of sign are equivalent, and I use the positive sign henceforth). This removes the $O(a)$ term (as it must since it is a special case of mass averaging where

the average is automatic[43]), and, in this case though not in general, also the $O(a^2)$ term.

I note two further features of the result (76). The $O(a)$ errors are determined throughout the twisted-mass plane by the combination $2\delta_W - \delta_{\widetilde{W}}$. In particular, on the Wilson axis, this term predicts an asymmetry in the slopes on the two sides of m_c. This has recently been observed numerically, as part of the initial studies of the properties of tmLQCD. I show an example of the results in Fig. 9. Defining the asymmetry in a quantity Q as[59]

$$\mathrm{AS}(Q) \equiv \frac{Q(m',\mu) - (-)^p Q(-m',\mu)}{Q(m',\mu) + (-)^p Q(-m',\mu)}\,, \tag{77}$$

(with p a parity which is $+1$ for most quantities), one finds

$$\mathrm{AS}(m^2_{\pi^\pm}) = (m'/m_q)(2\delta_W - \delta_{\widetilde{W}}) \stackrel{\mu=0}{\longrightarrow} (2\delta_W - \delta_{\widetilde{W}})\,. \tag{78}$$

The observed asymmetry in Fig. 9 is ~ 0.3, consistent with the expected size ($a^{-1} \approx 1$ GeV and $\Lambda_{\mathrm{QCD}} \approx 0.3$ GeV).

The second feature of (76) that I want to emphasize is that the w' term (proportional to a^2) gives an *additive* correction to m^2_π. This is indicative of the breaking of chiral symmetry. It can be of either sign. If $w' > 0$ then one expects a minimum pion mass, as observed in Fig. 9, while if $w' < 0$ the pion mass-squared can become negative, indicating an instability. This leads to the well-known Aoki phase[61].[VIII]

A graphic demonstration of the impact of discretization errors is provided by the contours of m^2_π in the twisted mass plane. Examples are shown later in Fig. 14. At LO the contours are circles (as in the continuum), indicating that ω_0 is redundant, but they are significantly deformed by the discretization errors that enter at NLO.

Discretization errors linear in a can be separated from the continuum parts of the predictions by using the asymmetries defined above. These are pure discretization effects, while the corresponding symmetries are $O(a)$ improved because symmetrization is the essentially the same as mass averaging. The asymmetries for quantities accessible with simulations are[55,59]

$$2\,\mathrm{AS}(\langle 0|P^\mp|\pi^\pm\rangle) = AS(\langle\pi|S^0|\pi\rangle) = \mathrm{AS}(m^2_{\pi^\pm}) = (m'/m_q)(2\delta_W - \delta_{\widetilde{W}}) \tag{79}$$

$$\mathrm{AS}(f_\pi) = (m'/m_q)\delta_{\widetilde{W}}/2\,, \qquad \mathrm{AS}(m_{\mathrm{PCAC}}) = (m_q/m')\delta_W\,. \tag{80}$$

[VIII]Since these phenomena occur when $m_q \sim a^2\Lambda^3_{\mathrm{QCD}}$, i.e. in the Aoki regime, there are corrections to (76), which will be discussed below.

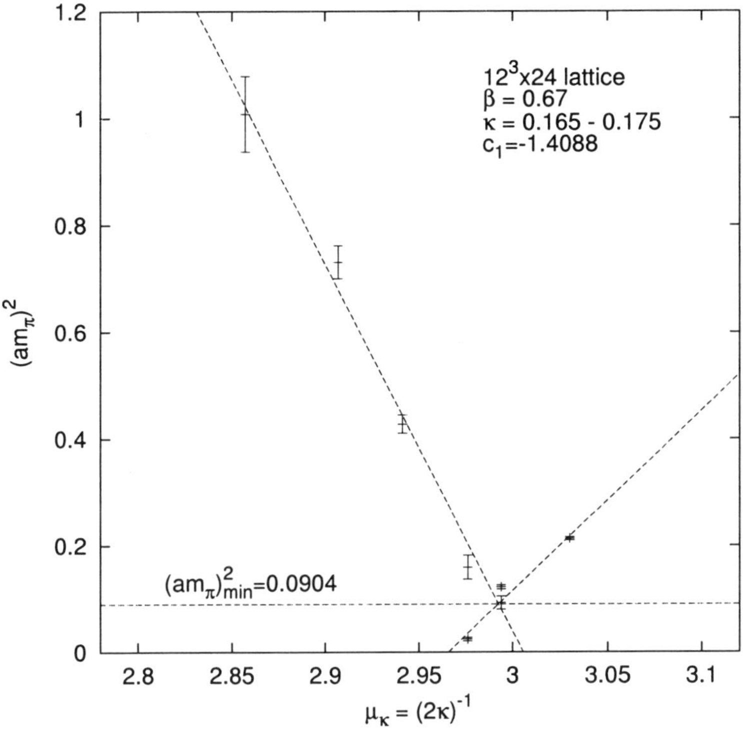

Figure 9. Unquenched results for $(am_\pi)^2$ as a function of $(2\kappa)^{-1} = m_0 + 4$ for $\mu = 0$ and with $a^{-1} \approx 0.2$ fm[60]. Straight lines are to guide the eye.

What is predicted here are the form of the mass dependence, and the relations between asymmetries in different quantities. The PCAC mass is defined below in Eq. (89). Note that the asymmetries on the first line vanish for a non-perturbatively improved action, while those on the second do not unless the axial current is also non-perturbatively improved.

TmLQCD explicitly breaks flavor and parity symmetries, and tmχPT can be used to see how such breaking effects physical quantities. Of particular interest is flavor breaking, e.g. the splitting of the pion multiplet or the Δ-baryon multiplet. The good news is that such breaking is automatically of $O(a^2)$ for any twist angle. This is because $\mu\gamma_5\tau_3$ structure of the $O(a)$ flavor breaking implies that it contributes at linear order only to parity-violating matrix elements. To obtain a contribution to masses one needs two parity-breaking insertions (to bring one back to the original parity). An example which can be studied using tmχPT at NLO is

$$m_{\pi^0}^2 - m_{\pi^\pm}^2 = -2w'(\sin\omega_0)^2 = -2w'\frac{\mu^2}{m'^2+\mu^2}\,. \tag{81}$$

(Recall that $w' \sim a^2$.) The splitting must vanish on the Wilson axis (as there is then no flavor breaking), and is necessarily even in μ^2 from the argument above. Not surprisingly, the splitting is maximized for $\omega_0 = \pi/2$. Similar results hold for Δ baryons[62]. Which pion is heavier depends on the sign of w', which, as already noted, also determines the nature of the phase structure in the Aoki regime[58].

Flavor-parity breaking of $O(a)$ does occur for unphysical parity-violating matrix elements. This is true for any non-zero twist, including maximal twist—the argument for automatic $O(a)$ improvement[43] does not hold for such quantities. Results for the following parity-flavor violating form factors are available in tmχPT[55]: $\langle\pi_b|A^b_{ph,\mu}, P^b_{ph}|\pi_3\rangle$, $\langle\pi_b|A^3_{ph,\mu}, P^3_{ph}|\pi_b\rangle$ and $\langle\pi_3|A^3_{ph,\mu}, P^3_{ph}|\pi_3\rangle$, where $b = 1, 2$. Here is one example:

$$\langle\pi_b(p_2)|P^3_{ph}|\pi_b(p_1)\rangle = iB_0\sin\omega_0\left[\delta_W - \delta_{\widetilde{W}} - \frac{(2\delta_W - \delta_{\widetilde{W}})q^2}{2(q^2+m_{\pi_3}^2)}\right] + \frac{iB_0\sin(2\omega_0)w'}{q^2+m_{\pi_3}^2}\,, \tag{82}$$

where $q = p_1 - p_2$ is the momentum transfer. These quantities provide an interesting window into the workings of tmLQCD, and are predicted once one has determined the LECs, but are difficult to study numerically because they involve quark-disconnected contractions.

I am aware of one detailed comparison of the results from simulations of dynamical tmLQCD with tmχPT formulae[63]. (There is also a detailed fit to quenched data[64] which I discuss below.) The simulations are for relatively coarse lattices ($a \approx 0.13$ and 0.18 fm), yet find reasonable agreement with the NLO forms sketched above, with the continuum LECs consistent with continuum results, and the magnitudes of the "lattice" LECs consistent with expectations (although poorly determined). The asymmetries discussed above are clearly present as a function of m'. The authors do their fits versus the PCAC mass, however, and for these the asymmetries are reduced, leading to the poor determination of the lattice LECs.

These fits give one confidence that tmχPT is a useful tool, and is likely to become more so as lattice spacings and quark masses are reduced. Thus it can aid extrapolations, and, perhaps more importantly, guide the investigation of the properties of tmLQCD, in particular its phase diagram and the issue of defining maximal twist. I return to these issues shortly.

Other simulations have studied the pion mass splitting[65,66,67]. This involves both quark-connected and disconnected contributions, with the latter hard to calculate accurately. Nevertheless, this is a key quantity to determine as it both sets the scale for isospin breaking and the size and nature of the phase structure. TmLQCD has some similarity to staggered fermions (although not the need to use rooting) because a desired continuum symmetry is broken (flavor for twisted mass fermions, taste for staggered), in both cases at $O(a^2)$. Fits to staggered fermions require this taste-breaking to be treated at *leading order* in "staggered" χPT (i.e. the equivalent of the Aoki regime here), because taste-splittings are large. Numerically, the splitting are $\sim a^2\Lambda^4$ with $\Lambda \approx 1$ GeV. The hope for tmLQCD is that the flavor-breaking is smaller, and can be treated as a NLO effect (the GSM regime). One can also tune its size by varying the quark and gluon actions. Quenched results for the mass splitting (with Wilson fermion and gluon actions and $a \approx 0.1$ fm) find $\Lambda \approx 0.7$ GeV [65] (i.e. four times smaller splittings than with staggered fermions), and there are indications of a significant reduction if one uses dynamical quarks and improved gluon actions[66] or the non-perturbatively improved quark action[67]. Thus the situation is promising.

3.8. *Defining m_c and the twist angle*

The critical mass plays a central role in tmLQCD, providing the origin about which one defines the twist angle (see the figure in Sec. 3.7.1). As noted above, it must be determined non-perturbatively as part of the simulation (and recalculated for each lattice spacing and choice of actions). The questions I address in this section are these: How accurately does one need to determine m_c in order that automatic $O(a)$ improvement holds at maximal twist? What methods allow this accuracy to be achieved? I will discuss these questions using the framework of tmχPT, rather than than use the Symanzik EFT as in the original treatment[43] and subsequent extensions[56,68,57]. TmχPT is less general (referring only to the vacuum and pion sectors), but more powerful in its domain of applicability (as it starts from the Symanzik EFT and includes further non-perturbative information).

The form of the $O(a)$ correction in the result for m_π^2, Eq. (76), is generic, namely that it is proportional to $a \cos\omega_0$. To obtain automatic $O(a)$ improvement one needs $\cos\omega_0 = O(a)$ and thus $\omega_0 = \pi/2 + O(a)$. In other words, maximal twist can mean "maximal up to $O(a)$". What does this imply for the required accuracy in the determination of m_c? This depends on the relative size of quark masses and discretization effects. In the GSM regime, with $\mu \sim a\Lambda_{\rm QCD}^2$, an $O(a)$ accuracy in ω_0 requires $m' = O(a^2)$ (and thus a determination of the dimensionless critical mass m_c with an accuracy of $O(a^3)$). This is relatively straightforward to achieve, as we will see. Once one enters the Aoki regime, $\mu \sim a^2\Lambda_{\rm QCD}^3$, the required accuracy increases to $m' = O(a^3)$. Since simulations are likely to need to enter this regime, it is important to know how to achieve this greater precision.

The traditional definition of m_c used with Wilson fermions is to extrapolate m_0 to the point where $m_\pi^2 = 0$. This method might be adequate in the GSM regime, but fails in the Aoki regime[56]. This failure has its origin in the a^2 (w') term in Eq. (76) and will be discussed in detail in the next section. Either the pion mass does not vanish but reaches a non-zero minimum at m_c (at a first-order phase boundary), in which case extrapolating to $m_\pi^2 = 0$ overshoots, or it vanishes over a *range* of m' of width a^2 (the Aoki phase), with the correct choice of m_c being in the middle of the range[56] but the extrapolation giving one of the end-points. In either case, $m' \sim O(a^2)$ at the putative critical point, which is not accurate enough for the Aoki regime. There are also practical issues with this method, reflecting the difficulty in doing accurate extrapolations, but I will not belabor them as this method is no longer being used in practice.

To do better one can adapt the method used to determine the normalization of currents and improvement coefficients, i.e. enforce the symmetries that are broken by discretization. Here, parity and flavor are broken explicitly, but are restored in the continuum limit. Enforcing this restoration in particular correlators for $a \neq 0$ gives a non-perturbative determination of the twist angle, which can, if desired, be tuned to maximal twist. Since flavor and parity *are* broken, different choices of correlator lead to $O(a)$ differences in the twist angle, but all choices lead to automatic $O(a)$ improvement.

I will need to use the relation between twisted and physical bases discussed in Sec. 3.7.1. Using this, a simple calculation finds the following relations between currents and densities in the two bases:

$$A^b_{ph,\mu} = \cos\omega\, A^b_\mu + \epsilon^{3bc}\sin\omega\, V^c_\mu\,, \quad A^3_{ph,\mu} = A^3_\mu\,, \tag{83}$$

$$V^b_{ph,\mu} = \cos\omega\, V^b_\mu + \epsilon^{3bc}\sin\omega\, A^c_\mu\,, \quad V^3_{ph,\mu} = V^3_\mu\,, \tag{84}$$

$$P^3_{ph} = \cos\omega\, P^3 + i\sin\omega\, S^0/2\,, \quad P^b_{ph} = P^b\,, \tag{85}$$

$$S^0_{ph} = \cos\omega\, S^0 + 2i\sin\omega\, P^3\,, \tag{86}$$

where the flavor label $b = 1, 2$. The flavor non-singlet scalar density vanishes at LO in $SU(2)$ χPT and I do not discuss it. What these relations mean is that, if you are working in the twisted basis (as one usually does on the lattice), then to construct the physical axial current with, say, $b = 1$, you must take a linear combination of the lattice currents A^1_μ and V^2_μ. These relations assume that the currents have been correctly normalized by multiplying by their corresponding Z-factors. Note that A^3_μ and P^b do not rotate, and so are good choices to create physical pions.

The idea is now to take either (83,84) or (85,86) as a *definition* of ω, and enforce parity-flavor restoration in a particular correlator. Two examples are [method (ii) will be explained later]:

- Method (i)[69] ("ω_A method") $\langle V^2_{ph,\mu}(x) P^1_{ph}(y)\rangle \propto \langle 0|V^2_{ph,\mu}|\pi^1\rangle = 0$.
- Method (iii)[55] ("ω_P method") $\langle S^0_{ph}(x) A^3_{ph,\mu}(y)\rangle \propto \langle 0|S^0_{ph}|\pi^3\rangle = 0$.

The correlators are to be evaluated for $x \neq y$, and the long-distance contribution is as indicated. Using (83-86) one can manipulate these criteria into results for the twist angle in terms of correlators in the twisted basis:

$$\tan\omega_A \equiv \frac{\langle V^2_\mu(x) P^1(y)\rangle}{\langle A^1_\mu(x) P^1(y)\rangle}\,, \quad \tan\omega_P \equiv \frac{i\langle S^0(x) A^3_\mu(y)\rangle}{2\langle P^3(x) A^3_\mu(y)\rangle}\,. \tag{87}$$

Maximal twist occurs when the denominators vanish, i.e

$$\omega_A = \pi/2 \Rightarrow \langle A^1_\mu(x) P^1(y)\rangle = 0\,, \quad \omega_P = \pi/2 \Rightarrow \langle A^3_\mu(x) P^3(y)\rangle = 0\,. \tag{88}$$

While superficially similar, the two criteria differ because of flavor breaking. The correlator in method (iii) includes quark-disconnected contractions and is much more difficult to calculate in practice. I include it for illustrative reasons. Method (i) is used in practice.[IX] One fixes μ_0 and varies m_0 until $\omega_A = \pi/2$. The resulting $m_0(\mu_0)$ depends on the choice of discretization of the axial current (e.g. $O(a)$ improved or not), and, in general, upon the separation $x - y$. At large distances, which are used in practice, the pion contribution dominates and the resulting m_0 becomes independent of

[IX] It has also been called the parity-violating method, or the PCAC method.

separation. At such distances, method (i) is equivalent to the vanishing of the PCAC mass:

$$m_{PCAC} = \langle \partial_\mu A^b_\mu(x) P^b(y) \rangle / 2 \langle P^b(x) P^b(y) \rangle = 0 \,. \tag{89}$$

One nice feature of either criterion is that knowledge of Z-factors is not required, unlike the determination of ω at non-maximal twist.[X]

Both methods (i) and (iii) can be implemented in tmχPT. With the technology developed above, we can work at NLO in the GSM regime, and calculate the correlators at long distances, for then the pion contribution dominates. I quote only the results at maximal twist:[55]

$$\text{(i): } \omega_A = \pi/2 \ \Rightarrow\ \omega_0 = \pi/2 + \delta_W \,, \quad \text{(iii): } \omega_P = \pi/2 \ \Rightarrow\ \omega_0 = \pi/2 \,, \tag{90}$$

[with δ_W defined in Eq. (75)]. These results are shown in Fig. 10. In method (i) one finds a line at an angle $\delta_W \sim a$ to the vertical. Thus one is not at maximal twist in terms of ω_0, but $\cos\omega_0 \sim a$ so automatic improvement still holds. It turns out that method (iii) leads to a vertical approach to the Wilson axis. The methods come together at the critical mass. Thus one could implement method (iii) by using method (i) for $\mu \neq 0$, extrapolating to $\mu = 0$ to determine m_c, and then working at fixed $m_0 = m_c$.

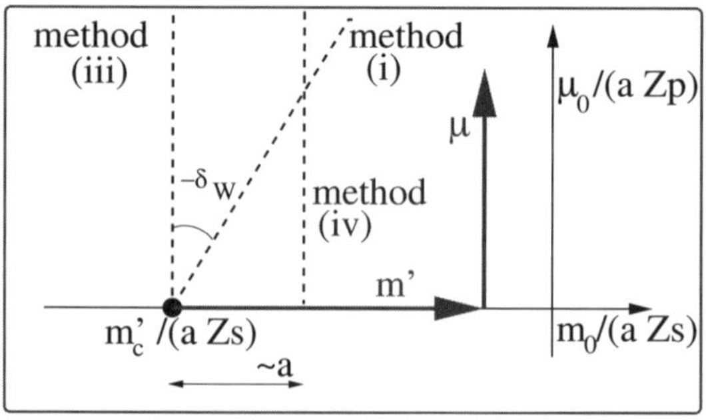

Figure 10. Results from different methods of defining maximal twist.

Quenched simulations at $1/a \approx 2$ GeV with unimproved Wilson fermions find an offset angle $\delta_W \approx -0.35$[70,71]. This is a direct measure

[X] A generalization of method (i) allows the determination of ω_A for any twist angle without *a priori* knowledge of $Z_{A,V}$[69].

of discretization errors. Written as $a\Lambda^2$ this gives $\Lambda \approx 0.7$ GeV, a large but not unreasonable value.

As noted above, the pion mass method can also be used, in principle, within the GSM regime. In practice the extrapolation can miss the critical mass, so it is useful to define a "straw-man" method (iv) in which m' is held fixed at a value of $O(a)$. As is clear from the figure, ω_0 varies from ~ 1 to 0 as μ varies from $\sim a$ to 0. Thus one is never near maximal twist in the GSM regime, and automatic improvement is lost.

3.9. *Results for* $m_q \sim a^2\Lambda^3_{\rm QCD}$ *(Aoki regime)*

As discussed above, to describe simulations we need to extend the analysis into the Aoki regime. This requires a change in the power counting:[64,59,57] at LO we keep terms of size $m' \sim \mu \sim a^2$,

$$\mathcal{L}^{\rm LO}_{\chi,Aoki} = \frac{f^2}{4}\mathrm{tr}(D_\mu\Sigma D_\mu\Sigma^\dagger) - \frac{f^2}{4}\mathrm{tr}(\chi'^\dagger\Sigma + p.c.) - W'\big[\mathrm{tr}(\hat{A}^\dagger\Sigma + p.c.)\big]^2, \quad (91)$$

while at NLO we keep those of size $ma \sim \mu a \sim a^3$,

$$\begin{aligned}\mathcal{L}^{\rm NLO}_{\chi,Aoki} &= -\frac{W_{3,1}}{f^2}\mathrm{tr}(\hat{A}^\dagger\hat{A})\mathrm{Tr}(\hat{A}^\dagger\Sigma + p.c.) - \frac{W_{3,3}}{f^2}\Big[\mathrm{tr}(\hat{A}^\dagger\Sigma)^3 + p.c.\Big] \\ &+ \widetilde{W}\mathrm{tr}(D_\mu\Sigma^\dagger D_\mu\Sigma)\mathrm{tr}(\hat{A}^\dagger\Sigma + p.c.) - W\mathrm{tr}(\chi'^\dagger\Sigma + p.c.)\mathrm{tr}(\hat{A}^\dagger\Sigma + p.c.). \quad (92)\end{aligned}$$

The two a^3 terms (those on the first line) are new, while previous NLO terms $\propto m^2$ become of NNLO. In fact, the $W_{3,1}$ term can be absorbed into χ' by a further shift of $O(a^4)$ in m_c. This leaves one new LEC at NLO, $W_{3,3}$. Note that only the source parts of the W and $\widetilde{W}$ terms enter at NLO.

In the following I will describe the analysis of the phase structure at LO, and then mention some modifications caused by NLO terms.

The orientation of $\langle\Sigma\rangle$ is now determined by an *equal* competition between the mass and a^2 terms in $\mathcal{L}^{\rm LO}_{\chi,Aoki}$. The former is minimized when $\langle\Sigma\rangle \propto \chi' \propto \exp(i\omega_0\tau_3)$, while the latter either favors $\langle\Sigma\rangle = \pm 1$ ($W' > 0$) or $\langle\Sigma\rangle = \exp[i(\pi/2)\hat{n}\cdot\vec{\tau}]$ with $\hat{n}^2 = 1$ ($W' < 0$). The analysis along the Wilson axis is simple[41], and one finds two cases. Either the condensate jumps discontinuously from $\langle\Sigma\rangle = +1$ to $\langle\Sigma\rangle = -1$ at a first-order transition ($W' > 0$), or it swings between these values continuously, as is possible within $SU(2)$ ($W' < 0$). In the latter case, the condensate breaks flavor, so there are exact lattice Goldstone bosons. This is the Aoki phase[61]. The presence of two possible phase structures was also predicted by Creutz[72].

Moving into the twisted-mass plane the analysis becomes more complicated as one must minimize a quartic[53,58,55]. I show below how the two scenarios on the Wilson axis extend into the mass plane.

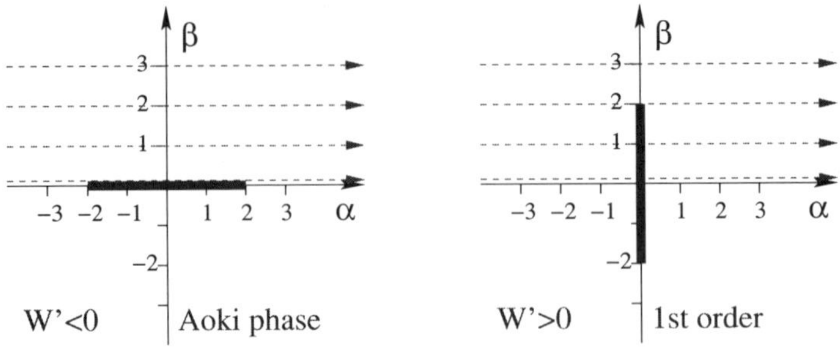

Here I use mass variables rescaled by $\sim a^2$:

$$\alpha = \frac{2B_0 m'}{(16|W'|\hat{a}^2/f^2)} = \frac{\hat{m'}}{|w'|}, \quad \beta = \frac{2B_0 \mu}{(16|W'|\hat{a}^2/f^2)} = \frac{\hat{\mu}}{|w'|}. \tag{93}$$

Both scenarios have a first-order transition boundary, indicated by the solid line, with second-order end-points. Note again that the parameter w', whose sign determines which scenario applies, and whose magnitude gives the size of the phase boundaries, is the same parameter as appears in the pion mass splitting[58], Eq. (81). Thus a calculation of this splitting *in the GSM regime* (which, as noted above, has been attempted) predicts the phase structure *in the Aoki regime.*

Wu and I have given detailed plots of the condensate and pion masses along the dashed horizontal lines in the phase diagrams above[55]. I show here only a sample. I write the condensate as $\langle\Sigma\rangle = A_m + iB_m\tau_3$, and plot the scalar component A_m, as well as the charged and neutral pion mass-squareds. Figure 11 shows results for $W' < 0$. The "swinging" of the condensate between ± 1, described in words above, is here the $\beta = 0$ curve. The charged pion masses vanish when $|A_m| < 1$, for then $|B_m| > 0$ and flavor is spontaneously broken. Moving away from the Wilson axis ($\beta = 1, 2, 3$), the Aoki-phase is washed out by the explicit breaking of flavor. The neutral pion (not shown) is always heavier than the charged pions.

If the mass is purely twisted ($\alpha = m' = 0$) then one finds for all β that $A_m = 0$, so the condensate points in the τ_3 direction. Thus, despite the large discretization errors, the condensate ends up pointing in the same direction as it would without them. This is one way of understanding why automatic $O(a)$ improvement continues to work, as discussed further below.

I illustrate the results for the first-order scenario ($W' > 0$) in Fig. 12. The left panel shows the discontinuity in the condensate for $|\beta| < 2$, trans-

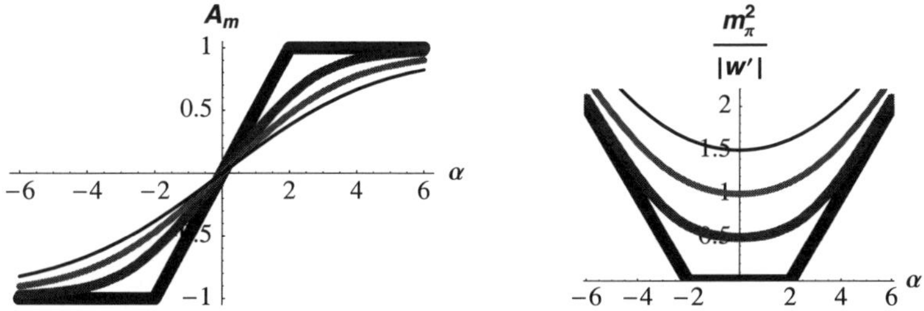

Figure 11. Condensate and charged pion masses for $W' < 0$, for four choices of twisted mass ($\beta = 0, 1, 2, 3$). The thickness of the lines decreases with increasing β.

forming into a smooth crossover for $|\beta| > 2$. Above the transition, note again that $A_m = 0$ when $\alpha = 0$. The right panel shows that the pion mass on the Wilson axis ($\beta = 0$) has a non-zero minimum, while for $\beta = 2$ the neutral mass goes down to zero at the second-order end-point. In this scenario, the charged pions are always heavier than the neutral once one moves off the Wilson axis. Unlike the Aoki-phase scenario, however, no pions are massless along the phase boundary away from the end-points. This is because no lattice symmetry is broken along this transition.

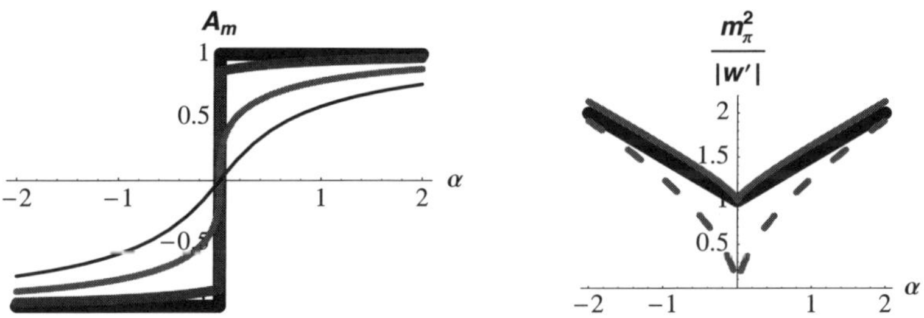

Figure 12. Condensate and pion masses for $W' > 0$. Notation as in Fig. 11, except that pion masses are shown only for $\beta = 0, 2$. For $\beta = 0$ charged and neutral pions are degenerate. For $\beta = 2$, the neutral mass is shown dashed.

3.9.1. *Applications to lattice simulations*

(I) The most important lesson we learn from tmχPT is to expect non-trivial phase structure when $m_q \sim a^2$, with two possible scenarios. The prediction of an Aoki-phase was actually made long before the tmχPT analysis[61], in order to understand how m_π^2 could vanish without an underlying chiral symmetry, and quenched studies in the 1980's and 90's found evidence for such a phase. What has happened in the last few years is the beginning of detailed dynamical studies of the twisted-mass plane. Those with unimproved Wilson gauge and fermion actions found evidence for the first-order scenario. I think it is fair to say this was a surprise, and that it was predicted by χPT I view as a significant success. It is amusing that for decades we have been assuming that m_π^2 extrapolates all the way to zero, whereas (at least with the simplest actions) it actually never makes it.

I show an example of the evidence for the first-order scenario in Fig. 13. This shows scans at fixed twisted mass (with μ roughly fixed in physical units) for three lattice spacings (a decreasing as β increases). The plaquette and $m_{\rm PCAC}$ both have a discontinuity, and show hysteresis. Both effects decrease as one approaches the continuum limit, qualitatively consistent with expectations. The fact that $m_{\rm PCAC}$ has a minimum away from zero is a manifestation of the non-zero minimum in the pion masses.

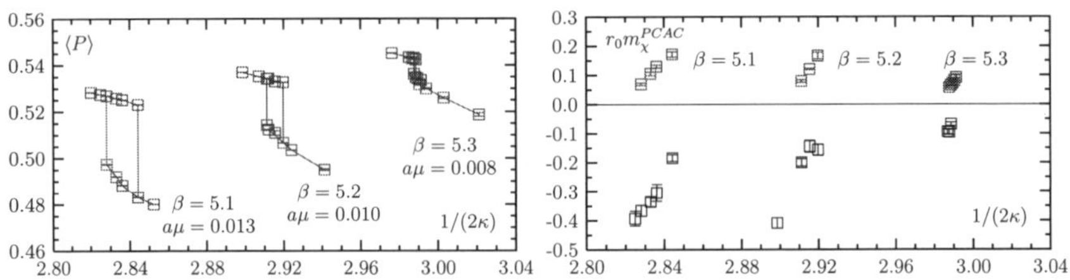

Figure 13. Average plaquette and $m_{\rm PCAC}$ plotted vs. $(2\kappa)^{-1} = m_0 + 4$ for fixed μ_0[73].

(II) The predictions for physical quantities that I sketched above in the GSM regime have been extended to NLO in the Aoki regime[55,64,59,57]. This is not trivial to implement, since minimization of the potential (with the $W_{3,3}$ term included) now involves a sextic equation. Detailed fits to quenched data have, however, been done, and find reasonable values for the resulting LECs[64].

I cannot resist showing one NLO prediction. Discretization effects distort the contours of constant $m_{\pi^\pm}^2$ from circles into the forms shown in Fig. 14. Here m'' is just the shifted mass m' with the $O(a^3)$ term proportional to $W_{3,1}$ absorbed as well. These plots are illustrative of the size of the expected effects, not the result of details fits. I have taken standard LO continuum LECs, and set the NLO continuum LECs and chiral logs to zero for simplicity. For the lattice LECs I use $\delta_W = \delta_{\widetilde{W}} = -0.3$, $|w'| = (250\,\mathrm{MeV})^2$ and $W_{3,3} = 0$, values which are not unreasonable for $a^{-1} \approx 2$ GeV. Quark masses range up to $\sim m_s/2$. Clearly the impact of discretization errors and phase structure could be very significant for actual simulation parameters.

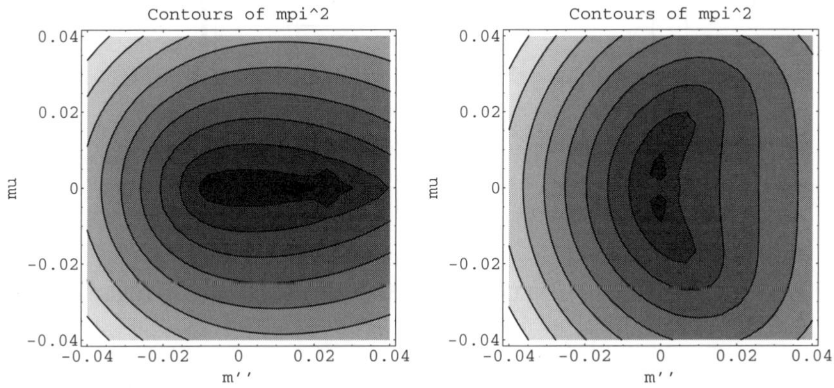

Figure 14. Contour plots of $m_{\pi^\pm}^2$ in twisted mass plane, for Aoki-phase (left) and first-order (right) scenarios, using a representative parameter set for $1/a \approx 2$ GeV. Quark masses are in GeV. Raggedness in contours is due to numerical errors.

(III) Another important question that can be studied using tmχPT in the Aoki regime is whether automatic $O(a)$ improvement at maximal twist still holds using the criteria introduced in Sec. 3.8. The answer is positive[56,55,59,57], aside from a caveat I will explain. In Fig. 15 I show the impact of NLO corrections on the phase boundaries and the lines of maximal twist[59]. These plots allow one to understand the qualitative features of the contours in Fig. 14.

The lines of maximal twist shown in the figure are those for methods (i) and (iii). Along either line the physical quark mass is proportional to μ. If one simulates on one of these lines, reducing a while holding μ fixed (i.e. working at fixed physical quark mass), then the dependence of physical quantities on a will be quadratic or of higher order. This is automatic $O(a)$ improvement. In the Aoki-phase scenario it holds even for $\mu \ll a^2$, i.e. all the way to the Wilson axis. For the first-order scenario the result breaks down, however, when one runs into the end-point with $\mu \approx |w'| \sim a^2$. This is the caveat mentioned above.

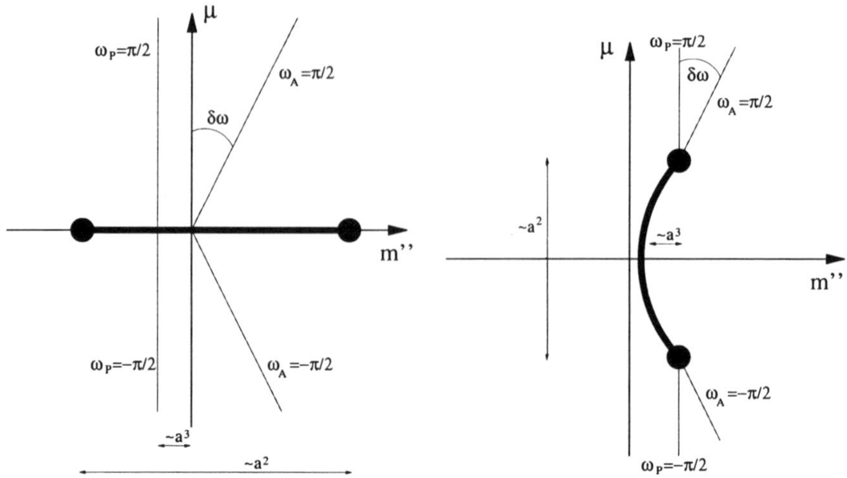

Figure 15. Phase diagram at NLO for Aoki-phase (left) and first-order (right) scenarios. Method (i) corresponds to $\omega_A = \pi/2$, method (iii) to $\omega_P = \pi/2$.

It is perhaps surprising that one can work in a regime where the quark mass is much smaller than the leading $O(a)$ effects of discretization, and yet be able to tune parameters so these effects do not enter into physical quantities. It is worth understanding this qualitatively. The point is that the tuning has to be *very* fine. If $\mu \sim a^2$, then, as Fig. 15 shows, one must tune m'' to an accuracy of a^3. This is what the criteria in Eq. (88) accomplish. In terms of the Symanzik Lagrangian, this amounts to canceling the untwisted mass (so that it is $O(a^3)$ or smaller). The result is

$$\mathcal{L}^{(4+5)} = \overline{\psi}\not{D}\psi + \mu\overline{\psi}i\gamma_5\tau_3\psi + ac\overline{\psi}i\sigma_{\mu\nu}F_{\mu\nu}\psi + O(a^2) \tag{94}$$

$$= \overline{\psi}_{\rm phys}\not{D}\psi_{\rm phys} + \mu\overline{\psi}_{\rm phys}\psi_{\rm phys} + ac\overline{\psi}_{\rm phys}\gamma_5\tau_3\sigma_{\mu\nu}F_{\mu\nu}\psi_{\rm phys}\,, \tag{95}$$

where in the second line I have rotated to the physical basis. The absence of a $\overline{\psi}\psi$ term on the first line means that the $O(a)$ term is purely flavor-parity breaking in the physical basis, and does not contribute to physical matrix elements however large it is.[XI] This is nothing more than a summary of the original argument for automatic improvement[43] (although simplified, as I am not including operators). The only new point here is that the argument

[XI] A similar argument can be made directly at the chiral Lagrangian level, and relies on the result, noted above, that the condensate points near the τ_3 direction at maximal twist [53,55].

goes through for arbitrarily small μ as long as m'' is tuned accurately enough.

It is now time to unveil what I call "method (ii)", which works as follows. One first defines a critical mass by extrapolating results from method (i) to the Wilson axis, and then keeps m_0 fixed at this critical mass for all μ. It turns out to correspond to working at $m'' = 0$. In the GSM regime (recall Fig. 10) method (ii) is equivalent to method (iii), but in the Aoki regime the line it defines is offset from that of method (iii) by $O(a^3)$. Such a difference might seem too small to be significant, but it need not be[57]. To see this, consider the Aoki-phase scenario. Then from Fig. 15 one sees that method (ii) interpolates between method (i) ($\mu < a^2$) and method (iii) ($\mu \gg a^2$). Now, it turns out that the condensate has a fixed orientation as one moves along the lines of either method, but the orientations for the two methods differ by an angle of $O(a)$. Thus for method (ii) the condensate varies direction rapidly for $\mu \lesssim a^2$, with a gradient of $\sim a/a^2$. This can disrupt extrapolations as the resulting form is not a simple polynomial[57]. I raise this worry because method (ii) is being used in practice by some groups, and it would be preferable to use method (i).

(IV) A further lesson from tmχPT is that both the size of the phase boundaries and the isospin splitting for pions are determined by the same parameter, w'. It thus makes sense to try and tune the gauge and fermion actions to reduce $|w'|$. Note that this tuning is not the same as a systematic improvement program, but it is nevertheless very important. Initial results in this direction are very encouraging. For example, Fig. 16 shows that the discontinuities in the plaquette are significantly reduced by alternative gauge actions.[XII] It is also found that using an improved *fermion* action reduces the pion isospin splitting and thus w'[67]. This is an active area of present research.

(V) It might appear from Fig. 15 that, if we had to choose, we would prefer the Aoki-phase scenario because we can work at maximal twist down to $\mu = 0$. I think, however, that the choice is not so clear. To illustrate why, I compare the results for pion masses in the two scenarios in Fig. 17, using the same "reasonable" choices of LECs as above. The role of the charged and neutral pions is interchanged in the two scenarios, and there is no clear advantage to using one over the other. The figure also emphasizes the

[XII] In fact, the reduction is greater than appears, as the results with the Wilson action are for $\mu \neq 0$, where the transition is weaker than for $\mu = 0$, the value used for the other actions.

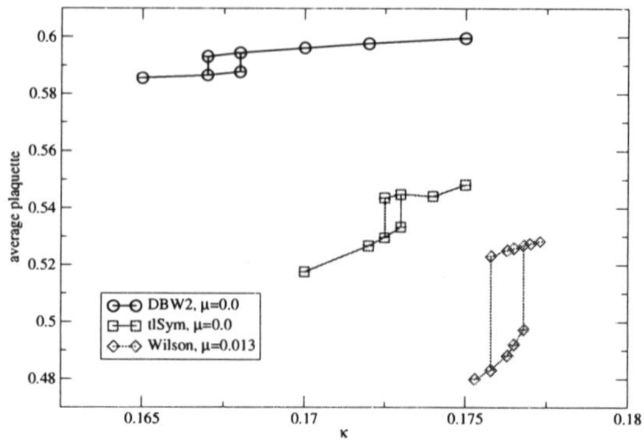

Figure 16. Hysteresis of plaquette at $a \approx 0.17$ fm for DBW2 (top), tree-level improved Symanzik (middle) and Wilson (bottom) gauge actions.[74].

significance of the $O(a^2)$ effects. I conclude that it is much more important to reduce the magnitude of w' than it is to end up in one or other scenario.

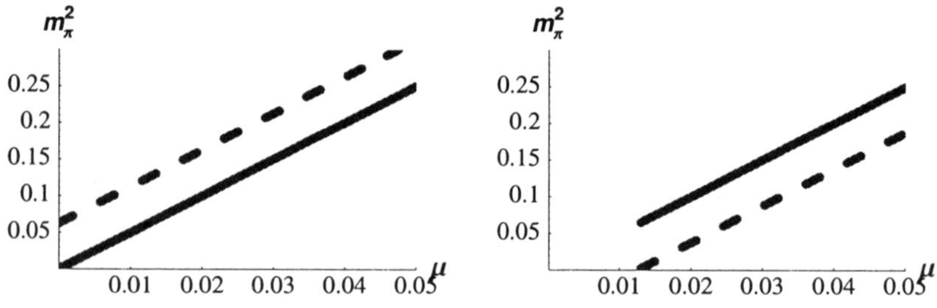

Figure 17. $m^2_{\pi^\pm}$ (solid) and $m^2_{\pi^0}$ (dashed) in GeV^2 versus μ in GeV for method (i). Left panel: Aoki-phase scenario; right panel: first-order scenario.

3.9.2. *Bending near maximal twist*

The discussion to this point makes clear the importance of accurate tuning to maximal twist. Initial studies of tmLQCD did not always attain this accuracy and observed a phenomenon called "bending". Although this is largely a closed chapter in the history of tmLQCD, it is worth learning the appropriate lessons, particularly as tmχPT played an important role in elucidating the phenomenon.

The sketch to the right shows the Aoki-phase scenario with three choices of the approach to the chiral limit shown by arrows. They are all nominally at maximal twist. These are method (i) [thick lines]; fixing m_c at an endpoint where $m_\pi = 0$ [medium thickness], and missing the endpoint by $\sim a^2$ [thin lines].

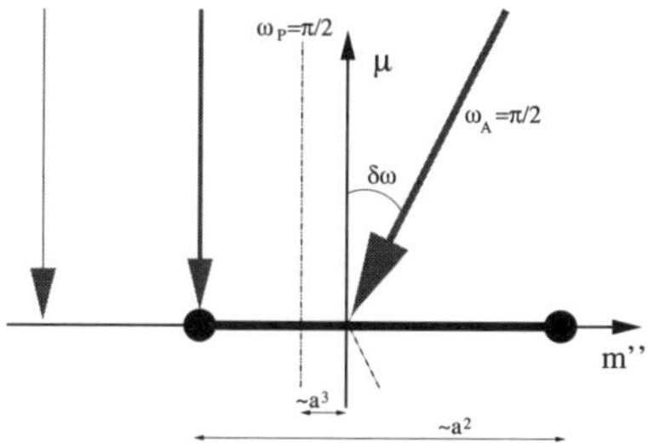

I show in Fig. 18 the NLO tmχPT results (using the same parameter set and assumptions as above) for $m^2_{\pi^\pm}/\mu$ for the three methods (with the thickness of the lines matching those above), along with results from simulations. The horizontal axis is μ, which is the quark mass when working at maximal twist. For my parameters (i.e. physical LECs and chiral logs dropped) the correct result is a constant, and is illustrated by the $O(a)$ improved result from method (i) [thick line]. The results from the other two choices show a noticeable bending away from the continuum result at small masses, i.e. a breakdown of $O(a)$ improvement. The numerical results are for a similar range of quark masses, but the axes are in different units and so the two plots can only be compared qualitatively. The numerical results are for method (i) ("PCAC definition"—triangles) and the "$m_\pi = 0$ definition" (circles), and are qualitatively consistent with tmχPT. A detailed fit of these and other quenched data to quenched tmχPT has been done, and finds quantitative agreement[64].

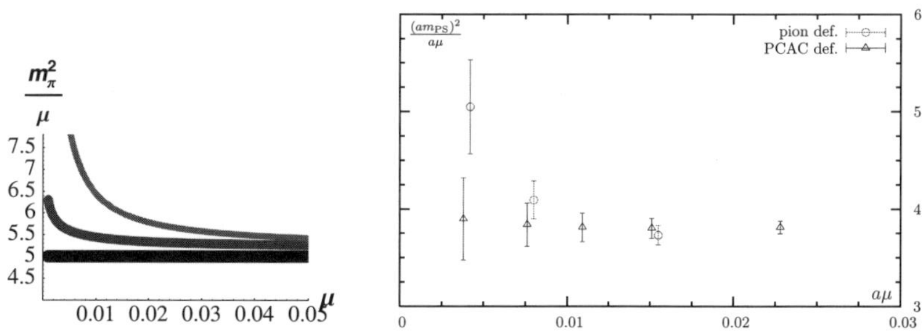

Figure 18. "Bending" in m^2_π/μ vs. μ from tmχPT[59] (left—both axes in GeV) and quenched simulations[44] (right—both axes in lattice units at $1/a \approx 2$ GeV using bare μ).

The most important conclusion from this figure is that one should use a good definition of maximal twist, such as method (i). In particular, the failure of the $m_\pi = 0$ method is apparent. This can be understood qualitatively as follows. The condensate at maximal twist should lie in the τ_3 direction, $\langle\Sigma\rangle = i\tau_3 + O(a)$. This is the case with method (i). In the other cases, however, the condensate starts in this direction for $\mu \gg a^2$ but rotates away when $\mu \sim a^2$ and ends up at $\langle\Sigma\rangle = -1$ when $\mu = 0$. Thus automatic $O(a)$ improvement is lost.[XIII]

3.9.3. *Does tmLQCD work in practice?*

TmχPT has been a very useful guide to the exploratory numerical studies of tmLQCD, but the most important question is whether working at maximal twist is practical. In particular, is the requisite fine tuning possible, and cheap enough, in practice? Are results for physical quantities really improved? These questions are being actively studied and the answers appear to be positive[44,74,75].

A different question is how best to include the strange quark, as one needs an even number of quarks to implement maximal twisting. One could just add a single non-perturbatively $O(a)$ improved untwisted strange quark to a maximally twisted up-down pair, but this would require non-perturbative improvement of each new operator involving the strange quark, which is exactly what we are trying to avoid with automatic improvement at maximal twist. An alternative is to introduce the charm quark as well, and treat the strange-charm pair as a twisted doublet[44]. Non-degenerate masses can be incorporated while maintaining maximal twist[76]. The disadvantages are that flavor breaking is more complicated (occurring in both τ_3 and τ_1 directions in the strange-charm sector), and that $(am_c)^2$ effects may not be small. This option is being actively studied[77].

Finally, I think it likely that isospin breaking in pion masses will need to be included in loop calculations, requiring one to work at NNLO in the GSM power counting. First steps in this direction have been taken[78].

For further possible uses of tmLQCD, see the lectures by Sint.

[XIII] If one treats deviations from $\langle\Sigma\rangle = i\tau_3$ as small perturbations, then discretization errors have π^0 poles and are infrared enhanced[68]. This appears superficially like a breakdown of the Symanzik expansion, but, in fact, the divergences are summed automatically in tmχPT if one expands about the correct vacuum[59].

4. Partial quenching and PQχPT

My final lecture is devoted to introducing partially quenched QCD and the corresponding EFT, PQχPT. I will consider PQQCD only in the continuum, and focus on conceptual issues. I will not discuss the extension to include discretization errors—this is straightforward following the methods of the previous lecture—nor the related topic of mixed action χPT. Nor will I discuss quenched QCD and quenched χPT (QχPT), except in passing, as it is not useful for obtaining quantitative physical information about QCD.

4.1. *What is PQQCD and why is it useful?*

I have given an overview of the meaning of partial quenching in the introduction. Here I give a concrete example for the pion correlator in QCD. If we write out how this is calculated in detail,

$$C_\pi(\tau) = -\left\langle \sum_{\vec{x}} \overline{u}\gamma_5 d(\vec{x},\tau)\; \overline{d}\gamma_5 u(0) \right\rangle \tag{96}$$

$$\equiv -\frac{1}{Z}\int DU \prod_q Dq D\overline{q} e^{-S_{\rm gauge} - \int_x \sum_q \overline{q}(\not{D}+m_q)q} \sum_{\vec{x}} \overline{u}\gamma_5 d(\vec{x},\tau)\; \overline{d}\gamma_5 u(0)$$

$$= \frac{1}{Z}\int DU \prod_q \det(\not{D}+m_q) e^{-S_{\rm gauge}} \sum_{\vec{x}} \mathrm{tr}\left[\gamma_5 \left(\frac{1}{\not{D}+m_d}\right)_{x0} \gamma_5 \left(\frac{1}{\not{D}+m_u}\right)_{0x}\right]$$

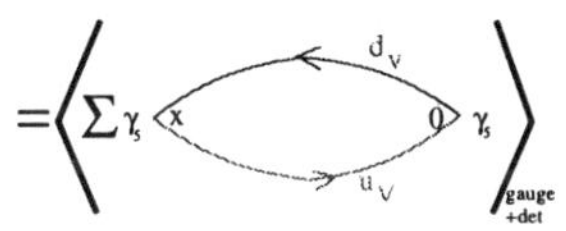

$$\propto f_\pi^2 e^{-m_\pi \tau} + \text{exponentially suppressed.}$$

we see that, in practice, we are free to use different masses in the determinant ("sea" or "dynamical" quark masses) and in the propagators ("valence" quark masses). Doing so is called partial quenching. It is relatively cheap to do so, since the calculation of propagators is usually a small overhead on the generation of configurations, thanks to the determinant, and because the rate of increase of CPU time as m_q decreases is less severe for propagators. The big question is whether we can make use of the extra information provided by working with $m_{\rm val} \neq m_{\rm sea}$.

At this point it is appropriate to interject a comment on nomenclature. In the bad old days we used the quenched approximation, in which $m_{\rm sea} \to \infty$, so the quark determinant could be ignored. This allowed simulations to be done, but at the heavy cost of ignoring quark loops and thus considering a theory which is not unitary and had various sicknesses.

QQCD is at best a model of QCD—there is no quantitative relation between the two theories. Partial quenching is in one sense a less extreme version of quenching, and thus the name. If $m_{\rm sea} \gg \Lambda_{\rm QCD}$ then PQQCD, like QQCD, is only qualitatively related to QCD, and the name is appropriate. What I consider here, however, is PQQCD with sea quarks light enough, $m_{\rm sea} \ll \Lambda_{\rm QCD}$, that (as I will argue) we can use χPT to relate the results *quantitatively* to physical QCD. For such theories a more appropriate name, with less negative connotations, would be "partially unquenched". But I will not try and create a naming revolution and will stick with the canonical nomenclature.

One thing that having light sea quarks does not change is that PQQCD is unphysical. To see this consider a contribution to $\pi^0\pi^0$ scattering in QCD: the two pions first annihilate into glue, from which is created, say, a $\pi^+\pi^-$ intermediate state, which subsequently annihilates back into glue, and then finally a new π^0 pair is created. In the corresponding process in PQQCD, the intermediate $\pi^+\pi^-$ pair must be composed of sea quarks, and thus have different masses from the valence π^0 pairs. Having different intermediate and external states means that the theory is not unitary. Another unphysical feature is the appearance of double poles in propagators. This is not possible if one can insert a complete set of physical states. The fact that there are double poles can be seen in PQχPT[79], but can also be seen more generally from the properties of the quark-level theory[80].

Because PQQCD is unphysical, it is essential to have a quantitative method relating its properties to those of QCD. This is provided by PQχPT. Thus we must simulate in the regime of quark masses where χPT is valid if PQQCD is to be a useful tool. This is illustrated above in Fig. 1. Note that the presence of the physical subspace with $m_{\rm val} = m_{\rm sea}$ within PQQCD is crucial in order to pin down the relation to QCD.

4.2. *A field theoretic formulation of PQQCD*

As we have seen, the construction of an EFT makes essential use of symmetries. Thus to develop PQχPT we need a formulation of PQQCD which makes its symmetries manifest. One way to do this is to use Morel's trick[81] of introducing commuting spin-1/2 fields (ghost quarks labeled $\widetilde{q}$), whose determinant can cancel that from the valence quarks:

$$\int D\bar{q} Dq e^{-\bar{q}(\not{D}+m_q)q} = \det(\not{D}+m_q)\,, \quad \int D\widetilde{q}^\dagger D\widetilde{q} e^{-\widetilde{q}^\dagger(\not{D}+m_q)\widetilde{q}} = \frac{1}{\det(\not{D}+m_q)}\,. \tag{97}$$

The formulation then includes three types of quark: valence quarks q_{V1}, $q_{V2}, \ldots q_{VN_V}$, ($N_V = 2, 3, \ldots$), sea quarks $q_{S1}, q_{S2}, \ldots q_{SN}$ ($N = 2, 3$), and ghost quarks $\widetilde{q}_{V1}, \widetilde{q}_{V2}, \ldots \widetilde{q}_{VN_V}$ ($N_V = 2, 3, \ldots$). The ghosts are degenerate with corresponding valence quarks. It is notationally convenient to package these fields into $(N + 2N_V)$-dim vectors:

$$Q^{tr} = \Big(\underbrace{q_{V1}, \ldots, q_{VN_V}}_{\text{valence}}, \underbrace{q_{S1}, \ldots, q_{SN}}_{\text{sea}}, \underbrace{\widetilde{q}_{V1}, \ldots, \widetilde{q}_{VN_V}}_{\text{ghost}} \Big), \tag{98}$$

$$\overline{Q} = \Big(\underbrace{\overline{q}_{V1}, \ldots, \overline{q}_{VN_V}}_{\text{valence}}, \underbrace{\overline{q}_{S1}, \ldots, \overline{q}_{SN}}_{\text{sea}}, \underbrace{\widetilde{q}^{\dagger}_{V1}, \ldots, \widetilde{q}^{\dagger}_{VN_V}}_{\text{ghost}} \Big), \tag{99}$$

and similarly for the masses

$$\mathcal{M} = \Big(\underbrace{m_{V1}, \ldots, m_{VN_V}}_{\text{valence}}, \underbrace{m_{S1}, \ldots, m_{SN}}_{\text{sea}}, \underbrace{m_{V1}, \ldots, m_{VN_V}}_{\text{ghost=valence}} \Big). \tag{100}$$

The action of PQQCD is a simple extension of that of QCD

$$S_{\text{PQ}} = S_{\text{gauge}} + \int \overline{Q}(\not{D} + \mathcal{M})Q \tag{101}$$

$$\overline{Q}(\not{D} + \mathcal{M})Q = \sum_{i=1}^{N_V} \overline{q}_{Vi}(\not{D} + m_{Vi})q_{Vi} + \sum_{j=1}^{N} \overline{q}_{Sj}(\not{D} + m_{Sj})q_{Sj}$$
$$+ \sum_{k=1}^{N_V} \widetilde{q}^{\dagger}_{Vk}(\not{D} + m_{Vk})\widetilde{q}_{Vk} \,. \tag{102}$$

If we similarly define an extended measure as

$$D\overline{Q}DQ \equiv \prod_{i=1}^{N_V} \Big(D\overline{q}_{Vi} Dq_{Vi} D\widetilde{q}^{\dagger}_{Vi} D\widetilde{q}_{Vi} \Big) \prod_{j=1}^{N} \left(D\overline{q}_{Sj} Dq_{Sj} \right) , \tag{103}$$

then we can write down a partition function containing valence quarks that nevertheless reproduces that of QCD:

$$Z_{\text{PQ}} = \int DU D\overline{Q}DQ \; e^{-S_{\text{PQ}}} \tag{104}$$

$$= \int DU e^{-S_{\text{gauge}}} \prod_{i=1}^{N_V} \left(\frac{\det(\not{D} + m_{Vi})}{\det(\not{D} + m_{Vi})} \right) \prod_{j=1}^{N} \det(\not{D} + m_{Sj}) \tag{105}$$

$$= \int DU e^{-S_{\text{gauge}}} \prod_{j=1}^{N} \det(\not{D} + m_{Sj}) = Z_{\text{QCD}} \tag{106}$$

So far this is rather trivial. The power of the method is that it provides field-theoretic expressions for PQ correlation functions, e.g.

$$C_\pi^{\rm PQ}(\tau) \equiv -Z_{\rm PQ}^{-1}\int DU D\overline{Q} DQ\ e^{-S_{\rm PQ}} \sum_{\vec{x}} \overline{u}_V\gamma_5 d_V(\vec{x},\tau)\ \overline{d}_V\gamma_5 u_V(0) \tag{107}$$

$$= Z_{\rm PQ}^{-1}\int DU \prod_{j=1}^{N} \det(\not{D} + m_{Sj}) e^{-S_{\rm gauge}}$$

$$\times \sum_{\vec{x}} {\rm tr}\left[\gamma_5 \left(\frac{1}{\not{D} + m_{Vd}}\right)_{x0} \gamma_5 \left(\frac{1}{\not{D} + m_{Vu}}\right)_{0x}\right] . \tag{108}$$

This is exactly the pion correlator with which we started, Eq. (96), but with differing valence and sea quark masses.[XIV]

As noted above, PQQCD is "anchored" to QCD (or, more precisely, to physical, QCD-like, theories). Let us see how this works in our new formulation. Consider the PQ pion correlator, but now set the valence quark masses equal to two of the sea quark masses, i.e. $m_{Vu} = m_{Sj}$ and $m_{Vd} = m_{Sk}$. Then the PQ correlator is equal to a physical QCD correlator:

$$C_\pi^{\rm PQ}(\tau) = Z_{\rm PQ}^{-1}\int DU D\overline{Q} DQ\ e^{-S_{\rm PQ}} \sum_{\vec{x}} \overline{u}_V\gamma_5 d_V(\vec{x},\tau)\ \overline{d}_V\gamma_5 u_V(0) \tag{109}$$

$$= Z_{\rm PQ}^{-1}\int DU D\overline{Q} DQ e^{-S_{\rm PQ}} \sum_{\vec{x}} \overline{q}_{Sj}\gamma_5 q_{Sk}(\vec{x},\tau)\ \overline{q}_{Sk}\gamma_5 q_{Sj}(0) \tag{110}$$

$$= Z_{\rm QCD}^{-1}\int DU \prod_{i=1}^{N} D\overline{q}_{Si} Dq_{Si}\ e^{-S_{\rm QCD}}$$

$$\times \sum_{\vec{x}} \overline{q}_{Sj}\gamma_5 q_{Sk}(\vec{x},\tau)\ \overline{q}_{Sk}\gamma_5 q_{Sj}(0) \tag{111}$$

$$= C_\pi^{\rm QCD}(\tau) . \tag{112}$$

To obtain the second line I used the result that the propagators obtained by doing the Wick contractions are the identical to those from the first line. This is an example of the enlarged symmetry of the PQ theory. Having removed the valence quarks from the operators, the valence and ghost integrals cancel, leaving a QCD correlation function. This analysis generalizes to any correlation function containing N or less different valence quark-antiquark pairs (recall that we can add any number of valence quarks). If there are more than N such pairs, e.g. a two point function containing

[XIV] This formulation works as well for quenched QCD—one just omits the sea quarks[81].

$\overline{q}_{V1}q_{V2}\overline{q}_{V3}q_{V4}$ and its conjugate in the presence of 3 light sea quarks, then there is no corresponding QCD correlator even if valence masses are all equal to sea masses. This is an example, developed further below, of how PQQCD gives one access to combinations of Wick contractions that do not occur in QCD itself.

The field theoretic formulation shows that PQQCD is a well-defined statistical system. In particular, the ghosts do not present a theoretical problem in the Euclidean functional integral. As long as quark masses are positive, the functional integrals over the ghost quarks converge (since $\not{D}$ has imaginary eigenvalues). Of course, the theory remains unphysical, and indeed we can now see this more directly. If we rotate to Minkowski space we will violate the spin-statistics theorem, and thus have an unphysical theory. Put another way, PQQCD does not satisfy reflection positivity (as can be seen, for example, from the fact that the ghost pion correlator has the opposite sign to that for the normal pion), and so one cannot construct a physical Hilbert space with a positive Hamiltonian. But the unphysical nature need not be a problem if we use PQQCD in Euclidean space as a tool to gain information about QCD.

4.3. *Developing PQχPT*

Before diving into the theoretical details let me give some qualitative motivation for what follows. We will need to assume that PQQCD is described by an effective theory close to that for QCD. I think it is fair to say that our confidence in this assumption is based in part on results from simulations. In particular, the charged pion correlator, $C_\pi^{\rm PQ}(\tau)$, has essentially the same form at long distances in PQQCD (and also QQCD) as in QCD: it falls as $\exp(-m_\pi \tau)$ at long times, and $m_\pi^2 \propto (m_{Vu} + m_{Vd})$ to good approximation. There is no sign of unphysical effects, e.g. double poles [which fall as $t\exp(-m_\pi t)$], or negative residues. It is only when one looks in detail that one finds deviations from QCD expectations, such as the enhanced chiral logarithms in QQCD. There are also other correlators where partial quenching has a dramatic effect: double poles do appear in the η' correlator, and there are negative contributions to the scalar-isovector correlator. Nevertheless, the apparent closeness of the infrared physics of PQQCD to that of QCD provides important motivation for the development of (P)QχPT.

This development has been done using two methods. The first is the "graded-symmetry" method[79] based on Morel's trick and which I use here. This is an extension of earlier work on the quenched theory[82]. The main issue, as we will see, is the whether the usual "derivation" of an EFT goes through when the theory is unphysical. The second method uses the "replica-trick"[83], in which one removes the valence determinant by sending $N_V \to 0$ rather than using ghosts.[XV] This has the advantage that the theories with integer N_V, from which one is extrapolating, are physical, and so one expects their long-distance physics to be described by χPT. Its disadvantage is that extrapolating in N_V is, in general, theoretically uncontrolled.

For the purposes of the present lecture I could use either method, since they are known to give the same results at one-loop in PQχPT, and it is plausible that this holds to all orders[83]. I choose the graded-symmetry method as I am more familiar with it.[XVI]

4.3.1. *Symmetries of PQQCD*

In the notation developed above, $S_{\rm PQQCD} = S_{\rm gauge} + \overline{Q}(\not{D} + \mathcal{M})Q$ looks just like $S_{\rm QCD}$, and appears to have a graded extension of chiral symmetry when $\mathcal{M} \to 0$, involving rotation of quarks into ghosts and vica-versa:

$$Q_{L,R} \longrightarrow U_{L,R} Q_{L,R}\,,\ \overline{Q}_{L,R} \longrightarrow \overline{Q}_{L,R} U^\dagger_{L,R}\,,\ U_{L,R} \in SU(N_V + N|N_V). \tag{113}$$

The apparent symmetry is $SU(N_V + N|N_V)_L \times SU(N_V + N|N_V)_R \times U(1)_V$. In fact, there are subtleties in the ghost sector: the transformations are inconsistent with the requirement that $\overline{Q}$ contain $\tilde{q}_V^\dagger$ in the ghost sector (and thus is related to Q, unlike in the quark sectors). This is necessary for convergence of the ghost functional integral[86]. I do not have space to discuss this technical detail, and I refer the interested reader to the literature[86,87,80,88]. The bottom line is that, for perturbative calculations in the EFT, one gets the same answer using the apparent symmetry group.

4.3.2. *Brief primer on graded Lie groups*

Since these are lectures, I recall a few basic properties of graded Lie groups. Graded means that the group matrices, U, contain both commuting and

[XV] This method gives a formalization of the "quark-line" method which I used to develop quenched χPT[84].

[XVI] The equivalence is less well established in contexts where EFT calculations are non-perturbative, such as in the ϵ−regime. Here calculations are harder in both approaches, but agree where they overlap[85].

anticommuting elements

$$U = \begin{pmatrix} A & B \\ \underbrace{C}_{N_V+N} & \underbrace{D}_{N_V} \end{pmatrix} \qquad A, D \text{ commuting; } B, C \text{ anticommuting} . \tag{114}$$

U is unitary [i.e. $U \in U(N_V + N|N_V)$] if $UU^\dagger = U^\dagger U = 1$, as for normal matrices, as long as one complex conjugates anticommuting variables as $(\eta_1\eta_2)^* \equiv \eta_2^*\eta_1^*$. Trace is generalized to "supertrace", defined to maintain cyclicity:

$$\text{str}\, U \equiv \text{tr}A - \text{tr}D \quad \Rightarrow \quad \text{str}(U_1U_2) = \text{str}(U_2U_1) . \tag{115}$$

Determinants generalize to "superdeterminants",

$$\text{sdet}\, U \equiv \exp[\text{str}(\ln U)] = \det(A - BD^{-1}C)/\det(D) , \tag{116}$$

which satisfy $\text{sdet}(U_1U_2) = \text{sdet}(U_1)\text{sdet}(U_2)$. Using this one can define $U \in SU(N_V + N|N_V)$ as unitary graded matrices with unit superdeterminant.

To get a feel for the subtleties of the graded groups it is useful to consider examples of $SU(N_V + N|N)$ matrices:

$$U_I = \begin{pmatrix} SU(N_V + N) & 0 \\ 0 & SU(N_V) \end{pmatrix} \quad \Rightarrow \quad \text{sdet}\, U_I = 1 , \tag{117}$$

$$U_{II} = \begin{pmatrix} e^{i\theta N_V} & 0 \\ 0 & e^{i\theta(N+N_V)} \end{pmatrix} \quad \Rightarrow \quad \text{sdet}\, U_{II} = \frac{(e^{i\theta N_V})^{N+N_V}}{(e^{i\theta(N+N_V)})^{N_V}} = 1 . \tag{118}$$

U_I looks just like an $SU(2N_V + N)$ matrix, while U_{II} does not (having a *determinant* differing from unity). Its superdeterminant is unity thanks to the $\det(D)$ in the denominator of (116).

One feature of $U(N_V+N|N_V)$ which is unchanged from ungraded groups is that one can pull out a commuting $U(1)$ factor, $U(N_V + N|N_V) = [SU(N_V + N|N_V) \otimes U(1)]/Z_N$, with the $U(1)$ being a phase rotation:[XVII]

$$U_{III} = \begin{pmatrix} e^{i\theta} & 0 \\ 0 & e^{i\theta} \end{pmatrix} \quad \Rightarrow \quad \text{sdet}\, U_{III} = \frac{e^{i\theta(N+N_V)}}{e^{i\theta N_V}} = e^{i\theta N} . \tag{119}$$

4.3.3. *Chiral symmetry breaking*

We now follow the same steps as we did for QCD in Sec. 2.2, noting differences along the way. We expand PQQCD about $\mathcal{M} = 0$, where the chiral

---

[XVII] Note that $\text{sdet}(U_{III}) = 1$ in the quenched theory ($N = 0$), so that U_{III} lies in $SU(N_V|N_V)$, indicating that the quenched group structure is more complicated.

symmetry group $G = SU(N_V+N|N_V)_L \times SU(N_V+N|N_V)_R$ is exact.[XVIII] We know that this symmetry is broken, because it is broken in the massless QCD contained in massless PQQCD. To study the symmetry breaking we introduce a graded generalization of the order parameter (6) used for QCD:

$$\Omega_{ij} = \langle Q_{L,i,\alpha,c}\overline{Q}_{R,j,\alpha,c}\rangle_{\rm PQ} \xrightarrow{G} U_L\,\Omega\,U_R^\dagger\,. \tag{120}$$

Next we assume that the graded vector symmetry, $SU(N_V+N|N_V)_V$, is not spontaneously broken. If $\mathcal{M}$ is diagonal, real and positive, this follows from an extension[80] of the Vafa-Witten theorem for QCD[23]. The assumption is thus that nothing singular happens as $\mathcal{M} \to 0$. Given this, we know that $\Omega = \omega \times \mathbf{1}$, and, furthermore, we know from the QCD sub-theory that $\omega = \langle q_S\overline{q}_S\rangle \neq 0$. Thus the symmetry breaking is $G \to H = SU(N_V+N|N_V)_V$, a simple graded generalization of that in QCD.

We can now derive Goldstone's theorem using Ward identities for two-point Euclidean correlators, which remain exact in the PQ theory[80]. The result is that there are massless, spinless poles which couple to $\overline{Q}\gamma_\mu\gamma_5T^aQ$ for all $(N+2N_V)^2-1$ traceless generators, T^a, of $SU(N_V+N|N_V)$. Of these, $2(N+N_V)N_V$ are fermionic (quark-ghost particles), $(N+N_V)^2-1$ are bosonic with normal sign two-point functions (quark-quark), and N_V^2 are bosonic with unphysical sign two-point functions (ghost-ghost).

4.3.4. *Constructing the EFT*

It would appear that we have the ingredients to construct an EFT, just as in QCD: we know the symmetries, and we have a separation of scales. In fact, we know much less than in QCD, because PQQCD is not physical. In QCD, the poles in two-point functions correspond to particle states in a physical Hilbert space, and thus we know that correlation functions of arbitrary order will also have these poles, and from their residues we can extract the S-matrix, which will be unitary. Furthermore, since QCD involves local interactions, the S-matrix will satisfy cluster decomposition. Then we can invoke Weinberg's "theorem" and write an effective, local field theory for the light particles. In PQQCD, by contrast, we *do not know* that the

[XVIII] *A posteriori* we will find that we must take the chiral limit with m_V and m_S in fixed ratio, because there are divergences if $m_V \to 0$ at fixed m_S[89]. This non-analyticity is not a barrier to the construction of the EFT. Non-analyticities are present in physical quantities in continuum χPT as well (e.g. the chiral logs), but arise from infrared physics, just as the divergences here. There is no reason to think that the coefficients in the EFT, which result from integrating out ultraviolet physics, are non-analytic in m_V, m_S.

poles we have found in two-point functions also appear in higher order correlators, and we *do know* that there is neither a physical Hilbert space nor S-matrix. Indeed, it can be shown that neutral correlators have double-poles if $m_V \neq m_S$[80]. So we cannot rely on Weinberg's argument. Instead we must simply *assume* that there is a local EFT containing the Goldstone modes and satisfying the symmetries. In other words we assume a minimal change from the EFT for QCD.

This is not as *ad hoc* as it might sound. Let me give three arguments in favor of this assumption. First, we know that there is a local EFT for the QCD sub-theory, and that this describes the long distance behavior of correlators. What PQQCD allows one to do is to separate individual Wick contractions contributing to QCD correlators. It seems implausible that, for example, the description of these individual contractions would require a non-local interaction (leading to a different pole structure) whose effects cancel when one adds them to form QCD correlators. Second, one can derive chiral Ward identities in PQQCD for arbitrary order correlation functions, that are generalizations of those in QCD. PQχPT satisfies these identities by saturating them with Goldstone pole contributions. In this regard, there is numerical evidence from simulations that in, say, four-point functions the Goldstone poles dominate when one pulls one of the operators far from the others, as predicted by PQχPT. Finally, one can imagine carrying out a Wilsonian renormalization group program in a Euclidean theory, in which one successively integrates out "shells" of high-momentum modes. This automatically leads to a local interaction, and symmetries are preserved. One can think of Symanzik's EFT for lattice QCD as an example. We cannot actually do the calculation here, given the non-perturbative physics of QCD, but if we could, and if the two-point functions correctly tell us the appropriate low-energy degrees of freedom, it is plausible that we would end up with PQχPT.

Having assumed the nature of the EFT we continue following the same steps as for QCD. We "promote" the condensate into a field,

$$\Omega/\omega \to \Sigma(x) \in SU(N_V+N|N)\,, \qquad \Sigma \xrightarrow{G} U_L\,\Sigma\,U_R^\dagger\,, \tag{121}$$

and, assuming standard masses so that $\langle\Sigma\rangle = 1$, we define NG particles by

$$\Sigma = \exp\left[\frac{2i}{f}\Phi(x)\right]\,, \qquad \Phi(x) = \begin{pmatrix} \phi(x) & \eta_1(x) \\ \eta_2(x) & \widetilde{\phi}(x) \end{pmatrix}\,. \tag{122}$$

Here ϕ are the quark-quark "normal" NG bosons, $\widetilde{\phi}$ are the ghost-ghost NG bosons, and $\eta_{1,2}$ are quark-ghost NG fermions. The constraint sdet $\Sigma = 1$

implies str $\Phi = \text{tr}\ \phi - \text{tr}\ \widetilde{\phi} = 0$. The QCD part of Σ is

$$\Phi(x) = \begin{pmatrix} 0 & 0 & 0 \\ 0 & \pi(x) & 0 \\ \underbrace{0}_{N_V} & \underbrace{0}_{N} & \underbrace{0}_{N_V} \end{pmatrix} \quad \Rightarrow \quad \Sigma = \begin{pmatrix} 1 & 0 & 0 \\ 0 & \Sigma_{\text{QCD}} & 0 \\ 0 & 0 & 1 \end{pmatrix} . \tag{123}$$

Next we construct the most general local, Euclidean-invariant, G-invariant Lagrangian built out of Σ, $D_\mu\Sigma$ and χ. Here the covariant derivative and χ are graded generalizations of the corresponding terms in χPT. In particular $\chi = 2B_0(s+ip) \to U_L\chi U_R^\dagger$, with the sources set to $s = \mathcal{M}$ and $p = 0$ at the end. We can use generalizations of the same building blocks as in χPT, e.g.

$$L_\mu = \Sigma D_\mu \Sigma^\dagger \to U_L L_\mu U_L^\dagger\,, \quad \text{str}(L_\mu) = 0\,, \tag{124}$$

The power counting (which is independent of the nature of the fields) is the same as in χPT.

In this way we arrive at the PQ chiral Lagrangian through NLO:

$$\mathcal{L}_{\text{PQ}}^{(2)} = \frac{f^2}{4}\text{str}\left(D_\mu\Sigma D_\mu\Sigma^\dagger\right) - \frac{f^2}{4}\text{str}(\chi\Sigma^\dagger + \Sigma\chi^\dagger) \tag{125}$$

$$\begin{aligned} \mathcal{L}_{\text{PQ}}^{(4)} = & -L_1\left[\text{str}(D_\mu\Sigma D_\mu\Sigma^\dagger)\right]^2 - L_2\,\text{str}(D_\mu\Sigma D_\nu\Sigma^\dagger)\text{str}(D_\mu\Sigma D_\nu\Sigma^\dagger) \\ & +L_3\,\text{str}(D_\mu\Sigma D_\mu\Sigma^\dagger D_\nu\Sigma D_\nu\Sigma^\dagger) \\ & +L_4\,\text{str}(D_\mu\Sigma^\dagger D_\mu\Sigma)\text{str}(\chi^\dagger\Sigma+\Sigma^\dagger\chi) + L_5\,\text{str}(D_\mu\Sigma^\dagger D_\mu\Sigma[\chi^\dagger\Sigma+\Sigma^\dagger\chi]) \\ & -L_6\left[\text{str}(\chi^\dagger\Sigma+\Sigma^\dagger\chi)\right]^2 - L_7\left[\text{str}(\chi^\dagger\Sigma-\Sigma^\dagger\chi)\right]^2 - L_8\,\text{str}(\chi^\dagger\Sigma\chi^\dagger\Sigma+p.c.) \\ & +iL_9\,\text{str}(L_{\mu\nu}D_\mu\Sigma D_\nu\Sigma^\dagger+p.c.) + L_{10}\,\text{str}(L_{\mu\nu}\Sigma R_{\mu\nu}\Sigma^\dagger) \\ & +H_1\,\text{str}(L_{\mu\nu}L_{\mu\nu}+p.c.) - H_2\,\text{str}(\chi^\dagger\chi) + \mathcal{L}_{\text{WZW,PQ}} \\ & +L_{\text{PQ}}\mathcal{O}_{PQ}\,. \end{aligned} \tag{126}$$

These are almost carbon copies of the corresponding results in χPT [Eqs. (28,31)], except that tr $\to$ str, and there is an additional term in $\mathcal{L}_{\text{PQ}}^{(4)}$ (the $\mathcal{O}_{\text{PQ}}$ term)[90]. To my knowledge, no-one has worked out the structure of the PQ WZW term in detail, though its contributions to $\pi^0 \to \gamma\gamma$ vertices have been analyzed[91].

Thus we find that the number of LECs in PQχPT is the same as in χPT at LO, and that there is only one more, L_{PQ}, at NLO. But how are these LECs related to those of χPT? The answer is simple[6]: *they are the same!* This can be seen by considering correlation functions created by sources s, p, l_μ, r_μ restricted to the QCD sub-space. At the quark level, these are QCD correlators, since valence and ghost contributions cancel identically. Thus they are described by χPT. At the EFT level, one can show diagramatically in PQχPT that a similar cancellation occurs, and that one can do the calculation using Σ restricted to the QCD subspace as in (123). Inserting this form into $\mathcal{L}_{\rm PQ}^{(2,4)}$ one finds ($i = 1, 10$, HEC ignored):

$$\mathcal{L}_{\rm PQ}^{(2,4)}(\Sigma, L_i, L_{\rm PQ}) = \mathcal{L}^{(2,4)}(\Sigma_{\rm QCD}, L_i)\,. \tag{127}$$

In words, the calculation one does with QCD sources in PQχPT is exactly that one would do in χPT. For the results to be equal it must be that the L_i are equal. This is the key result in PQχPT, for it means that the predictions of PQχPT involve only slightly more LECs than those of χPT.

4.3.5. *What about $\mathcal{O}_{\rm PQ}$?*

Back when we were constructing $\mathcal{L}^{(4)}$ in χPT, I noted that one possible four-derivative term was not independent for $N \leq 3$. This is due to Cayley-Hamilton relations between traces of finite matrices. Such relations do not hold for graded matrices, and so the term is independent in PQχPT. It is convenient to write it as

$$\begin{aligned}\mathcal{O}_{\rm PQ} &= {\rm str}(D_\mu\Sigma D_\nu\Sigma^\dagger D_\mu\Sigma D_\nu\Sigma^\dagger) + 2\,{\rm str}(D_\mu\Sigma D_\nu\Sigma^\dagger D_\mu\Sigma D_\nu\Sigma^\dagger)\\ &\quad - {\rm str}(D_\mu\Sigma D_\mu\Sigma^\dagger)^2/2 - {\rm str}(D_\mu\Sigma D_\nu\Sigma^\dagger){\rm str}(D_\mu\Sigma D_\nu\Sigma^\dagger)\,,\end{aligned} \tag{128}$$

for then it vanishes if Σ is restricted to its QCD subspace as in (123). This is why $L_{\rm PQ}$ does not appear on the right-hand side of Eq. (127). $\mathcal{O}_{\rm PQ}$ does not vanish for general Σ, however, and thus appears in $\mathcal{L}_{\rm PQ}^{(4)}$ with a new LEC. This additional LEC also appears in standard χPT if $N \geq 4$, when the Cayley-Hamilton relations become less restrictive. As one goes to higher order, the number of additional such operators in PQχPT increases[90].

How does this new operator enter into results for measurable quantities in PQQCD? It can only contribute to PQ quantities, for the considerations above show that it vanishes when restricted to the QCD sub-space. It turns out to contribute to PQ $\pi\pi$ scattering at NLO, but to PQ m_π and f_π only at NNLO[90]. Thus its practical impact is small.

It is worthwhile, however, understanding its origin more deeply. As I have repeatedly mentioned, PQQCD allows one to separate individual Wick contractions, unlike QCD. For example, π^+K^0 scattering in QCD has two contractions (thin [blue] is u, medium [red] is d, and thick [brown] is s):

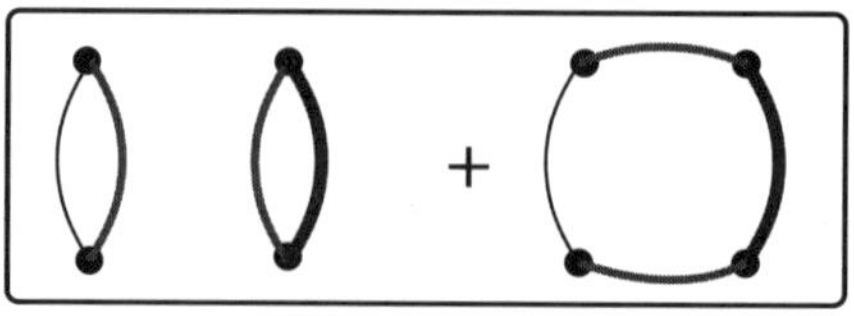

and $\mathcal{O}_{\rm PQ}$ makes no contribution to this process. We can separate these contractions in PQQCD using, for example, scattering involving ghost quarks:

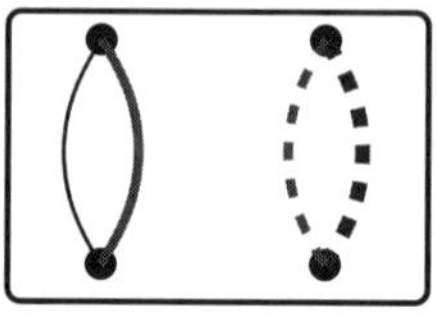

This is the same as the first contraction contributing to QCD scattering, up to a sign. $\mathcal{O}_{\rm PQ}$ does contribute to this correlator. Thus $L_{\rm PQ}$ contains information about the relative size of two contractions in the QCD process.

4.4. *PQχPT at LO*

With the Lagrangian in hand it is straightforward to develop perturbation theory. Inserting the expansion (122) into $\mathcal{L}^{(2)}_{\rm PQ}$, we find

$$\mathcal{L}^{(2)}_{\rm PQ} = \mathrm{str}(\partial_\mu\Phi\partial_\mu\Phi) + \mathrm{str}(\chi\Phi^2) + \dots \tag{129}$$

$$= \mathrm{tr}(\partial_\mu\phi\partial_\mu\phi + \partial_\mu\eta_1\partial_\mu\eta_2 - \partial_\mu\eta_2\partial_\mu\eta_1 - \partial_\mu\widetilde{\phi}\partial_\mu\widetilde{\phi})$$
$$+ \mathrm{tr}\left[(\phi^2 + \eta_1\eta_2)\begin{pmatrix}\chi_V & 0\\ 0 & \chi_S\end{pmatrix}\right] - \mathrm{tr}(\widetilde{\phi}^2\chi_V) - \mathrm{tr}(\eta_2\eta_1\chi_V). \tag{130}$$

Here $\chi_{V,S}$ are the mass matrices in the valence and sea sectors, respectively, multiplied by $2B_0$. ϕ is like the pion field in χPT, except that it includes both valence and sea quarks. The propagator for "charged" mesons with flavor $\overline{q}_1 q_2$ (which can be VV, VS or SS) is

$$(p^2 + m_{12}^2)^{-1}, \qquad m_{12}^2 = (\chi_1 + \chi_2)/2\,. \tag{131}$$

On the other hand, the terms involving the "ghost-ghost" boson $\widetilde{\phi}$ have unphysical signs. It appears that we are expanding the ghost-ghost sector of Σ about the wrong point, since the potential is maximized. Furthermore, the kinetic term will not give a convergent functional integral. Both these problems result from our earlier decision to use the symmetry group G, even though it was inconsistent with convergence. A more careful treatment[86,80,88] shows that one should have changed $\widetilde{\phi} \to i\widetilde{\phi}$, which solves the convergence problem, at the cost of introducing i's into vertices.[XIX] In perturbation theory we can reshuffle the i's by hand, and work with the naive propagator one gets from (130). For "charged" ghost mesons with flavor $\bar{\tilde{q}}_1\tilde{q}_2$ one has

$$-(p^2+m_{12}^2)^{-1}\,,\qquad m_{12}^2=(\chi_1+\chi_2)/2\,. \tag{132}$$

Finally, the NG fermion propagators can have either sign. There are no convergence issues for fermions, but signs are important for cancellations.

What about the "neutral" fields ($\bar{q}_1 q_1$, etc.)? Here we have to implement the constraint $\mathrm{str}(\Phi)=\mathrm{tr}(\phi)-\mathrm{tr}(\tilde{\phi})=0$. There are two ways to do this. The first is simply use a basis of generators which is straceless: $\Phi=\sum_a \Phi_a T^a$ with $\mathrm{str}(T^a)=0$. This is analogous to excluding the η' in χPT, but is more complicated in PQχPT. In the second method, we remove the constraint by including a singlet field, $\Phi\to\Phi+\Phi_0/\sqrt{N}$, adding a mass term to the action,

$$\mathcal{L}_{\mathrm{PQ}\chi}\to\mathcal{L}_{\mathrm{PQ}\chi}+m_0^2\,\mathrm{str}(\Phi)^2/N\,, \tag{133}$$

and then integrating out Φ_0 by sending $m_0^2\to\infty$. This is just a trick to project out the singlet. To make it formally correct, we must regularize the theory with a cut-off so that m_0^2 always exceeds any loop momenta. This is the method mostly used in practice as it is simple to implement.

Using this method, the neutral propagator is obtained from:

$$\mathcal{L}^{(2)}=\sum_{j=1}^{N+2N_V}\epsilon_j(\partial_\mu\Phi_{jj}\partial_\mu\Phi_{jj}+m_j\Phi_{jj}^2)+(m_0^2/N)(\sum_j\epsilon_j\Phi_{jj})^2+\ldots \tag{134}$$

$$\epsilon_j=\begin{cases}+1 & \text{valence or sea quarks}\\ -1 & \text{ghosts}\end{cases} \tag{135}$$

The m_0 term couples all the Φ_{jj}, so that, in particular, neutral sea-quark states can contribute to neutral valence propagators. The inversion of the kernel is not trivial for general quark masses, but can be accomplished using linear algebra tricks[79,6]. I show an example of the result for $N=3$ non-degenerate sea quarks, after having sent $m_0^2\to\infty$

$$\langle\Phi_{ii}\Phi_{jj}\rangle=\frac{\epsilon_i\delta_{ij}}{p^2+\chi_i}-\frac{1}{N}\frac{1}{(p^2+\chi_i)(p^2+\chi_j)}\frac{(p^2+\chi_{S1})(p^2+\chi_{S2})(p^2+\chi_{S3})}{(p^2+M_{\pi^0}^2)(p^2+M_\eta^2)}\,. \tag{136}$$

[XIX] Obtaining a positive ghost propagator does not imply a restoration of reflection positivity. This is now violated by the i's in the vertices.

Here M_{π^0} and M_η are the masses of the sea-sector neutrals, after inclusion of $\pi^0 - \eta$ mixing evaluated at LO in χPT (as discussed in the first lecture). If we take i, j to be valence labels, and set $\chi_i = \chi_j$ (or simply consider $i = j$), we see the infamous double pole in the second term. It is reduced to a single pole if the valence mass equals that of one of the sea quarks, $\chi_i = \chi_{Sj}$. The form simplifies if the sea quarks are degenerate,

$$\langle \Phi_{ii} \Phi_{jj} \rangle = \frac{\epsilon_i \delta_{ij}}{p^2 + \chi_i} - \frac{1}{N} \frac{(p^2 + \chi_S)}{(p^2 + \chi_i)(p^2 + \chi_j)} \,. \tag{137}$$

The residue of the double pole for $\chi_i = \chi_j$ is then $(\chi_i - \chi_S)/N$, showing how it vanishes in the physical subspace. Setting $\chi_i = \chi_j = \chi_S$ we obtain:

$$\langle \Phi_{SS} \Phi_{SS} \rangle = \frac{1}{p^2 + \chi_S} \left(1 - \frac{1}{N} \right) . \tag{138}$$

This is the correct result in the sea sector, with the $1/N$ term projecting out the η'.

Introducing Φ_0 has allowed us to use the basis $\Phi_{ij} \sim Q_i \overline{Q}_j$ for neutral as well as charged states. This means that one can follow the flavor indices in an unambiguous way through any PQχPT diagram, resulting in "quark-line diagrams". This is a useful qualitative tool in thinking about calculations. Charged particles propagators are simple:[XX]

$$\langle \Phi_{ij} \Phi_{ji} \rangle = \pm \tfrac{1}{p^2 + (\chi_i + \chi_j)/2} =$$ 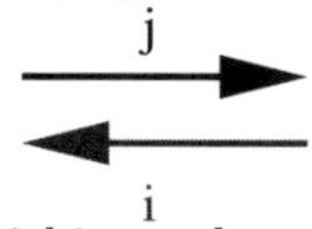

Typically one uses solid lines for quarks (distinguishing valence and sea by a label), and dashed lines for ghosts. In the diagram above I chose i, j to be quark labels. With this notation the neutral propagator is, schematically,

where the "hairpin" is the m_0 vertex. The first term corresponds to the single pole in (137), while the remaining diagrams all have a double pole (if $\chi_i = \chi_j$), and their summation gives the second term in (137). Note that the valence and ghost contributions cancel exactly between two hairpin vertices.

[XX]The sign of the propagator is to be determined from Eq. (130).

4.5. *NLO calculations in PQχPT and outlook*

With the machinery developed above, perturbative calculations in PQχPT are straightforward extensions of those in standard χPT. (I stress again that this is not true for non-perturbative calculations like those in the $\epsilon-$regime.) Let me sketch an example, that of the mass of a pion composed of valence quarks $V1, V2$. The types of quark-line diagrams corresponding to the one-loop diagram are:

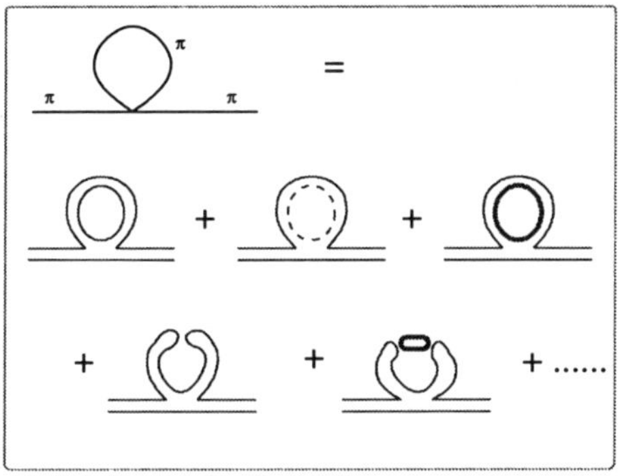

Here thin (blue) lines are valence quarks, dashed lines are ghosts, and thick (brown) lines are sea quarks. I have used the result that the four-pion vertices from $\mathcal{L}^{(2)}_{\rm PQ}$ involve a single supertrace and so give "connected" quark-line vertices. The contributions of loops of valence-valence bosons cancel those of valence-ghost fermions, as expected from the underlying theory. This leaves only loops of valence-sea bosons on the second line (which turn out to cancel for m_π but not for f_π), while the third line shows the hairpin contributions.

The result (simplified by assuming degenerate sea quarks and $m_{V1} = m_{V2} \equiv m_V$) is[89]

$$m^2_{VV} = \chi_V \Bigg(1 + \frac{1}{N} \frac{2\chi_V - \chi_S}{\Lambda^2_\chi} \ln(\chi_V/\mu^2) + \frac{\chi_V - \chi_S}{N\Lambda^2_\chi} + \frac{8}{f^2} \left[(2L_8 - L_5)\chi_V + (2L_6 - L_4)N\chi_S \right] \Bigg) . \tag{139}$$

The terms proportional to $1/N$ arise from the loops involving hairpins, while the second line shows the analytic terms from the NLO Lagrangian. The unphysical nature of the double poles in the hairpin contribution gives rise to the "enhanced logarithm", proportional to $\chi_S \ln(\chi_V)$, which diverges when $m_V \to 0$ at fixed m_S, and leads to a breakdown of PQχPT. This is the divergence noted above which requires that one take the chiral limit with m_V/m_S fixed. The divergence occurs only at extremely small valence quark masses, however, and should not be a problem in practice[6].

The analytic terms in (139) show the utility of partial quenching. By varying m_V and m_S, and including the chiral logs of the first line in the fit, one can separately determine $2L_8 - L_5$ and $2L_6 - L_4$. This should be compared to the setting $\chi_V = \chi_S$, so that one is in the physical subspace. Then the chiral logs are physical and not enhanced, but varying χ_V only allows a determination of $2L_8 - L_5 + 2L_6 - L_4$. In fact, PQ simulations have been used by various groups to determine $2L_8 - L_5$, which is of particular interest as its value determines whether it is possible for $m_u = 0$ (which would solve the strong CP problem). The answer is clearly negative[3].

4.5.1. *Status of PQχPT calculations*

It is now standard to extend to PQχPT any χPT calculation relevant for extrapolating lattice results. Many quantities have been considered at one-loop,[XXI] including the masses and form-factors of pions, baryons, vector mesons, and heavy-light hadrons, structure functions of baryons, the scalar two-point function, weak matrix elements (B_K, $K \to \pi\pi$), the neutron electric dipole moment, and pion and nucleon scattering amplitudes. Similarly standard are partially quenched extensions of calculations including discretization effects: tmχPT, (rooted) staggered χPT, and mixed-action χPT. In most cases this extension is straightforward. The cases where it is not involve non-trivial generalizations of normal representations to those of graded groups. Two notable examples are octect baryons[92,93] and four-fermion matrix elements[94]. Particularly striking examples of the unphysical nature of PQQCD are the negative contributions to correlators that are strictly positive in QCD, e.g. the scalar two-point function[95], and the appearance of non-unitary effects in two-pion correlators[96,97]. Another interesting application of PQQCD is to the calculation of the spectrum of the continuum[98] and lattice[99] Dirac operators.

The MILC studies of pion and kaon properties show the potential power of using partial quenching[3], while at the same time exposing the challenges of extrapolating using χPT. With very precise numerical results, and masses ranging from $m_s/10$ to m_s, an accurate description of their data requires

[XXI] I have chosen not to give references for the subsequent list—there is not space for the $O(100)$ that would be needed.

not only terms of NLO but also of NNLO and, at the higher masses, of NNNLO. The loop contributions are not known beyond NLO in staggered χPT, so only analytic NNLO and NNNLO terms are kept. This is clearly a phenomenological approach, but the key point to keep in mind is that these higher order terms have essentially no impact on the resulting extrapolated results for physical quantities: they are very small corrections for the physical up and down quark masses, and the strange quark is anyway at, or close to, its physical value. The MILC fits would not be possible without the large number of partially quenched points, and their success provides *a posteriori* justification for the assumptions needed to develop PQχPT.

The need to work beyond NLO in practice has spurred some heroic work from continuum χPT experts: there are now full NNLO (i.e. two-loop) calculations for pion and kaon properties in PQχPT[31]!

4.5.2. *A final example:* L_7

I close these lectures with a final example of which I am particularly fond and which nicely illustrates the power of PQQCD[6]. This concerns the LEC L_7, which multiplies a "two (s)trace term",

$$\mathcal{L}^{(4)}_{\chi,\mathrm{PQ}} = \cdots - L_7 \mathrm{str}\,(\chi\Sigma^\dagger - \Sigma\chi^\dagger)^2 + \ldots, \tag{140}$$

and which contributes to PGB masses only for non-degenerate quarks. Its most significant contribution in QCD is to m_η, and this leads to violations of the GMO relation:

$$4m_K^2 - m_\pi^2 - 3m_\eta^2 = \frac{32(m_K^2 - m_\pi^2)^2}{3f^2}(L_5 - 6L_8 - 12L_7) + \text{known chiral logs}\,. \tag{141}$$

Obtaining the physical result for m_η would be a highly non-trivial check of the lattice methodology, as it involves quark-disconnected diagrams with intermediate glue. For the same reason, this is a challenging calculation. Partial quenching can help in the usual way by providing more data to fit, but also by obviating the need to actually do the extrapolation to the physical η'. In particular, since L_5 and L_8 are known quite accurately, as are the chiral logs, η' physics is tested, using Eq. (141), by any method that allows a calculation of L_7. One such method is to calculate the residue of the double pole in a disconnected valence-valence correlator:

$$\left.\frac{\int d^3x \langle \Phi_{V1,V1}(t,\vec{x})\Phi_{V2,V2}(0)\rangle}{\int d^3x \langle \Phi_{V1,V2}(t,\vec{x})\Phi_{V2,V1}(0)\rangle}\right|_{m_{V1}=m_{V2}} \overset{t\to\infty}{\longrightarrow} \frac{\mathcal{D}t}{2M_{VV}}\,. \tag{142}$$

With $N = 3$ degenerate sea quarks one finds

$$\mathcal{D} = \frac{\chi_V - \chi_S}{N} - \frac{16}{f^2}\left(L_7 + \frac{L_5}{2N}\right)(\chi_V - \chi_S)^2 + \text{known chiral logs}\,, \quad (143)$$

so L_7 can be determined from the term quadratic (and thus even) in the deviation from the unquenched theory. The generalization of this formula to a $2+1$ theory has not been worked out, but should be straightforward.

I like this formula as it gets to the essence of partial quenching: using unphysical phenomena (in this case the double pole) to obtain physical results (here L_7). Of course, we have not removed the need to do the challenging calculation of quark-disconnected correlators, but rather have packaged the calculation in a way that is more flexible because there are more knobs to turn. The result has been very recently extended to NNLO[100].

Acknowledgments

I thank Oliver Bär for a helpful correspondence, Will Detmold, Maarten Golterman and André Walker-Loud for comments, and Y. Kuramashi and co-organizers for a delightful school.

References

1. K. G. Wilson, *Phys. Rev.* **D10**, 2445 (1974).
2. M. Creutz, *Phys. Rev. Lett.* **45**, 313 (1980).
3. C. Aubin *et al.* [MILC Collaboration], *Phys. Rev.* **D70**, 114501 (2004) [arXiv:hep-lat/0407028].
4. A. S. Kronfeld *et al.* [Fermilab Lattice Collaboration], *PoS* **LAT2005**, 206 (2006) [*Int. J. Mod. Phys.* **A21**, 713 (2006)] [arXiv:hep-lat/0509169].
5. J. Heitger and R. Sommer [ALPHA Collaboration], *JHEP* **0402**, 022 (2004) [arXiv:hep-lat/0310035].
6. S. R. Sharpe and N. Shoresh, *Phys. Rev.* **D62**, 094503 (2000) [arXiv:hep-lat/0006017].
7. C. Bernard, M. Golterman and Y. Shamir, *Phys. Rev.* **D73**, 114511 (2006) [arXiv:hep-lat/0604017].
8. W. J. Lee and S. R. Sharpe, *Phys. Rev.* **D60**, 114503 (1999) [arXiv:hep-lat/9905023].
9. C. Bernard [MILC Collaboration], *Phys. Rev.* **D65**, 054031 (2002) [arXiv:hep-lat/0111051];
   C. Aubin and C. Bernard, *Phys. Rev.* **D68**, 034014 (2003) [arXiv:hep-lat/0304014].
10. C. Bernard, *Phys. Rev.* **D73**, 114503 (2006) [arXiv:hep-lat/0603011].
11. J.F. Donoghue, E. Golowich and B.R. Holstein, "*Dynamics of the Standard Model*", (Cambridge, 1992).

12. G. Ecker, "*Chiral Perturbation Theory*", [arXiv:hep-ph/9608226,9805300].
13. H. Georgi, "*Weak Interactions and Modern Particle Theory*", (Benjamin-Cummings, 1984).
14. D.B. Kaplan, "*5 lectures on Effective Field Theory*", [arXiv:nucl-th/0510023].
15. A. S. Kronfeld, "*Uses of effective field theory in lattice QCD*", arXiv:hep-lat/0205021.
16. A.V. Manohar, "*Effective Field Theories*", [arXiv:hep-ph/9606222].
17. A. Pich, "*Introduction to Chiral Perturbation Theory*", [arXiv:hep-ph/9502366].
18. S. Weinberg, *Physica* **A96**, 327 (1979).
19. M. Lüscher, S. Sint, R. Sommer and P. Weisz, *Nucl. Phys.* **B478**, 365 (1996) [arXiv:hep-lat/9605038].
20. K. Symanzik, *Commun. Math. Phys.* **45**, 79 (1975); *Nucl. Phys.* **B226**, 187 and 205 (1983).
21. M. Knecht, B. Moussallam, J. Stern and N. H. Fuchs, *Nucl. Phys.* **B457**, 513 (1995) [arXiv:hep-ph/9507319]; *ibid* **B471**, 445 (1996) [arXiv:hep-ph/9512404].
22. G. Colangelo, J. Gasser and H. Leutwyler, *Phys. Rev. Lett.* **86**, 5008 (2001) [arXiv:hep-ph/0103063].
23. C. Vafa and E. Witten, *Nucl. Phys.* **B234**, 173 (1984).
24. S.R. Coleman, J. Wess and B. Zumino, *Phys. Rev.* **177**, 2239 (1969).
25. C.G. Callan, S.R. Coleman, J. Wess and B. Zumino, *Phys. Rev.* **177**, 2247 (1969).
26. J. Gasser and H. Leutwyler, *Annals Phys.* **158**, 142 (1984); *Nucl. Phys.* **B250**, 465 (1985).
27. J. Wess and B. Zumino, *Phys. Lett.* **B37**, 95 (1971); E. Witten, *Nucl. Phys.* **B223**, 422 (1983).
28. J. Bijnens, G. Colangelo and G. Ecker, *JHEP* **9902**, 020 (1999) [arXiv:hep-ph/9902437] and *Annals Phys.* **280**, 100 (2000) [arXiv:hep-ph/9907333].
29. M. Creutz, *Phys. Rev. Lett.* **92**, 201601 (2004) [arXiv:hep-lat/0312018].
30. D. B. Kaplan and A. V. Manohar, *Phys. Rev. Lett.* **56**, 2004 (1986).
31. J. Bijnens, N. Danielsson and T. A. Lahde, *Phys. Rev.* **D73**, 074509 (2006) [arXiv:hep-lat/0602003].
32. H. Leutwyler, *Annals Phys.* **235**, 165 (1994) [arXiv:hep-ph/9311274].
33. D. Becirevic and G. Villadoro, *Phys. Rev.* **D70**, 094036 (2004) [arXiv:hep-lat/0408029].
34. G. Colangelo, S. Durr and C. Haefeli, *Nucl. Phys.* **B721**, 136 (2005) [arXiv:hep-lat/0503014].
35. J. Gasser and H. Leutwyler, *Phys. Lett.* **B188**, 477 (1987).
36. L. Giusti, C. Pena, P. Hernandez, M. Laine, J. Wennekers and H. Wittig, *PoS* **LAT2005**, 344 (2006) [arXiv:hep-lat/0510033].
37. J. Bijnens, *AIP Conf. Proc.* **768**, 153 (2005) [arXiv:hep-ph/0409068] and *Prog. Part. Nucl. Phys.* **58**, 521 (2007) [arXiv:hep-ph/0604043].
38. G. Amoros, J. Bijnens and P. Talavera, *Nucl. Phys.* **B585**, 293 (2000) [Erratum, *ibid.* **B598**, 665 (2001)] [arXiv:hep-ph/0003258].

39. O. Bär, G. Rupak and N. Shoresh, *Phys. Rev.* **D67**, 114505 (2003) [arXiv:hep-lat/0210050].
40. O. Bär, *Nucl. Phys. Proc. Suppl.* **140**, 106 (2005) [arXiv:hep-lat/0409123].
41. S. R. Sharpe and R. L. Singleton, *Phys. Rev.* **D58**, 074501 (1998) [arXiv:hep-lat/9804028].
42. R. Frezzotti, P. A. Grassi, S. Sint and P. Weisz, *Nucl. Phys. Proc. Suppl.* **83**, 941 (2000) [arXiv:hep-lat/9909003]; *JHEP* **0108**, 058 (2001) [arXiv:hep-lat/0101001].
43. R. Frezzotti and G. C. Rossi, *JHEP* **0408**, 007 (2004) [arXiv:hep-lat/0306014].
44. A. Shindler, *PoS* **LAT2005**, 014 (2006) [arXiv:hep-lat/0511002].
45. R. Frezzotti and G. C. Rossi, *Nucl. Phys. Proc. Suppl.* **153**, 250 (2006) [arXiv:hep-lat/0511035].
46. T. Reisz, *Commun. Math. Phys.* **117**, 79 (1988); *Nucl. Phys.* **B318**, 417 (1989).
47. K. Jansen *et al.*, *Phys. Lett.* **B372**, 275 (1996) [arXiv:hep-lat/9512009];
48. S. R. Sharpe and J. M. S. Wu, *Phys. Rev.* **D70**, 094029 (2004) [arXiv:hep-lat/0407025].
49. M. Lüscher and P. Weisz, *Commun. Math. Phys.* **97**, 59 (1985) [Erratum, *ibid.* **98**, 433 (1985)].
50. B. Sheikholeslami and R. Wohlert, *Nucl. Phys.* **B259**, 572 (1985).
51. O. Bär, G. Rupak and N. Shoresh, *Phys. Rev.* **D70**, 034508 (2004) [arXiv:hep-lat/0306021].
52. G. Rupak and N. Shoresh, *Phys. Rev.* **D66**, 054503 (2002) [arXiv:hep-lat/0201019].
53. G. Münster and C. Schmidt, *Europhys. Lett.* **66**, 652 (2004) [arXiv:hep-lat/0311032].
54. T. Bhattacharya, R. Gupta, W. Lee, S. R. Sharpe and J. M. S. Wu, *Phys. Rev.* **D73**, 034504 (2006) [arXiv:hep-lat/0511014].
55. S. R. Sharpe and J. M. S. Wu, *Phys. Rev.* **D71**, 074501 (2005) [arXiv:hep-lat/0411021].
56. S. Aoki and O. Bär, *Phys. Rev.* **D70**, 116011 (2004) [arXiv:hep-lat/0409006].
57. S. Aoki and O. Bär, Phys. Rev. **D74**, 034511 (2006) [arXiv:hep-lat/0604018].
58. L. Scorzato, *Eur. Phys. J.* **C37**, 445 (2004) [arXiv:hep-lat/0407023].
59. S. R. Sharpe, *Phys. Rev.* **D72**, 074510 (2005) [arXiv:hep-lat/0509009].
60. F. Farchioni *et al.*, *Eur. Phys. J.* **C42**, 73 (2005) [arXiv:hep-lat/0410031].
61. S. Aoki, *Phys. Rev.* **D30**, 2653 (1984); *Phys. Rev. Lett.* **57**, 3136 (1986).
62. A. Walker-Loud and J. M. S. Wu, *Phys. Rev.* **D72**, 014506 (2005) [arXiv:hep-lat/0504001].
63. F. Farchioni *et al.*, *Eur. Phys. J.* **C47**, 453 (2006) [arXiv:hep-lat/0512017].
64. S. Aoki and O. Bär, *PoS* **LAT2005**, 046 (2006) [arXiv:hep-lat/0509002].
65. K. Jansen *et al.* [XLF Collaboration], *Phys. Lett.* **B624**, 334 (2005) [arXiv:hep-lat/0507032].
66. F. Farchioni *et al.*, *PoS* **LAT2005**, 033 (2006) [arXiv:hep-lat/0509036].
67. D. Becirevic, P. Boucaud, V. Lubicz, G. Martinelli, F. Mescia, S. Simula

and C. Tarantino, *Phys. Rev.* **D74**, 034501 (2006) [arXiv:hep-lat/0605006].
68. R. Frezzotti, G. Martinelli, M. Papinutto and G. C. Rossi, *JHEP* **0604**, 038 (2006) [arXiv:hep-lat/0503034].
69. F. Farchioni *et al.*, *Eur. Phys. J.* **C42**, 73 (2005) [arXiv:hep-lat/0410031].
70. A. M. Abdel-Rehim, R. Lewis and R. M. Woloshyn, *Phys. Rev.* **D71**, 094505 (2005) [arXiv:hep-lat/0503007].
71. K. Jansen, M. Papinutto, A. Shindler, C. Urbach and I. Wetzorke [XLF Collaboration], *Phys. Lett.* **B619**, 184 (2005) [arXiv:hep-lat/0503031].
72. M. Creutz, arXiv:hep-ph/9608216.
73. F. Farchioni *et al.*, *Phys. Lett.* **B624**, 324 (2005) [arXiv:hep-lat/0506025].
74. F. Farchioni *et al.*, *PoS* **LAT2005**, 072 (2006) [arXiv:hep-lat/0509131].
75. K. Jansen, M. Papinutto, A. Shindler, C. Urbach and I. Wetzorke [XLF Collaboration], *JHEP* **0509**, 071 (2005) [arXiv:hep-lat/0507010].
76. R. Frezzotti and G. C. Rossi, *Nucl. Phys. Proc. Suppl.* **128**, 193 (2004) [arXiv:hep-lat/0311008].
77. T.Chiarappa, *et al.*, arXiv:hep-lat/0606011
78. S. Aoki, *Phys. Rev.* **D68**, 054508 (2003) [arXiv:hep-lat/0306027];
S. Aoki, O. Bär, S. Takeda and T. Ishikawa, *Phys. Rev.* **D73**, 014511 (2006) [arXiv:hep-lat/0509049];
S. Aoki, O. Bär and S. Takeda, *Phys. Rev.* **D73**, 094501 (2006) [arXiv:hep-lat/0601019].
79. C. W. Bernard and M. F. Golterman, *Phys. Rev.* **D49**, 486 (1994) [arXiv:hep-lat/9306005].
80. S. R. Sharpe and N. Shoresh, *Phys. Rev.* **D64**, 114510 (2001) [arXiv:hep-lat/0108003].
81. A. Morel, *J. Phys. (Paris)* **48**, 111 (1987)
82. C. W. Bernard and M. F. Golterman, *Phys. Rev.* **D46**, 853 (1992) [arXiv:hep-lat/9204007].
83. P. H. Damgaard, *Phys. Lett.* **B476**, 465 (2000) [arXiv:hep-lat/0001002];
P. H. Damgaard and K. Splittorff, *Phys. Rev.* **D62**, 054509 (2000) [arXiv:hep-lat/0003017].
84. S. R. Sharpe, *Nucl. Phys. Proc. Suppl.* **17**, 146 (1990); *Phys. Rev.* **D41**, 3233 (1990) and *Phys. Rev.* **D46**, 3146 (1992) [arXiv:hep-lat/9205020].
85. P. H. Damgaard, *Nucl. Phys. Proc. Suppl.* **106**, 29 (2002) [arXiv:hep-lat/0110192].
86. P. H. Damgaard, J. C. Osborn, D. Toublan and J. J. M. Verbaarschot, *Nucl. Phys.* **B547**, 305 (1999) [arXiv:hep-th/9811212].
87. M. Zirnbauer, *J. Math. Phys. (N.Y.)* **37**, 4986 (1996).
88. M. Golterman, S. R. Sharpe and R. L. . Singleton, *Phys. Rev.* **D71**, 094503 (2005) [arXiv:hep-lat/0501015].
89. S. R. Sharpe, *Phys. Rev.* **D56**, 7052 (1997) [Erratum, *ibid.* **D62**, 099901 (2000)] [arXiv:hep-lat/9707018].
90. S. R. Sharpe and R. S. Van de Water, *Phys. Rev.* **D69**, 054027 (2004) [arXiv:hep-lat/0310012].
91. W. Detmold, B. C. Tiburzi and A. Walker-Loud, *Phys. Rev.* **D73**, 114505 (2006) [arXiv:hep-lat/0603026].

92. J. N. Labrenz and S. R. Sharpe, *Phys. Rev.* **D54**, 4595 (1996) [arXiv:hep-lat/9605034].
93. J. W. Chen and M. J. Savage, *Phys. Rev.* **D65**, 094001 (2002) [arXiv:hep-lat/0111050].
94. M. Golterman and E. Pallante, *JHEP* **0110**, 037 (2001) [arXiv:hep-lat/0108010] and *Phys. Rev.* **D74**, 014509 (2006) [arXiv:hep-lat/0602025].
95. W. A. Bardeen, A. Duncan, E. Eichten, N. Isgur and H. Thacker, *Phys. Rev.* **D65**, 014509 (2002) [arXiv:hep-lat/0106008];
S. Prelovsek, C. Dawson, T. Izubuchi, K. Orginos and A. Soni, *Phys. Rev.* **D70**, 094503 (2004) [arXiv:hep-lat/0407037].
96. C. W. Bernard and M. F. L. Golterman, *Phys. Rev.* **D53**, 476 (1996) [arXiv:hep-lat/9507004].
97. C. J. D. Lin, G. Martinelli, E. Pallante, C. T. Sachrajda and G. Villadoro, *Phys. Lett.* **B553**, 229 (2003) [arXiv:hep-lat/0211043].
98. J. C. Osborn, D. Toublan and J. J. M. Verbaarschot, *Nucl. Phys.* **B540**, 317 (1999) [arXiv:hep-th/9806110].
99. S. R. Sharpe, *Phys. Rev.* **D74**, 014512 (2006) [arXiv:hep-lat/0606002].
100. J. Bijnens and N. Danielsson, *Phys. Rev.* **D74**, 054503 (2006) [arXiv:hep-lat/0606017].

LATTICE QCD WITH A CHIRAL TWIST*

S. SINT
School of Mathematics,
Trinity College Dublin,
Dublin 2, Ireland
E-mail: sint@maths.tcd.ie

In these lectures I explain how chiral symmetry of continuum QCD naturally leads to a class of lattice regularisations known as twisted mass QCD (tmQCD). As compared to standard Wilson quarks, its advantages are the absence of unphysical zero modes, the possibility to circumvent lattice renormalisation problems and automatic O(a) improvement. On the other hand, the physical parity and flavour symmetries are explicitly broken. I discuss these aspects and then turn to the theory in a finite space-time volume with Schrödinger functional boundary conditions. Again, chiral transformations of the continuum theory may be used as a guide to formulate an alternative lattice regularisation of the Schrödinger functional, with interesting applications to renormalization problems in QCD.

1. Introduction

In recent years, twisted mass QCD (tmQCD) has become a popular variant of lattice QCD with Wilson-type quarks [1,2,3]. Initially designed to render the (partially) quenched approximation well-defined through the elimination of unphysical zero modes, it was soon realised that tmQCD could also be used to circumvent some notorious lattice renormalization problems [1,2]. Later, Frezzotti and Rossi [4] observed that scaling violations in tmQCD can be reduced to O(a^2) without the need for all the O(a) counterterms required with standard Wilson quarks (a being the lattice spacing). This property, referred to as "automatic O(a) improvement", has attracted further attention and a number of groups have started large scale numerical simulations using tmQCD. In these lectures I do not attempt to review this work in progress[a]. Here I would rather like to give an introduction to the basic con-

*based on lectures given at the school "Perspectives in Lattice QCD", Nara, Japan, October 31- November 14, 2005

[a]See Ref. 5 for a review and further references.

cepts. This includes in particular a discussion of O(a) improvement and the question whether it is compromised by currently used non-perturbative renormalization procedures based on the QCD Schrödinger functional (SF-schemes). In fact, the standard Schrödinger functional boundary conditions turn out to be difficult to reconcile with automatic O(a) improvement and the construction of an alternative set-up for the Schrödinger functional may therefore be advantageous.

This writeup is organised as follows: I start with the interplay between the choice of the quark mass term and the form taken by parity, flavour and chiral symmetry transformations (Sect. 2). After a reminder of standard Wilson quarks and the problem of unphysical zero modes (Section 3), lattice tmQCD is introduced in Sect. 4. Based on the formal continuum theory a dictionary between tmQCD and QCD correlation functions is readily established, which is expected to hold between properly renormalised correlation functions. It then becomes clear how to by-pass certain renormalization problems of standard Wilson quarks (Sect. 5), and the computation of B_K is discussed in some detail. In Sect. 6 automatic O(a) improvement of tmQCD is analysed using Symanzik's effective theory. Potential problems of tmQCD associated with flavour and parity breaking are shortly mentioned in Sect. 7. In Sect. 8, the properties of Schrödinger functional renormalisation schemes (SF schemes) are discussed. Motivated by the clash of the standard set-up with automatic O(a) improvement and by the slow decoupling of heavy quarks in mass-dependent SF schemes, a modified definition of the Schrödinger functional is proposed, and its effectiveness regarding O(a) improvement is illustrated in an example taken from perturbation theory. Section 9 contains some conclusions.

2. Continuum QCD and chiral transformations

Let us consider the continuum action of QCD with $N_{\rm f} = 2$ massless quarks[b]. Decomposing the action into a pure gauge and a fermionic part, $S = S_g + S_f$, we here focus on the fermionic part,

$$S_f = \int \mathrm{d}^4x \; \bar{\psi}(x) \not{D} \psi(x), \qquad \not{D} = \gamma_\mu D_\mu. \tag{1}$$

The quark and antiquark fields $\psi, \bar{\psi}$ are flavour doublets, interacting minimally with the gluon field A_μ via the covariant derivative $D_\mu = \partial_\mu + A_\mu$.

---

[b]Conventions used for Euclidean γ-matrices in 4 dimensions: $\{\gamma_\mu, \gamma_\nu\} = 2\delta_{\mu\nu}$, $\gamma_\mu^\dagger = \gamma_\mu$, where $\mu, \nu = 0, 1, 2, 3$, and $\gamma_5 = \gamma_0\gamma_1\gamma_2\gamma_3$, $\sigma_{\mu\nu} = \frac{i}{2}[\gamma_\mu, \gamma_\nu]$.

The massless fermionic action has a global chiral-flavour SU(2)×SU(2) invariance, corresponding to the transformations,

$$\begin{aligned}\psi \to \psi' &= \exp(i\omega_V^a \tau^a/2)\exp(i\omega_A^b \gamma_5 \tau^b/2)\psi,\\ \bar\psi \to \bar\psi' &= \bar\psi \exp(i\omega_A^b \gamma_5 \tau^b/2)\exp(-i\omega_V^a \tau^a/2),\end{aligned} \tag{2}$$

where τ^a $(a = 1,2,3)$ are Pauli matrices, $\omega_{V,A}^a$ are transformation parameters[c]. This notation distinguishes the axial from the vector generators (corresponding to the flavour or isospin SU(2) subgroup) in a standard way.

A quark mass term breaks the chiral flavour symmetry explicitly, leaving only the vector or isospin symmetry intact. The above notation for the symmetry transformations was introduced with the standard quark mass term $\bar\psi\psi$ in mind, but e.g. the choice

$$\bar\psi'\psi' = \bar\psi \exp(i\omega_A^a \gamma_5 \tau^a)\psi = \cos(\omega_A)\bar\psi\psi + i\sin(\omega_A)u_A^a \bar\psi\gamma_5\tau^a\psi, \tag{3}$$

would be completely equivalent. Here, ω_A denotes the modulus of $(\omega_A^1, \omega_A^2, \omega_A^3)$ and $u_A^a = \omega_A^a/\omega_A$ is a unit vector. In fact, it is only after the introduction of the quark mass term that the distinction between axial and vector symmetries acquires a meaning. By definition, the vector symmetry transformations are those which leave the quark mass term invariant. Similarly, the quark mass term is supposed to be invariant under parity transformations. As a consequence, the form of a symmetry transformation depends on the choice of the mass term. While a standard mass term implies that a parity transformation can be realised as

$$\psi(x_0,\mathbf{x}) \to \gamma_0\psi(x_0,-\mathbf{x}), \qquad \bar\psi(x_0,\mathbf{x}) \to \bar\psi(x_0,-\mathbf{x})\gamma_0, \tag{4}$$

the alternative choice of (3) for the mass term means that a parity transformation will look more complicated, for instance

$$\begin{aligned}\psi(x_0,\mathbf{x}) &\to \gamma_0 \exp(i\omega_A^a\gamma_5\tau^a)\psi(x_0,-\mathbf{x}),\\ \bar\psi(x_0,\mathbf{x}) &\to \bar\psi(x_0,-\mathbf{x})\exp(i\omega_A^a\gamma_5\tau^a)\gamma_0.\end{aligned} \tag{5}$$

Similarly, the isospin transformation obtained with a standard mass term corresponds to (2) with all axial transformation parameters set to zero, $\omega_A^a = 0$ (whence the notation), whereas the mass term (3) leads to the much less intuitive formula

$$\begin{aligned}\psi &\to \exp(-i\omega_A^a\gamma_5\tau^a/2)\exp(i\omega_V^b\tau^b/2)\exp(i\omega_A^c\gamma_5\tau^c/2)\psi,\\ \bar\psi &\to \bar\psi\exp(i\omega_A^a\gamma_5\tau^a/2)\exp(-i\omega_V^b\tau^b/2)\exp(-i\omega_A^c\gamma_5\tau^c/2),\end{aligned} \tag{6}$$

[c] Summation over repeated indices $a, b = 1,2,3$ is understood.

where ω_V^b $(b = 1, 2, 3)$ are transformation parameters while ω_A^a $(a = 1, 2, 3)$ are again fixed. The situation is reminiscent of the choice of a coordinate system, and our intuition about the form of symmetry transformations is thus based on a particular choice of "field coordinates". Of course, this raises the question why one should deviate from the standard choice of the mass term. In the continuum and for regularisations preserving chiral symmetry there is indeed no point in introducing a twisted mass term, for any non-standard choice could be brought into the standard form by using an axial rotation, which, being a symmetry of the massless theory, has no further effects. The situation is different in regularisations which break chiral symmetry, such as lattice regularisations with Wilson type quarks. One may then obtain different regularisations of QCD which have equivalent continuum limits but differ at the cutoff level. This will be made more precise a bit later.

3. Standard Wilson quarks

Standard Wilson quarks are characterised by the fermionic lattice action,

$$S_f = a^4 \sum_x \bar{\psi}(x)(D_{\rm W} + m_0)\psi(x), \tag{7}$$

$$D_{\rm W} = \sum_{\mu=0}^{3} \left\{ \tfrac{1}{2}(\nabla_\mu + \nabla_\mu^*)\gamma_\mu - a\nabla_\mu^*\nabla_\mu \right\}. \tag{8}$$

Here, m_0 is a bare mass parameter and the covariant lattice derivatives in the Wilson-Dirac operator are defined as usual (see Ref. 6 for unexplained notation). Assuming $N_{\rm f}$ quark flavours the lattice action has an exact $\mathrm{U}(N_{\rm f})$ vector symmetry, and is invariant under axis permutations, reflections such as parity and charge conjugation. Furthermore, unitarity of lattice QCD with Wilson quarks has been rigorously established [7]. These nice properties of standard Wilson quarks come with a price: all axial symmetries are explicitly broken by the last term in Eq. (8), called the Wilson term. This has a number of consequences:

(1) Linear mass divergence: the quark mass term is not protected against additive renormalization, i.e. any renormalized quark mass is of the form $m_{\rm R} = Z_m(m_0 - m_{\rm cr})$, where the critical mass is linearly divergent, i.e. $m_{\rm cr} \propto 1/a$.
(2) Axial current renormalization: since axial transformations are not an exact symmetry, there is no exact current algebra, and the non-

singlet axial current requires a non-trivial multiplicative renormalization to restore current algebra up to $\mathrm{O}(a)$ effects.

(3) Definition of the chiral condensate as expectation value of a local operator: the renormalised iso-singlet scalar density has the structure,

$$(\bar{\psi}\psi)_{\mathrm{R}} = Z_{\mathrm{S}^0}\{\bar{\psi}\psi + c_{\mathrm{S}}a^{-3}\}. \quad (9)$$

In a regularisation which respects chiral symmetry, the additive renormalization constant c_{S} would be proportional to am, with m being a multiplicatively renormalisable bare quark mass. This means that the chiral condensate is well-defined in the chiral limit once its *multiplicative* renormalisation has been carried out. In contrast, with Wilson quarks one first needs to subtract the cubic power divergence, even in the chiral limit.

(4) Cutoff effects: the leading cutoff effects with Wilson-type fermions are proportional to a, rather than a^2. Again, this is a consequence of chiral symmetry breaking. This is easily seen by looking at the structure of the counterterms which are to be included for the on-shell $\mathrm{O}(a)$ improvement of the theory à la Symanzik [6].

From a field theoretical point of view this illustrates the proliferation of additional counterterms in a case where the regularisation breaks a continuum symmetry. One should note, however, that there is no remaining theoretical or conceptual problem.

3.1. *Wilson quarks and unphysical fermionic zero modes*

Nevertheless, technical problems may arise within the current practice of numerical simulations with Wilson-type quarks. This is related to the fact that, for a given gauge background field, the massive Wilson-Dirac operator $D_W + m_0$ is not protected against zero modes unless the bare mass parameter m_0 is positive. However, due to additive quark mass renormalisation, the masses of the light quarks typically correspond to negative bare mass parameters, which leaves the Wilson-Dirac operator unprotected against zero modes in the physically interesting region. These modes are considered unphysical, since one expects from the continuum theory that any non-zero value of the renormalised quark mass prohibits zero modes of the Dirac operator.

It is instructive to look at a typical fermionic correlation function, such as the pion propagator given by

$$G^{ab}(x-y) = -\left\langle \bar{\psi}(x)\gamma_5\frac{\tau^a}{2}\psi(x)\bar{\psi}(y)\gamma_5\frac{\tau^b}{2}\psi(y)\right\rangle$$
$$= -\mathcal{Z}^{-1}\int D[U,\psi,\bar{\psi}]\mathrm{e}^{-S}\bar{\psi}(x)\gamma_5\frac{\tau^a}{2}\psi(x)\bar{\psi}(y)\gamma_5\frac{\tau^b}{2}\psi(y), \quad (10)$$

where τ^a, $a = 1,2,3$, are the Pauli matrices acting in flavour space and $\mathcal{Z} = \langle 1 \rangle$. It is convenient to introduce the operator,

$$Q = \gamma_5(D_\mathrm{W} + m_0), \qquad Q = Q^\dagger, \quad (11)$$

which acts in single flavour space. Integrating over the quark and anti-quark fields one obtains

$$G^{ab}(x-y) = \frac{1}{2}\delta^{ab}\mathcal{Z}^{-1}\int D[U]\mathrm{e}^{-S_g}\det\left(Q^2\right)\mathrm{tr}\left[Q^{-1}(x,y)Q^{-1}(x,y)\right], \quad (12)$$

where the flavour structure has been reduced analytically and the remaining trace is over colour and spin indices. The important point to notice is that the resulting expression is never singular. Denoting the eigenfunction of Q for a given eigenvalue λ_i by $\varphi_i(x)$, the pion propagator takes the form

$$G^{ab}(x-y) = \frac{1}{2}\delta^{ab}\mathcal{Z}^{-1}\int D[U]\mathrm{e}^{-S_g}\left(\prod_i \lambda_i^2\right)$$
$$\times\sum_{j,k}\lambda_j^{-1}\lambda_k^{-1}\varphi_j(x)\varphi_j^*(y)\varphi_k(x)\varphi_k^*(y). \quad (13)$$

In other words, the eigenvalues in the denominator are always compensated by corresponding factors from the determinant. The limit of vanishing eigenvalues is always regular and a strict lower bound on the eigenvalue spectrum is not required for the theory to be well-defined.

However, the absence of a lower bound on $|\lambda_i|$ may still lead to technical problems, either due to the use of unphysical approximations or due to the set-up of numerical simulations:

3.1.1. *Quenched and partially quenched approximations*

As the computational cost for the generation of a an ensemble of gauge field configurations is dominated by the inclusion of the quark determinant, a widely used approximation consists in omitting the determinant when taking the average over gauge fields. The quark propagators with the eigenvalues in the denominator may then become singular, and gauge field configurations where this happens are called "exceptional". The example in Figure 1 taken from Ref. 8 shows the ensemble average of the pion propagator over all gauge configurations but a a single exceptional one (dashed line), where the propagator deviates dramatically from the average (dots and solid line). The inclusion of the exceptional configuration in the ensemble average would lead to much larger errors, while its omission invalidates the Monte Carlo procedure. In principle one should say that the quenched approximation with Wilson type quarks is ill-defined, since zero modes are bound to occur if the ensemble of gauge configurations is large enough. However, the frequency of near zero modes depends very sensitively upon the bare quark mass and is in fact a function of the lattice size and all the other bare parameters in the lattice action. One may therefore think of the quenched approximation as being operationally defined, if for an ensemble of, say, a few hundred configurations the problem is typically absent. "Safe" parameter ranges may then be quoted for a given action, but this situation is clearly unsatisfactory. In particular, as the problem is not sharply defined, one may always be unlucky and encounter near zero modes even at parameter values which have previously been considered safe. In practice it is this problem which has limited the approach to the chiral limit, rather than finite volume effects due to the pions becoming too light.

Obviously, the problem is expected to disappear once the quark determinant is properly included. Usually this is done by including the complete determinant in the effective gauge field measure used in the importance sampling, and the probability for a gauge configuration to be included in the ensemble becomes proportional to the eigenvalues. Exceptional configurations are then never produced. However, even in this case, one is often interested in varying the valence quark masses independently of the sea quark masses, a situation which is referred to as the partially quenched approximation. One may also have different numbers of valence and sea quarks, or Wilson valence quarks and sea quarks of a different kind. In all these cases one expects similar problems with unphysical zero modes as in the quenched approximation.

3.1.2. *Potential problems in the Hybrid Monte Carlo algorithm*

Most numerical simulations use some variant of the Hybrid Monte Carlo algorithm [9]. Integrating the molecular dynamics trajectories in fictitious phase space then requires the evaluation of the fermionic force term and thus the inversion of the Dirac operator at each step in molecular dynamics

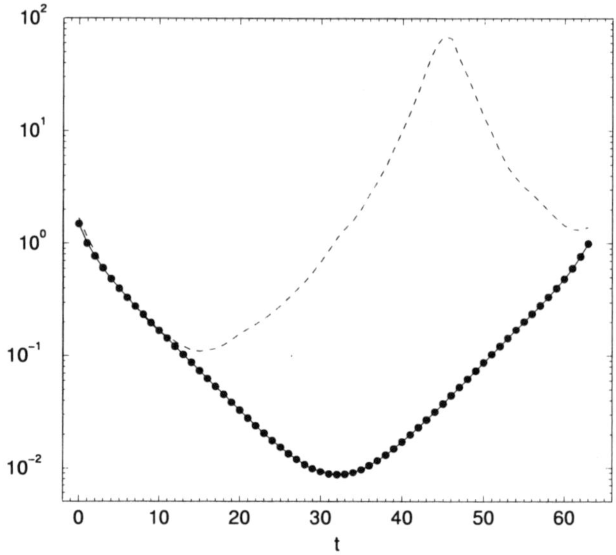

Figure 1. The pion propagator vs. time separation from a quenched simulation on a $32^3 \times 64$ lattice at $\beta = 6.2$ (cf. text for an explanation).

time. The force term may become very large if an exceptional configuration is encountered, and the molecular dynamics integrator tends to become unstable if the product of the force and the step size exceeds a certain critical value [10]. To avoid this situation one may hence be forced to decrease the step size to very small values thereby increasing the cost of the simulation. It is likely that this problem was at the heart of the difficulties encountered in the past with simulations of Wilson type quarks [11]. However, various developments over the past few years seem to have solved this problem (see Ref. 12 for a recent account of current simulation algorithms and cost estimates and Ref. 13 for further discussion).

4. Twisted mass lattice QCD

Initially the main motivation for introducing a twisted mass term was the problem with zero modes discussed in the previous section. The lattice action for a doublet ψ of $N_{\rm f} = 2$ mass degenerate quarks is now given by

$$S_f = a^4 \sum_x \bar{\psi}(x)(D_{\rm W} + m_0 + i\mu_{\rm q}\gamma_5\tau^3)\psi(x), \tag{14}$$

where $\mu_{\rm q}$ denotes the bare twisted mass parameter. It is easy to see that the presence of this parameter elimiates any unphysical zero modes, for

$$\det\left(D_{\rm W}+m_0+i\mu_{\rm q}\gamma_5\tau^3\right)=\det\begin{pmatrix}Q+i\mu_q & 0\\ 0 & Q-i\mu_q\end{pmatrix}$$
$$=\det(Q^2+\mu_q^2)>0. \qquad (15)$$

The difference in the determinant already shows that twisted mass and standard QCD cannot be the same regularisation. In fact, any attempt to perform an axial rotation so as to eliminate the twisted mass term will rotate the Wilson term in Eq. (8), too. the equivalence between both regularisations can therefore only be expected to hold in the continuum limit. We will discuss this more in detail below.

Here it suffices to say that the chiral flavour symmetry of twisted mass QCD is reduced to an exact U(1) symmetry with generator $\tau^3/2$. Furthermore, charge conjugation, axis permutations and reflections combined with a flavour permutation, e.g.

$$\psi(x_0,\mathbf{x})\rightarrow\gamma_0\tau^1\psi(x_0,-\mathbf{x}),\qquad \bar\psi(x_0,\mathbf{x})\rightarrow\bar\psi(x_0,-\mathbf{x})\gamma_0\tau^1, \qquad (16)$$

are exact symmetries. Finally, the construction of a positive and self-adjoint transfer matrix for standard Wilson quarks can be generalised to twisted mass QCD, provided $\mu_{\rm q}$ is real and the usual condition on the standard bare mass parameter, $|\kappa|<1/6$, with $\kappa=(2am_0+8)^{-1}$ is satisfied [3].

4.1. *Equivalence between tmQCD and QCD*

Taking the continuum limit, we see that the fermionic continuum action of tmQCD,

$$S_f=\int \mathrm{d}^4x\;\bar\psi(x)(\not{D}+m+i\mu_{\rm q}\gamma_5\tau^3)\psi(x), \qquad (17)$$

can be related to the standard action by a global chiral field rotation,

$$\psi'=R(\alpha)\psi,\quad \bar\psi'=\bar\psi R(\alpha),\quad R(\alpha)=\exp\left(i\alpha\gamma_5\frac{\tau^3}{2}\right). \qquad (18)$$

Choosing the angle α such that $\tan\alpha=\mu_{\rm q}/m$, the action for the primed fields takes the standard form,

$$S_f'=\int \mathrm{d}^4x\;\bar\psi'(x)(\not{D}+M)\psi'(x),\qquad M=\sqrt{m^2+\mu_{\rm q}^2}. \qquad (19)$$

In QCD all physical observables can be extracted from gauge invariant correlation functions of composite fields. We would therefore like to study the relationship between correlation functions in tmQCD and standard QCD. To this end we introduce polar mass coordinates,

$$m = M\cos(\alpha), \qquad \mu_q = M\sin(\alpha), \tag{20}$$

and consider the correlation functions labelled by (M,α),

$$\left\langle O[\psi,\bar{\psi}]\right\rangle_{(M,\alpha)} = \mathcal{Z}^{-1}\int D[U,\psi,\bar{\psi}]\; O[\psi,\bar{\psi}]\; \mathrm{e}^{-S[m,\mu_\mathrm{q}]}. \tag{21}$$

Treating the functional integral like an ordinary integral we change the variables to ψ' and $\bar{\psi}'$ of Eq. (18) and re-label these new integration variables to ψ and $\bar{\psi}$ afterwards. In this way we arrive at the identity,

$$\left\langle O[\psi,\bar{\psi}]\right\rangle_{(M,0)} = \left\langle O[R(\alpha)\psi,\bar{\psi}R(\alpha)]\right\rangle_{(M,\alpha)}. \tag{22}$$

To go a step further, we now assume that the functional $O[\psi,\bar{\psi}]$ consists of factors which are members of a chiral multiplet. Considering such a field $\phi_A^{(r)}[\psi,\bar{\psi}]$ in the representation r, the transformation of ψ and $\bar{\psi}$ by $R(\alpha)$ induces the transformation of $\phi_A^{(r)}$ by $R^{(r)}(\alpha)$ in the representation r,

$$\phi_A^{(r)}[R(\alpha)\psi,\bar{\psi}R(\alpha)] = R_{AB}^{(r)}(\alpha)\phi_B^{(r)}[\psi,\bar{\psi}]. \tag{23}$$

For n-point functions of such fields, one obtains the identity,

$$\langle\phi_{A_1}^{(r_1)}\cdots\phi_{A_n}^{(r_n)}\rangle_{(M,0)} = \left(\prod_{i=1}^{n} R_{A_iB_i}^{(r_i)}(\alpha)\right)\langle\phi_{B_1}^{(r_1)}\cdots\phi_{B_n}^{(r_n)}\rangle_{(M,\alpha)}. \tag{24}$$

The correlation functions in standard QCD labelled by $(M,0)$ are just linear combinations of those in twisted mass QCD, labelled by (M,α). The inverse relation can be obtained by inverting the matrices $R^{(r)}(\alpha)$. This is trivial, as the axial rotation (18) forms an abelian subgroup of the chiral flavour group, so that $[R^{(r)}(\alpha)]^{-1} = R^{(r)}(-\alpha)$. Examples of chiral multiplets are the non-singlet currents (A_μ^a, V_μ^a) or the non-singlet axial density combined with the singlet scalar density, $(\frac{1}{2}S^0, P^a)$. In terms of quark fields one has

$$\begin{aligned} A_\mu^a &= \bar{\psi}\gamma_\mu\gamma_5\tfrac{\tau^3}{2}\psi, & V_\mu^a &= \bar{\psi}\gamma_\mu\tfrac{\tau^3}{2}\psi, \\ P^a &= \bar{\psi}\gamma_5\tfrac{\tau^3}{2}\psi, & S^0 &= \bar{\psi}\psi, \end{aligned} \tag{25}$$

and one may then easily infer the transformation behaviour of these chiral multiplets:

$$\begin{aligned} A_\mu'^1 &= cA_\mu^1 + sV_\mu^2, & V_\mu'^1 &= cV_\mu^1 + sA_\mu^2, \\ A_\mu'^2 &= cA_\mu^2 - sV_\mu^1, & V_\mu'^2 &= cV_\mu^2 - sA_\mu^1, \\ A_\mu'^3 &= A_\mu^3, & V_\mu'^3 &= V_\mu^3, \\ P'^a &= P^a, \quad (a=1,2), & S'^0 &= cS^0 + 2isP^3, \\ P'^3 &= cP^3 + is\tfrac{1}{2}S^0. \end{aligned} \tag{26}$$

Here the notation $O' \equiv O[\psi', \bar{\psi}']$, $c \equiv \cos(\alpha)$, $s \equiv \sin(\alpha)$ was used. For a correlator of $A^1_\mu(x)$ and $P^1(y)$ in standard QCD this means

$$\left\langle A^1_\mu(x) P^1(y) \right\rangle_{(M,0)} = \cos(\alpha) \left\langle A^1_\mu(x) P^1(y) \right\rangle_{(M,\alpha)} + \sin(\alpha) \left\langle V^2_\mu(x) P^1(y) \right\rangle_{(M,\alpha)}. \tag{27}$$

In other words, Eqs. (26) relate an insertion of the primed fields into standard QCD correlators to the insertion of the corresponding r.h.s. into tmQCD correlators. In particular, we note that the PCAC and PCVC relations in the physical basis

$$\partial_\mu A'^a_\mu = 2MP'^a, \qquad \partial_\mu V'^a_\mu = 0, \tag{28}$$

are equivalent to linear combinations of their twisted counterparts,

$$\begin{aligned} \partial_\mu A^a_\mu &= 2mP^a + \delta^{3a} i\mu_q S^0, \\ \partial_\mu V^a_\mu &= -2\mu_q \epsilon^{3ab} P^b. \end{aligned} \tag{29}$$

In conclusion, the formal continuum theory provides us with a dictionary between correlation functions in standard and twisted mass QCD. However, all these considerations have been quite formal, and we need to specify how such a dictionary carries over to the renormalized theories.

4.2. *Beyond the formal continuum theory*

To clarify this question let us suppose that tmQCD is regularised on the lattice with Ginsparg-Wilson quarks, where chiral and flavour symmetries are the same as in the continuum. Identities such as Eqs. (24) may then be derived in the *bare* theory. If in addition, we start from a finite volume with, say, periodic boundary conditions for all fields, the functional integral becomes a finite dimensional Grassmann integral. Therefore, these identities are no longer formal, but on firm mathematical grounds, and all one has to show is that the renormalisation procedure can be carried out such that they continue to hold in the renormalised theory. This is straightforward, as one just has to make sure that all members of a given chiral multiplet are renormalised in the same way, and that the multiplicative renormalization constants do not depend on the twist angle α. This can be achieved e.g. by imposing renormalisation conditions in the massless limit. Hence, in this case, the dictionary introduced above holds between the renormalised correlation functions of both theories. Assuming universality to hold beyond perturbation theory, this establishes the equivalence of both versions of QCD at the non-perturbative level, since any other regularisation, chirally symmetric or not, will then lead to the same renormalised correlation functions up to cutoff effects. While there is no reason to doubt that universality holds generally, one should be aware that it has rigorously been established only in perturbation theory and for selected regularisations (e.g. lattice regularisations with Wilson type quarks [14]).

4.3. *Lattice tmQCD with Wilson quarks*

In tmQCD on the lattice with Wilson quarks the axial transformation relating continuum tmQCD to standard QCD is not an exact symmetry. Therefore, equivalence can only be expected to hold in the continuum limit, i.e. for properly renormalized correlation functions and up to cutoff effects. The lattice symmetries imply the counterterm structure, with the following result for the renormalised parameters,

$$g_{\mathrm{R}}^2 = Z_g g_0^2, \qquad m_{\mathrm{R}} = Z_m(m_0 - m_{\mathrm{cr}}), \qquad \mu_{\mathrm{R}} = Z_\mu \mu_{\mathrm{q}}. \tag{30}$$

It is a priori not obvious how the twist angle α should be defined from the mass parameters. The key observation is that chiral symmetry can be restored in the bare lattice theory up to cutoff effects, by imposing axial Ward identities as normalisation conditions [15]. This fixes the *relative* renormalization of all members of a chiral multiplet, such as $Z_{\mathrm{A}}/Z_{\mathrm{V}}$ for the symmetry currents[d], or Z_{S^0}/Z_P for the iso-triplet axial and the iso-inglet scalar densities. Note that such ratios are scale independent functions of g_0 only, which are expected to converge to 1 in the continuum limit with a rate $g_0^2 \propto -1/\ln a$. In particular, these ratios do not depend upon the quark mass parameters and may therefore be determined in the massless limit [16] where the tmQCD and standard QCD actions coincide. The connection between the mass parameters and chiral Ward identities is established by choosing renormalisation schemes such that the PCAC and PCVC relations hold, with the renormalised currents and and axial density, and the renormalised mass parameters. The renormalization constants may then be shown to satisfy the identities, $Z_m = Z_{\mathrm{S}^0}^{-1}$ and $Z_\mu = Z_{\mathrm{P}}^{-1}$. With these conventions it is clear that the ratio of renormalised mass parameters is known once the critical mass and the ratio Z_{S^0}/Z_P are given,

$$\tan\alpha = \frac{\mu_{\mathrm{R}}}{m_{\mathrm{R}}} = \frac{Z_{\mathrm{S}^0}}{Z_{\mathrm{P}}}\frac{\mu_{\mathrm{q}}}{m_0 - m_{\mathrm{cr}}}. \tag{31}$$

[d] $Z_{\mathrm{V}} = 1$ only holds if the (partially) conserved point-split vector current $\tilde{V}_\mu^a$ is used.

Besides the ratio of renormalization constants one thus needs to determine the critical mass. In practice this can be done by measuring a bare PCAC mass m from correlation functions with some external field O,

$$m = \frac{\langle \partial_\mu A^1_\mu(x) O\rangle}{\langle P^1(x) O\rangle}, \tag{32}$$

and by using the relation

$$m_{\rm R} = Z_P^{-1} Z_{\rm A} m, \qquad m = Z_{\rm A}^{-1} Z_m Z_P (m_0 - m_{\rm cr}). \tag{33}$$

Alternatively one may use the measured bare PCAC quark mass m to obtain α directly,

$$\tan\alpha = \mu_{\rm q}/(Z_{\rm A} m), \tag{34}$$

provided one has previously determined $Z_{\rm A}$. Already at this point one notes that the choice $\alpha = \pi/2$ is special, as in this case one merely needs to determine the critical mass. The choice $\alpha = \pi/2$ is referred to a full or maximal twist, because the physcial quark mass is then entirely defined by the twisted mass parameter $\mu_{\rm q}$.

Having determined the twist angle, and the relative renormalizations within chiral multiplets, chiral symmetry is restored up to cutoff effects for the correlation functions of members of these multiplets. In a second step one just needs to make sure that this property of the bare theory is not compromised by the renormalization procedure, i.e. one is in a similar situation as in the bare theory with Ginsparg-Wilson quarks. Proceeding in the same way, the formal identities of Subsect. 4.1 will hold in the renormalised theory.

An important point to notice is that the twist angle α is a new parameter which reflects the freedom to choose a direction in chiral flavour space for the explicit chiral flavour symmetry breaking. Our physical interpretation is such that by definition only the axial generators are broken by the mass term thus defining the residual vector symmetry. With Wilson quarks at non-zero α there is an additional breaking of flavour symmetry by the Wilson term. which is expected to disappear in the continuum limit, just like chiral symmetry is restored with standard Wilson quarks. In order to define the continuum limit properly one must make sure that cutoff effects are a smooth function of $\beta = 6/g_0^2$. In general this can be achieved by taking the continuum limit at constant physical conditions. For instance one may keep m_π/F_π constant as β is varied. However, in tmQCD this observable is a function of two mass parameters, or, equivalently of one mass parameter and the twist angle. It is crucial that the twist angle is kept constant as the continuum limit is taken, since the twist angle labels different lattice regularisations of two-flavour QCD. In particular, if α is changed from one β-value to the next, there is no reason to expect a smooth continuum approach and a continuum extrapolation may become impossible.

5. A few applications of tmQCD

The relations between tmQCD and standard QCD correlation functions can be used to by-pass certain lattice renormalization problems of standard Wilson quarks (cf. Sect. 3). As the different operators of a continuum chiral multiplet are not necessarily related by lattice symmetries, their renormalisation properties can be very different. Moreover, the renormalisation properties do not change in the presence of (twisted or non-twisted) mass terms except when power divergences are present. Excluding these cases it is thus sufficient to renormalise a given composite field in the chiral limit where the actions of tmQCD and standard Wilson quarks coincide. One may then choose the operator with the best renormalisation properties that can be related to the desired standard QCD operator by the dictionary established earlier. Moreover, it may not even be necessary to match the operators directly. In principle, it is enough to match the desired correlation function up to cutoff effects. Perhaps these remarks become clearer by going through a few examples:

5.1. *Computation of F_π*

Both the pion mass m_π and the pion decay constant F_π can be obtained from the long distance behaviour of the 2-point function

$$\begin{aligned}\left\langle (A_\mathrm{R})_0^1(x)(P_\mathrm{R})^1(y)a\right\rangle_{(M_\mathrm{R},0)} &= \cos(\alpha)\ \left\langle (A_\mathrm{R})_0^1(x)(P_\mathrm{R})^1(y)\right\rangle_{(M_\mathrm{R},\alpha)} \\ &\quad + \sin(\alpha)\ \left\langle \tilde{V}_0^2(x)(P_\mathrm{R})^1(y)\right\rangle_{(M_\mathrm{R},\alpha)} . \quad (35)\end{aligned}$$

The problem with the standard Wilson computation on the l.h.s. is that the axial current requires a non-trivial renormalisation, which needs to be determined from Ward identities, as done e.g. in Ref. 16. On the other hand the vector current $\tilde{V}_\mu^a$ is protected against such a rescaling since it is conserved at $\mu_\mathrm{q} = 0$. At $\alpha = \pi/2$ the axial current is mapped to the vector current and one may thus avoid the current renormalisation by computing the vector correlation function in tmQCD. It is in fact not necessary to set

$\alpha = \pi/2$; when inverting the relation (35),

$$\left\langle (\tilde{V}_0^2(x)(P_\mathrm{R})^1(y) \right\rangle_{(M_\mathrm{R},\alpha)} = \cos(\alpha) \left\langle (\tilde{V}_0^2(x)(P_\mathrm{R})^1(y) \right\rangle_{(M_\mathrm{R},0)} + \sin(\alpha) \left\langle (A_\mathrm{R})_0^1(x)(P_\mathrm{R})^1(y) \right\rangle_{(M_\mathrm{R},0)} . \quad (36)$$

one notices that the first term on the r.h.s violates both parity and flavour symmetry of standard QCD. On the lattice this correlation function therefore contributes at most an $\mathrm{O}(a)$ effect. One may thus obtain F_π at values $\alpha \neq \pi/2$ by computing the l.h.s of this equation. Finally, it should be mentioned that the exact PCVC lattice relation,

$$\partial_\mu^* \tilde{V}_\mu^a = -2\mu_\mathrm{q} \varepsilon^{3ab} P^b, \quad (37)$$

may be used to replace the vector current by the axial density. Summing over $\mathbf{x}$, translation invariance eliminates the spatial part of the divergence, and the time derivative reduces to a multiplication by m_π at large time separations [17]. The results of a quenched computation along these lines [18,19] are shown in Figures 2 and 5.

5.2. *Direct determination of the chiral condensate*

A computation of the chiral condensate from the local scalar density has never been performed with Wilson quarks, due to the cubic divergence (9) which persists in the chiral limit. In tmQCD the rôle of the scalar density is played by the axial density, i.e. one expects the relation

$$\left\langle (P_\mathrm{R})^3(x) \right\rangle_{(M_\mathrm{R},\alpha)} = \cos(\alpha) \left\langle (P_\mathrm{R})^3(x) \right\rangle_{(M_\mathrm{R},0)} - \frac{i}{2} \sin(\alpha) \left\langle (S_\mathrm{R})^0(x) \right\rangle_{(M_\mathrm{R},0)} \quad (38)$$

Again, the first term on the r.h.s. vanishes up to $\mathrm{O}(a)$ due to parity, so that the computation of the l.h.s. yields the chiral condensate up to the factor $(-i/2)\sin(\alpha)$. This is advantageous as the renormalised axial density is of the form,

$$(P_\mathrm{R})^3 = Z_\mathrm{P} \left(P^3 + \mu_\mathrm{q} c_\mathrm{P} \, a^{-2} \right), \quad (39)$$

i.e. the power divergence vanishes for $\mu_\mathrm{q} = 0$. Still, in order to determine the condensate one needs to perform first the infinite volume limit followed by the $\mu_\mathrm{q} = 0, m_0 = m_\mathrm{cr}$ limits (at fixed α) and the continuum limit, which remains a rather delicate task. In particular, the chiral limit is complicated by the fact that the uncertainty in $\sin(\alpha)$ increases as the quark mass is decreased, due to the intrinsic $\mathrm{O}(a)$ ambiguity of m_cr. In practice this means that one has to extrapolate to the chiral limit from some distance, but this is anyway required for finite volume effects to remain small.

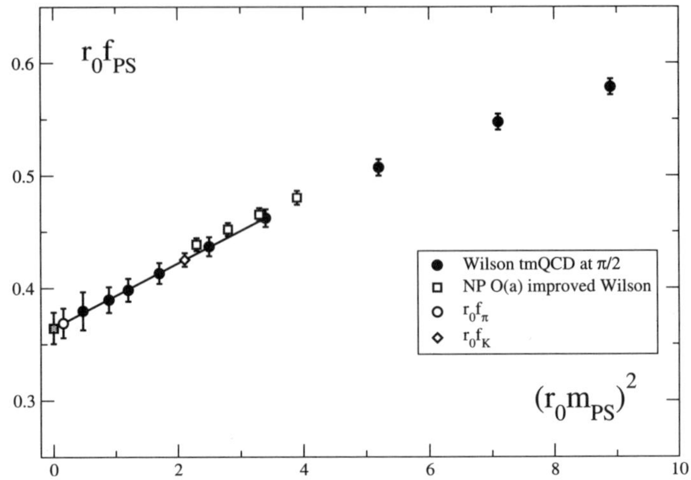

Figure 2. Quenched continuum results for F_π. The plot also illustrates the absence of the zero mode problem in tmQCD, as much smaller pion masses could be reached than with standard O(a) improved Wilson quarks.

5.3. *The computation of B_K*

Four-quark operators provide an interesting playground for mappings between tmQCD and standard QCD. We start with the B_K parameter which is defined in QCD with dynamical u, d, s quarks by,

$$\langle \bar{K}^0 \mid O^{\Delta S=2}_{(\mathrm{V-A})(\mathrm{V-A})} \mid K^0 \rangle = \tfrac{8}{3} F_K^2 m_K^2 B_K. \tag{40}$$

The local operator

$$O^{\Delta S=2}_{(\mathrm{V-A})(\mathrm{V-A})} = \sum_\mu [\bar{s}\gamma_\mu(1-\gamma_5)d]^2, \tag{41}$$

is the effective local interaction induced by integrating out the massive gauge bosons and t-,b- and c-quarks in the Standard Model. The transition between the pseudoscalar states K^0 and $\bar{K}^0$ does not change parity. Therefore, only the parity-even part in the effective operator,

$$O_{(\mathrm{V-A})(\mathrm{V-A})} = \underbrace{O_{\mathrm{VV+AA}}}_{\text{parity-even}} - \underbrace{O_{\mathrm{VA+AV}}}_{\text{parity-odd}}, \tag{42}$$

contributes to B_K. With Wilson type quarks, the operators $O_{\rm VV+AA}$ and $O_{\rm VA+AV}$ are renormalised as follows

$$(O_{\rm VV+AA})_{\rm R} = Z_{\rm VV+AA}\Big\{O_{\rm VV+AA} + \sum_{i=1}^{4} z_i\, O_i^{d=6}\Big\}, \tag{43}$$

$$(O_{\rm VA+AV})_{\rm R} = Z_{\rm VA+AV} O_{\rm VA+AV}. \tag{44}$$

While the parity-even component mixes with four other operators of dimension 6, the parity-odd component only requires multiplicative renormalisation, due to CP and flavour exchange symmetries [20]. This raises the question if one can by-pass the mixing problem by exchanging the rôles of both operators through the introduction of twisted mass terms. This is indeed possible, but one first needs to introduce the strange quark. The simplest possibility consists in adding a standard s-quark to a twisted quark doublet ψ of the light up and down quarks, which are thus taken to be degenerate. The corresponding continuum Lagrangian is given by

$$\mathcal{L} = \bar\psi(\not{D} + m + i\mu_q\gamma_5\tau^3)\psi + \bar s(\not{D} + m_s)s, \tag{45}$$

and, passing to the physical basis of primed fields, one finds

$$\begin{aligned} O'_{\rm VV+AA} &= \cos(\alpha) O_{\rm VV+AA} - i\sin(\alpha) O_{\rm VA+AV} \\ &= -iO_{\rm VA+AV} \quad (\alpha = \pi/2). \end{aligned} \tag{46}$$

At full twist, we thus get a direct mapping between both operators, i.e. $O_{\rm VA+AV}$ in twisted mass QCD at $\pi/2$ is interpreted as $O_{\rm VV+AA}$ in standard QCD. A second possibility consists in exchanging the rôles of up and strange quark, i.e. one considers a twisted doublet of strange and down quarks and a standard u-quark. In this case one finds

$$\begin{aligned} O'_{\rm VV+AA} &= \cos(2\alpha) O_{\rm VV+AA} - i\sin(2\alpha) O_{\rm VA+AV} \\ &= -iO_{\rm VA+AV} \quad (\alpha = \pi/4), \end{aligned} \tag{47}$$

i.e. the same mapping is obtained, but with the twist angle $\alpha = \pi/4$. Several comments are in order: while both options, referred to as $\pi/2$ and $\pi/4$ scenarios respectively, are possible, the second one is clearly more remote from reality, as it assumes mass degenerate down and strange quarks. However, this is precisely the limit in which most lattice calculations to date have been performed. The justification rests on chiral perturbation theory where a a weak dependence upon the strange-down mass difference is predicted. Moreover, in the quenched approximation, any deviation from the degenerate case leads to an unphysical logarithmic quark mass dependence [21,22].

5.3.1. *Renormalisation of* $O_{\rm VA+AV}$

Whatever the chosen strategy, the operator which requires renormalisation is $O_{\rm VA+AV}$. The renormalisation is multiplicative, and the general strategy of Ref. 23 can be applied. The scale evolution of the operator in a few Schrödinger functional schemes has been traced in the quenched approximation over a wide range of scales (for first results with $N_{\rm f} = 2$ sea quarks cf. Ref. 24). The result is shown in Figure 3. It thus remains to calculate the bare matrix element for B_K at various values of β, and, after multiplication with the Z-factor at the low energy scale, perform the continuum limit extrapolation. In the continuum limit one may then use the known scale evolution to reach the truly perturbative regime where contact is made with the perturbative renormalisation schemes of the continuum.

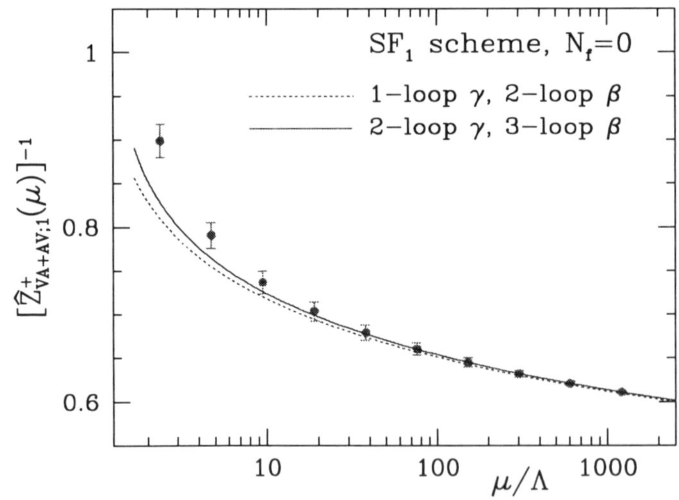

Figure 3. The data points show the non-perturbatively computed scale evolution of B_K in the SF scheme. Also shown are two perturbative approximations.

5.3.2. *Results for B_K in the quenched approximation*

Both scenarios have been implemented in the quenched approximation with lattice spacings $a = 0.05 - 0.1$ fm and lattice sizes up to $L/a = 32$ [25]. If one sticks to mass degenerate down and strange quarks the $\pi/2$ scenario requires some chiral extrapolation, due to the problem with unphysical zero modes (recall that the s-quark remains untwisted). In the $\pi/4$ scenario the zero mode problem is eliminated and the kaon mass can be reached by interpolation, provided the finite volume effects are small enough. This is the case with all lattice spacings except the finest one, where some extrapolation is required. A combined continuum extrapolation to both data sets, linear in a, leaving out the data at the coarsest lattice spacing led to the result $\hat{B}_K = 0.789(46)$ [25], where $\hat{B}_K$ denotes the renormalisation group invariant B-parameter. Unfortunately, the twist angle at $\beta = 6.1$ had not been tuned precisely enough, a fact that was only noticed after publication of Ref. 25. A new analysis indicates that higher than linear lattice artefacts are still significant at $\beta = 6.1$. As the data set is not sufficient to fit to both a and a^2 terms, it was decided to discard the data at $\beta = 6.1$, too, with the result [28]

$$\hat{B}_K = 0.735(71) \qquad \Leftrightarrow \qquad B_K^{\overline{\rm MS}}(2\,{\rm GeV}) = 0.534(52), \tag{48}$$

which is compatible with the earlier result, albeit with a larger uncertainty. In conclusion, the quenched result for B_K has a total error of almost 10 percent, which includes all systematic effects (renormalisation, chiral inter- or extrapolations, continuum extrapolation) except quenching and the fact that the valence quarks are mass degenerate. However a variation of the mass difference up to $(M_s - M_d)/(M_s + M_d) \approx 0.5$ did not show sizeable effects. While the error could still be improved by including data at a finer lattice spacing it seems fair to say that further progress requires the inclusion of sea quark effects.

5.4. *Further applications*

Twisted mass QCD does not provide a general recipe for by-passing the lattice specific renormalisation problems of Wilson quarks. Rather, one needs to discuss on a case by case basis whether it can be advantageous to use some variant of tmQCD. For further applications to four-quark operators and $K \to \pi$ transitions I refer the reader to Refs. 26, 27. While the first reference insists on an equal treatment of sea and valence quarks, the second paper explores a mixed action approach, where the valence quarks are

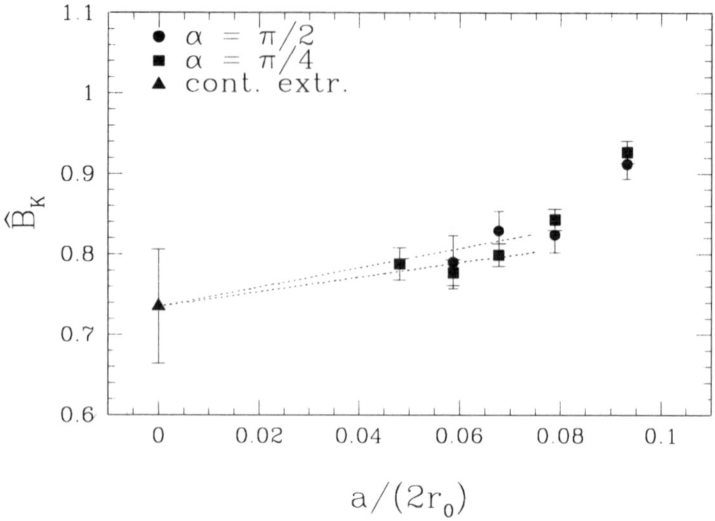

Figure 4. Quenched lattice data for both scenarios, $\alpha = \pi/2$ and $\alpha = \pi/4$. Also shown is the combined continuum extrapolation, leaving out the data at the two coarsest lattice spacings.

chirally twisted individually, independently of the sea quark action. This yields a much greater flexibility and allows for a complete elimination of lattice specific mixings and subtractions, even including O(a) improvement. Finally, similar considerations apply to QCD with static b-quarks, where the mixing of four-quark operators is considerably simplified by twisting the light quarks (see Ref. 28 for a recent review and further references).

6. $O(a)$ improvement and tmQCD

Given that the quenched approximation is currently being overcome, and the zero mode problem for algorithms can be alleviated, there remain essentially two arguments in favour of tmQCD as opposed to standard or O(a) improved Wilson quarks: the first consists in the possibility to by-pass renormalisation problems, as explained in the preceding section. The second, is the property of "automatic O(a) improvement" at maximal twist (i.e. $\alpha = \pi/2$), as first observed by Frezzotti and Rossi [4]. I will explain this point more in detail below, after a brief reminder of the situation with standard Wilson quarks.

6.1. *O(a) improvement of Wilson quarks*

In lattice QCD with Wilson quarks, results are typically affected by O(a) lattice effects, which is to be contrasted with staggered or Ginsparg-Wilson quarks where the leading cutoff effects are quadratic in a. As illustrated by the B_K determination described above, linear lattice artefacts render continuum extrapolations more difficult, and it would be nice to get rid of them altogether. This is possible by introducing O(a) counterterms to the action and the composite operators such that O(a) effects are cancelled in on-shell quantities. The basic idea goes back to Symanzik [29], while the restriction to on-shell quantities in gauge theories has been first advocated by Lüscher and Weisz [30]. When applied to Wilson quarks [31], it turns out that O(a) improvement of the spectrum (particle masses and energies) can be achieved by adding a single counterterm to the action, the so-called Sheikholeslami-Wohlert (SW) or clover term, $i\bar{\psi}\sigma_{\mu\nu}F_{\mu\nu}\psi$, where $F_{\mu\nu}$ is the gluon field tensor [e]. This term is of dimension five and therefore comes with an explicit factor a when included in the lattice action density.

While O(a) improvement of the spectral quantities is quite economical, one is often interested in matrix elements of composite operators, and each operator comes with its own set of O(a) counterterms, all of which have to be tuned in order to cancel the linear lattice artefacts. While this may still be possible for quark bilinear operators, the counterterms quickly proliferate in the case of 4-quark operators, and O(a) improvement becomes completely impractical if the quarks are taken to be mass non-degenerate (cf. Ref. 32).

6.2. *Automatic O(a) improvement of tmQCD in a finite volume*

The Symanzik effective theory can also be applied to tmQCD and a list of O(a) counterterms for the action and a few quark bilinear operators can be found in Ref. 3. The observation in Ref. 4 is, that at maximal twist,

---

[e] On the lattice the field tensor is usually discretised using four plaquette terms in the (μ, ν)-plane whence the name "clover term".

all the O(a) counterterms become irrelevant in the sense that they can at most contribute at O(a^2). The argument for automatic O(a) improvement can be made such that it only relies on Symanzik's effective continuum theory [33,5,34,35] To simplify the discussion, let us first assume that the space-time volume is finite, so that spontaneous symmetry breaking is excluded and all observables are analytic in the quark mass parameters. We furthermore assume that we have tuned some PCAC current quark mass $m_{\rm PCAC} = 0$, i.e. the renormalized standard mass parameter vanishes up to $O(a)$ effects. Then Symanzik's effective continuum action is given by

$$S_{\rm eff} = S_0 + aS_1 + O(a^2), \qquad S_0 = \int {\rm d}^4x\ \bar\psi(\not{D} + i\mu_q\gamma_5\tau^3)\psi, \tag{49}$$

where S_0 is the maximally twisted tmQCD continuum action. S_1 is given by

$$S_1 = \int {\rm d}x^4\ \{c\ i\bar\psi\sigma_{\mu\nu}F_{\mu\nu}\psi + b_\mu\mu^2\bar\psi\psi + \ldots\}, \tag{50}$$

where the dots stand for further operators of dimension 5 (possibly including explicit mass factors), which share the symmetries of the lattice action. The reason why I omitted them here is that they can be eliminated by the equations of motion. Furthermore, the second operator can be absorbed in an O(a) shift of the standard quark mass parameter, so that one is really left with the SW term as the only relevant operator for on-shell improvement. Renormalised (connected) lattice correlation functions can be analysed in the effective theory,

$$\langle O\rangle = \langle O\rangle^{\rm cont} - a\langle S_1 O\rangle^{\rm cont} + a\langle \delta O\rangle^{\rm cont} + O(a^2), \tag{51}$$

where the cutoff dependence is explicit. We are here only interested in the leading cutoff effects at O(a). To this order there are two contributions, first the insertion of the O(a) part of the effective action S_1, and second the field specific counterterms δO. For example, with the choice

$$O = V^1_\mu(x)P^2(y), \tag{52}$$

one finds the counterterm,

$$\delta O = \{c_{\rm V} i\partial_\nu T^1_{\mu\nu}(x) + \tilde b_{\rm V}\mu_{\rm q}A^2_\mu(x)\}P^2(y) + \ldots, \tag{53}$$

where the dots stand for further terms which vanish by the equations of motion It should be emphasised that the O(a) (and higher) corrections in the effective action are only treated as insertions, i.e. the expectation values $\langle\cdot\rangle^{\rm cont}$ are taken with respect to the continuum action S_0. In writing down the effective Symanzik theory there is thus an implicit assumption made, namely that one is working in the regime of continuum QCD where cutoff effects only appear as asymptotically small corrections. This assumption may certainly be wrong in some regions of parameter space, and particular care has to be taken in the presence of phase transitions.

To proceed I introduce the $\gamma_5\tau^1$-transformation,

$$\psi \to i\gamma_5\tau^1\psi, \qquad \bar{\psi} \to \bar{\psi}i\gamma_5\tau^1, \tag{54}$$

which is part of the vector symmetry of two-flavour QCD. Hence S_0 is invariant, but this is not the case for S_1, i.e. one finds

$$S_0 \to S_0, \qquad S_1 \to -S_1. \tag{55}$$

For gauge invariant fields the transformation (54) squares to the identity, so that one may define an associated parity. For fields O with a definite $\gamma_5\tau^1$-parity one then finds,

$$O \to \pm O \quad \Rightarrow \quad \delta O \to \mp\delta O. \tag{56}$$

By applying the $\gamma_5\tau^1$ transformation to the integration variables in the functional integral, one may derive identities between correlation functions, due to the invariance of the continuum action and functional measure. In particular, if we choose a $\gamma_5\tau^1$-even field O, we find for the correlation functions at O(a)

$$\begin{aligned} \langle S_1 O\rangle^{\rm cont} &= -\langle S_1 O\rangle^{\rm cont} = 0, \\ \langle \delta O\rangle^{\rm cont} &= -\langle \delta O\rangle^{\rm cont} = 0, \end{aligned} \tag{57}$$

and therefore

$$\langle O\rangle = \langle O\rangle^{\rm cont} + O(a^2). \tag{58}$$

For a $\gamma_5\tau^1$-odd O, one obtains

$$\begin{aligned} \langle O\rangle^{\rm cont} &= -\langle O\rangle^{\rm cont} = 0, \\ \langle S_1 O\rangle^{\rm cont} &= \langle S_1 O\rangle^{\rm cont}, \\ \langle \delta O\rangle^{\rm cont} &= \langle \delta O\rangle^{\rm cont}, \end{aligned} \tag{59}$$

which implies

$$\langle O\rangle = -a\langle S_1 O\rangle^{\rm cont} + a\langle \delta O\rangle^{\rm cont} + O(a^2). \tag{60}$$

We may thus conclude that, at least in a small finite volume lattice correlation functions of $\gamma_5\tau^1$-even fields are automatically $O(a)$ improved, while those of $\gamma_5\tau^1$-odd fields vanish up to $O(a)$ terms. As a corollary, one may state that standard Wilson quarks in a finite volume are automatically $\mathrm{O}(a)$ improved in the chiral limit. Although this is not the most interesting regime of QCD, it is somewhat surprising that this fact had not been noticed for more than 2 decades! To conclude this section note that in terms of the physical basis, (54) corresponds to the discrete flavour transformation,

$$\psi' \to -i\tau^2\psi', \qquad \bar{\psi}' \to \bar{\psi}' i\tau^2. \tag{61}$$

A very similar argument based on parity transformations has been given by Shindler in Ref. 5. In Ref. 35 a systematic analysis of the $\gamma_5\tau^1$ symmetry (called T_1 in this paper) can be found, showing that not only $\mathrm{O}(a)$ but all odd powers of a vanish in $\gamma_5\tau^1$-even correlators. This is not surprising, as this is implicit in the earlier analysis in Ref. 4, where the same conclusion was drawn.

6.2.1. *Uncertainty of the chiral limit*

If $\mathrm{O}(a)$ improvement is automatic one might think that it should be possible to determine the critical mass $m_{\rm cr}$ up to an intrinsic $\mathrm{O}(a^2)$ uncertainty. This is not so, as I will now explain. The critical mass can be determined by tuning some PCAC mass to zero, and there is no obstacle for doing this in a finite volume. Now, the PCAC relation involves the axial current and density, A_μ^a and P^a, which have opposite $\gamma_5\tau^1$-parities. According to the preceding discussion this means, for the first flavour components and with a $\gamma_5\tau^1$-even source field $O_{\rm even}$,

$$\langle \partial_\mu A_\mu^1(x) O_{\rm even}\rangle = 2\underbrace{m_{\rm PCAC}}_{\mathrm{O}(a)}\underbrace{\langle P^1(x) O_{\rm even}\rangle}_{\mathrm{O}(a)} = O(a^2). \tag{62}$$

The l.h.s. being $\gamma_5\tau^1$-even must vanish up to $\mathrm{O}(a^2)$, provided maximal twist is realised at least up to cutoff effects, i.e. $m_{\rm R} = \mathrm{O}(a)$. This implies that the PCAC mass is of $\mathrm{O}(a)$, too, multiplying a correlation function which is $\gamma_5\tau^1$-odd and therefore of $\mathrm{O}(a)$. Thus no contradiction arises, the $\mathrm{O}(a^2)$ of the l.h.s. is matched on the r.h.s. by two factors of $\mathrm{O}(a)$.

Another way to understand that an $\mathrm{O}(a)$ shift in the critical mass does not ruin $\mathrm{O}(a)$ improvement is to treat such a shift as an insertion of the standard mass operator $\bar{\psi}\psi$ into correlation functions. This operator is $\gamma_5\tau^1$-odd so that its insertion into a $\gamma_5\tau^1$-even correlator produces an $\mathrm{O}(a)$ effect, which together with the $\mathrm{O}(a)$ mass shift yields an $\mathrm{O}(a^2)$ effect.

6.3. *Automatic O(a) improvement in infinite volume*

When the infinite volume limit is taken, the basic difference is the presence of spontaneous symmetry breaking and the appearance of non-analyticities in the mass parameters near the chiral limit. As discussed earlier, twisted mass QCD is a valid regularisation of two-flavour QCD provided the continuum limit is taken at fixed twist angle. To maintain maximal twist, i.e. $\alpha = \pi/2$ one needs to tune the standard quark mass to $m_{\rm cr}$, which has an intrinsic $\mathrm{O}(a)$ ambiguity. As long as the twisted mass is much larger than the typical $\mathrm{O}(a)$ spread of $m_{\rm cr}$, the twist angle may be considered well-defined, and the continuum limit is reached with $\mathrm{O}(a^2)$ corrections. However, in practice one is interested in varying the quark mass at fixed cutoff, rather than studying the quark mass dependence only in the continuum limit. Approaching the chiral limit at fixed a by lowering the twisted mass one enters the regime where the twisted mass parameter becomes comparable to the $\mathrm{O}(a)$ ambiguity of $m_{\rm cr}$. One may debate at this point whether the relevant comparison is with the uncertainty of $m_{\rm cr}$ itself or rather with the size of typical $\mathrm{O}(a^2)$ effects in correlation functions generated by this uncertainty. In any case one reaches a point where the control over the twist angle is lost. When delivering my Nara lectures I interpreted this fact as a breakdown of the effective Symanzik theory. This is perhaps too rigid an interpretation. Rather one could say that for every definition of $m_{\rm cr}$, an effective twist angle is formed by the dynamics of the system, which may be far from the maximal twist one would like to maintain. Moreover, without further input is is impossible to know the effective twist angle for a given definition of $m_{\rm cr}$. This is a disaster, as the whole interpretation of the theory rests on the twist angle, and a change in the effective twist angle (which remains unnoticed!) might strongly affect some correlators even at O(1)! Fortunately this problem occurs close to the chiral limit, and thus in a region of parameter space where Chiral Perturbation Theory (χPT) is expected to describe the dynamics in terms of pion physics [36]. In particular, χPT is able to identify definitions of $m_{\rm cr}$ in terms of pionic observables, which lead to an effective twist angle of $\alpha = \pi/2$, so that the Symanzik effective theory for maximally twisted mass QCD remains applicable in this region. For instance, this should be the case if one requires parity or flavour symmetry restauration, e.g. by imposing that a $\gamma_5\tau^1$-odd pion correlation function vanishes. Note that the vanishing of the PCAC mass for a pion correlation function is a special case of such a condition. On the other hand, according to Refs. 37, 35 the condition of vanishing pion mass (calculated in the untwisted theory) does indeed lead to a O(1) variation of the effective twist angle. However, apart from larger cutoff effects of $O(a^2)$ this does not (yet?) seem to be a major problem in Ref. 19, cf. Figure 5. In any case, as the spontaneous symmetry breaking is closely related to the dynamics of pions, it seems that no statement can be made about generic definitions of m_{cr} in a small volume, either from axial current conservation, or from parity or flavour symmetry restauration.

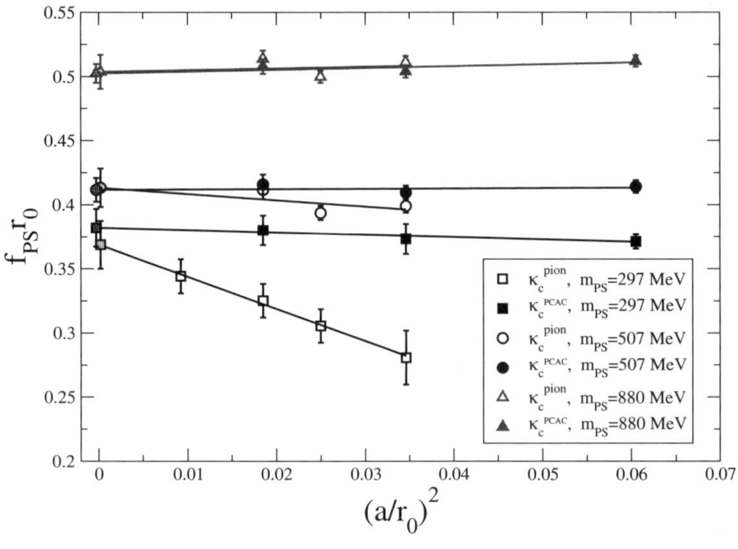

Figure 5. The continuum approach of F_π in quenched tmQCD for various pion masses vs. a^2/r_0^2.

7. Consequences of Parity and Flavour breaking

The exact symmetries of lattice QCD with standard Wilson quarks include parity and flavour symmetry which are used to classify the hadron spectrum. This is very convenient in any hadron analysis: even at fixed lattice spacing, the excited states which may occur in a given channel can be read from the Particle Data Book, with the exception of states with higher spin and/or angular momentum where the correspondence is spoilt by the lack of rotational symmetry on the lattice.

The situation is different in tmQCD since both parity and flavour symmetry are broken by the Wilson term. As a consequence, the classification in isospin multiplets fails by terms of $\mathrm{O}(a)$, or $\mathrm{O}(a^2)$ if $\mathrm{O}(a)$ improvement is at work. For instance, the neutral pion is not mass degenerate with the charged pions, or the nucleon Δ-resonances no longer form an exact isospin multiplet. Various simulations of quenched tmQCD have confirmed these expectations, and point to a restoration of flavour symmetry in the continuum limit [38,39,40], although the expected rate $\propto a^2$ for maximally twisted mass QCD has not in all cases been demonstrated convincingly.

However, the splittings of isospin multiplets by cutoff effects are not the most serious drawback of parity and flavour symmetry breaking. In the spectral analysis of a hadronic two-point function all excited states with the same lattice quantum numbers may contribute. Even though the states violating continuum symmetries are multiplied by coefficients proportional to a, these states need to be taken into account when working at fixed lattice spacing. Particularly annoying is the neutral pion, which shares all the lattice quantum numbers with the vacuum. One may thus add a neutral pion to any state without changing its lattice quantum numbers. The presence of additional relatively light states may require a multistate analysis just to identify and subtract states which are a pure lattice artefact. Moreover, correlation functions involving the light pion require the evaluation of disconnected diagrams. However, it should be emphasised that these problems are purely technical; conceptually tmQCD is on a very solid basis, and in contrast to staggered fermions there is no mixing between flavour and spin degrees of freedom.

7.1. *Non-degenerate quarks and additional flavours*

Twisted mass QCD was originally formulated for a single doublet of mass degenerate flavours. This can easily be generalised to include more mass degenerate doublets. However, such a spectrum is quite unrealistic unless a non-degeneracy can be introduced within a doublet. Moreover, this non-degeneracy should not cause too much damage to all the nice properties of tmQCD. In particular, one needs to maintain the reality and positivity of the quark determinant, if such an action is to be used for simulations of full tmQCD. This is indeed possible, by introducing a mass splitting term as follows [41],

$$\mathcal{L} = \bar{\psi}(\not{D} + m + i\mu_{\mathrm{q}}\gamma_5\tau^3 + \delta_m\tau^1)\psi, \tag{63}$$

where δ_m is the mass splitting parameter. The mass spectrum is easily obtained by going to the physical basis and diagonalising the mass matrix. Its eigenvalues are then found to be $M_{\pm} = \sqrt{m^2 + \mu_{\mathrm{q}}^2} \pm \delta_m$. Translating this continuum situation to Wilson quarks in the obvious way, one first notices that the determinant of the twisted Wilson-Dirac operator must be real due to the conjugation property,

$$\begin{aligned}\gamma_5\tau^1 \left(D_{\mathrm{W}} + m_0 + i\mu_q\gamma_5\tau^3 + \delta_m\tau^1\right) \gamma_5\tau^1 \\ = \left(D_{\mathrm{W}} + m_0 + i\mu_q\gamma_5\tau^3 + \delta_m\tau^1\right)^{\dagger} .\end{aligned} \tag{64}$$

Furthermore, the flavour structure of the determinant can again be reduced analytically, with the result,

$$\begin{aligned}\det(D_{\mathrm{W}} + m_0 + i\mu_q\gamma_5\tau^3 + \delta_m\tau^1) \\ = \det(Q^2 + \delta_m[\gamma_5, Q] + \mu_q^2 - \delta_m^2),\end{aligned} \tag{65}$$

and this determinant is non-zero provided $\mu_{\mathrm{q}}^2 > \delta_m^2$. The positivity of the determinant at $\delta_m = 0$ and continuity in δ_m then imply positivity of this determinant for non-zero δ_m.

The mass splitting parameter is renormalised multiplicatively $\delta_{m,\mathrm{R}} = Z_{\mathrm{S}}^{-1}\delta_m$, where Z_{S} is the renormalisation constant of the non-singlet scalar density. As the positivity of the determinant follows from a condition on the bare parameters μ_{q} and δ_m, the corresponding condition in terms of the renormalised parameters involves a ratio of renormalisation constants, i.e. $\delta_{m,\mathrm{R}} < (Z_{\mathrm{P}}/Z_{\mathrm{S}})\mu_{\mathrm{R}}$. The value of $Z_{\mathrm{P}}/Z_{\mathrm{S}}$ depends on details of the regularization, so that one cannot make a general statement about the ensuing limitations (if any). However, it is remarkable that one may use this action to perform numerical simulations with two non-degenerate light quark flavours, as needed for instance to study small isospin breaking effects. If used for strange and charm quarks [42,43], however, one potentially has to deal with a fine tuning problem for the strange quark mass: for instance, assuming $m_s = 100$ MeV and $m_c = 1300$ MeV, these values are obtained as (700 ± 600) MeV. Finally, it should be said that the presence of the additional flavour non-diagonal breaking term renders the relationship to standard QCD more complicated, and the flavour structure needs to be dealt with explicitly in numerical calculations of quark propagators.

8. A chiral twist to the QCD Schrödinger functional

In order to solve scale dependent renormalisation problems the introduction of an intermediate renormalisation scheme based on the Schrödinger functional (SF scheme) is an attractive possibility[23]. Here I start by summarising its basic features in order to prepare the discussion of possible improvements.

8.1. *The QCD Schrödinger functional*

The QCD Schrödinger functional[44,45](SF) is the functional integral for QCD where the Euclidean space-time manifold is taken to be a hyper cylinder. The quantum fields are periodic in space, and Dirichlet conditions are imposed at (Euclidean) times $x_0 = 0$ and $x_0 = T$.

$$\begin{aligned} P_+\psi(x)\,|_{x_0=0} &= \rho, \qquad P_-\psi(x)\,|_{x_0=T}= \rho', \\ \bar\psi(x)P_-\,|_{x_0=0} &= \bar\rho, \qquad \bar\psi(x)P_+\,|_{x_0=T}= \bar\rho', \\ A_k(x)\,|_{x_0=0} &= C_k, \qquad A_k(x)\,|_{x_0=T}= C_k', \quad k=1,2,3, \end{aligned} \tag{66}$$

with the projectors $P_\pm = \frac{1}{2}(1 \pm \gamma_0)$. Correlation functions are then defined as usual,

$$\langle O \rangle = \left\{ \mathcal{Z}^{-1} \int_{\text{fields}} O\ \mathrm{e}^{-S} \right\}_{\rho=\rho'=0;\bar\rho=\bar\rho'=0} . \tag{67}$$

O denotes some gauge invariant functional of the fields, possibly including the quark and antiquark boundary fields ζ and $\bar\zeta$, which are obtained by taking derivatives with respect to the quark boundary fields, viz.

$$\zeta(\mathbf{x}) \equiv P_-\zeta(\mathbf{x}) = \frac{\delta}{\delta\bar\rho(\mathbf{x})}, \qquad \bar\zeta(\mathbf{x}) \equiv \bar\zeta(\mathbf{x})P_- = -\frac{\delta}{\delta\rho(\mathbf{x})}. \tag{68}$$

The name "Schrödinger functional" derives from the fact that such wave functionals arise naturally in the Schrödinger representation of Quantum Field Theory [46], and the SF provides an example of a Quantum Field Theory defined on a manifold with a boundary.

Using correlation functions derived from the Schrödinger functional, it is possible to define renormalised QCD parameters (the strong coupling and the quark masses), as well as renormalised composite operators (e.g. four-quark operators). Such renormalization schemes based on the Schrödinger functional (SF schemes) are attractive for the following reasons:

- The finite volume is part of the scheme definition, i.e. all dimensionful quantities such as Euclidean time extent T, or boundary field parameters are scaled proportionally to L, the linear extent of the volume. As a consequence L remains the only scale in the system and can be identified with the renormalization scale by setting

$\mu = L^{-1}$. Running parameters and operators then run with the size of the space-time volume, and one may apply recursive finite size techniques to bridge large scale differences (cf. Subsect. 5.3.1)

- SF schemes are made quark mass independent by imposing the renormalisation conditions in the chiral limit. Fortunately, the SF boundary conditions introduce a gap in the spectrum of the Dirac operator, which persists as the quark mass is taken to zero. This means that numerical simulations can be performed in the chiral limit, and no chiral extrapolation is needed to evaluate the renormalisation conditions.
- SF schemes are gauge invariant, no gauge fixing is needed.
- Perturbation theory up to two loops is still feasable, due to the existence of a unique absolute minimum of the action[44]. This is to be contrasted with the situation on a hyper torus where perturbation theory becomes very intricate already at the one-loop level.
- A further technical advantage consists in the possibility to use correlators involving zero momentum boundary quark and anti-quark fields. This is convenient in perturbation theory, and it leads to good numerical signals and reduced cutoff effects as compared to gauge invariant correlators in a periodic setting.

All these nice properties come with a price: first of all, the presence of the boundary means that even the pure gauge theory suffers from $\mathrm{O}(a)$ cutoff effects, caused by effective local operators of dimension 4, such as $\mathrm{tr}\{F_{0k}F_{0k}\}$ and $\mathrm{tr}\{F_{kl}F_{kl}\}$, integrated over the boundary. When the quarks are included, there is even a dimension 3 operator, which can be absorbed in a multiplicative rescaling of the quark and antiquark boundary fields [47]. At order a, one expects dimension 4 operators like $\bar{\psi}\gamma_0 D_0 \psi$ and $\bar{\psi}\gamma_k D_k \psi$ to contribute additional $\mathrm{O}(a)$ effects [6]. It is important to note that these cutoff effects are, unlike the $\mathrm{O}(a)$ bulk effects of Wilson quarks, not due to the breaking of a continuum symmetry by the regularisation. Rather, such terms are to be expected with any regularisation of the Schrödinger functional. One may, however, write down a complete basis of $\mathrm{O}(a)$ counterterms which contribute to a given observable. After reduction via the equations of motion, one typically ends up with 2-3 $\mathrm{O}(a)$ boundary counterterms. In practice it is then possible to monitor the size of the boundary $\mathrm{O}(a)$ effects by varying the coefficients. Perturbative results for these coefficients are often known to one-loop or even two-loop order [48], and a non-perturbative determination may be conceivable. In summary, with some extra work, the $\mathrm{O}(a)$ boundary effects can be controlled and eventually eliminated. This is important, as otherwise the SF renormalisation procedure risks to introduce $\mathrm{O}(a)$ effects even in $\mathrm{O}(a)$ improved regularisations such as tmQCD at maximal twist or lattice QCD with Ginsparg-Wilson quarks.

8.2. *Decoupling of heavy quarks in SF schemes*

Quark mass independent schemes are very convenient to study the scale evolution for a theory with fixed quark flavour content. However, it also means that the decoupling of heavy quarks is not automatic, and one needs to match theories with different numbers of active flavours over quark thresholds. This is routinely done in perturbation theory, but it is not obvious that perturbation theory is adequate e.g. for matching the $N_\mathrm{f} = 4$ and $N_\mathrm{f} = 3$ effective theories over the charm quark threshold. One possibility to study decoupling consists in introducing a quark mass dependent SF scheme which would allow to study the non-perturbative evolution over the quark threshold until the heavy quark has decoupled. To define a mass dependent SF scheme it suffices to impose the renormalisation conditions at finite quark masses. Unfortunately, it turns out that the decoupling of a heavy quark in such a scheme is only *linear* in the inverse quark mass rather than quadratic. If the quark decouples very slowly, this means that it has to be kept longer in the evolution as an active degree of freedom, which could mean that widely different scales have to be accomodated on the same lattice.

An example from perturbation theory[49,50] is given in Figure 6. It shows the one-loop β-function of the running coupling in the SF scheme as a function of $z = mL$, where m is some renormalised quark mass (its precise definition is not required to one-loop order). As $z = mL$ is varied from 0 to infinity, one expects to see a smoothed out step function going from -1 to 0 around the threshold $z = 1$. The solid and dotted curves (from 2 different SF schemes) do indeed show this behaviour, but the decoupling is rather slow compared to the MOM scheme[51] (dashed line).

To understand this behaviour I propose a closer look at the Dirac operator for free quarks and its spectrum in the continuum limit.

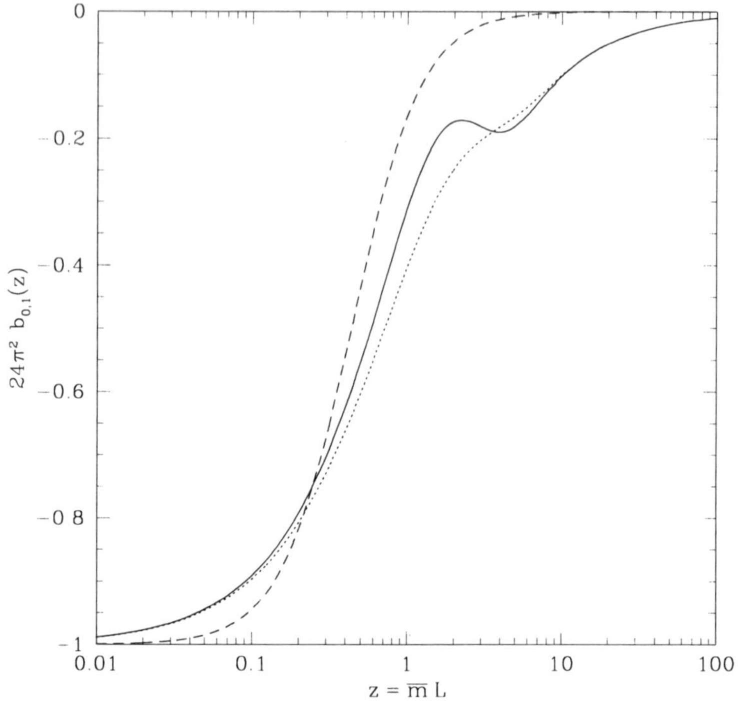

Figure 6. Decoupling of a heavy quark in the one-loop β function in two SF schemes and in the MOM scheme. See text for further explanation

8.2.1. *Free quarks with SF boundary conditions*

Let us consider a free quark ψ in the continuum with homogeneous SF boundary conditions,

$$P_+\psi(x)\,|_{x_0=0}= 0, \qquad P_-\psi(x)\,|_{x_0=T}= 0. \tag{69}$$

Then $\gamma_5(\not{\partial}+m)$ is a hermitian operator with smooth eigenfunctions and no zero modes [45]. Evaluating the eigenvalue equation for any of its eigenfunctions φ at the boundaries one finds,

$$\begin{aligned} P_+\gamma_5(\not{\partial}+m)\varphi\,|_{x_0=0}= 0 \quad &\Rightarrow \quad (\partial_0 - m)P_-\varphi\,|_{x_0=0}= 0, \\ P_-\gamma_5(\not{\partial}+m)\varphi\,|_{x_0=T}= 0 \quad &\Rightarrow \quad (\partial_0 + m)P_+\varphi\,|_{x_0=T}= 0. \end{aligned} \tag{70}$$

The complementary components thus satisfy Neumann conditions modified by the mass term m. The eigenvalues λ are of the form λ =

$\pm\sqrt{p_0^2+\boldsymbol{p}^2+m^2}$, where p_0 is determined as non-vanishing solution of $\tan(p_0T) = -p_0/m$. It is obvious from this equation that p_0 and thus λ are not symmetric under $m \to -m$. This is generic and can be understood as a consequence of chiral symmetry breaking by the boundary conditions. As a result one expects, for any observable in the SF, the asymptotic small mass behaviour $\propto m$ (rather than m^2), and similarly for heavy quarks the corrections $\propto 1/m$ (instead of $1/m^2$), as illustrated in Figure 6.

At least for even numbers of flavours a possible way out consists in adding a twisted mass term and setting $m = 0$. Then $\gamma_5\tau^1(\not{D} + i\mu_{\rm q}\gamma_5\tau^3)$ is again hermitian. With this Dirac operator, the complementary field components at the boundaries satisfy simple Neumann conditions, and the spectrum is symmetric under a change of sign of the twisted mass term. A physically equivalent solution is obtained by staying with the standard mass term and rotating the boundary projectors instead. This will be discussed in more detail below. However, a caveat remains as only the simultaneous decoupling of an even number of quarks can be studied in this formalism. On the other hand, it may be sufficient to compare to perturbative decoupling in this slightly unphysical setting, in particular if a perturbative treatment turns out to be satisfactory.

8.3. *SF boundary conditions and chiral rotations*

Let us consider flavour doublets ψ' and $\bar{\psi}'$ which satisfy homogeneous standard SF boundary conditions. Performing a chiral rotation,

$$\psi' = \exp(i\alpha\gamma_5\tau^3/2)\psi, \qquad \bar{\psi}' = \bar{\psi}\exp(i\alpha\gamma_5\tau^3/2), \tag{71}$$

one finds that the fields ψ and $\bar{\psi}$ satisfy the chirally rotated boundary conditions,

$$\begin{aligned} P_+(\alpha)\psi(x)\,|_{x_0=0} &= 0, & P_-(\alpha)\psi(x)\,|_{x_0=T} &= 0, \\ \bar{\psi}(x)\gamma_0 P_-(\alpha)\,|_{x_0=0} &= 0, & \psi(x)\gamma_0 P_+(\alpha)\,|_{x_0=T} &= 0, \end{aligned} \tag{72}$$

with the projectors,

$$P_\pm(\alpha) = \tfrac{1}{2}[1 \pm \gamma_0\exp(i\alpha\gamma_5\tau^3)]. \tag{73}$$

Special cases are $\alpha = 0$ and $\alpha = \pi/2$ where one obtains,

$$P_\pm(0) = P_\pm, \qquad P_\pm(\pi/2) \equiv Q_\pm = \tfrac{1}{2}(1 \pm i\gamma_0\gamma_5\tau^3). \tag{74}$$

We perform again a change of variables in the functional integral. Including mass terms as well, we label correlation functions by a subscript

$(m, \mu_{\rm q}, P_+(\alpha))$, i.e. we include the projector defining the Dirichlet component of the quark field at $x_0 = 0$. The generalisation of formula (22) then reads:

$$\langle O[\psi, \bar\psi]\rangle_{(m,\mu_{\rm q},P_\pm)} = \langle O[R(\alpha)\psi, \bar\psi R(\alpha)]\rangle_{(\tilde m,\tilde\mu_q,P_\pm(\alpha))}, \qquad (75)$$

with mass parameters $\tilde m$ and $\tilde{\mu_q}$ given by

$$\tilde m = m\cos\alpha - \mu_q \sin\alpha, \qquad \tilde{\mu_q} = m\sin\alpha + \mu_q\cos\alpha. \qquad (76)$$

The boundary quark fields are included in this transformation by replacing

$$\bar\zeta(\mathbf{x}) \leftrightarrow \bar\psi(0,\mathbf{x})P_+, \qquad \zeta(\mathbf{x}) \leftrightarrow P_-\psi(0,\mathbf{x}). \qquad (77)$$

This extends the equivalence between correlation functions of tmQCD and standard QCD to correlation functions derived from the Schrödinger functional. Simple examples are provided by purely gluonic observables $O[U]$, such as the SF coupling constant. Eq. (75) then implies,

$$\langle O[U]\rangle_{(0,\mu_{\rm R},P_+)} = \langle O[U]\rangle_{(\mu_{\rm R},0,Q_+)}. \qquad (78)$$

In other words, either the mass term is twisted and one stays with standard SF boundary conditions, or the mass term is standard and the boundary conditions are fully twisted. In both cases one expects a quadratic dependence on the mass parameter and hence a relatively fast decoupling of heavy quarks.

8.4. *SF schemes with Wilson quarks and O(a) improvement*

From the discussion of $\mathrm{O}(a)$ improvement in Section 6 one may conclude that $\gamma_5\tau^1$-even observables computed with Wilson quarks in a finite volume and with periodic boundary conditions are automatically $\mathrm{O}(a)$ improved at zero quark mass. As SF schemes are usually defined at zero quark mass, it seems natural to ask how the SF boundary conditions interfere with this property. It is useful to think of $\mathrm{O}(a)$ effects to arise from different sources. First there are the $\mathrm{O}(a)$ boundary effects, which are cancelled by introducing the $\mathrm{O}(a)$ boundary counterterms to the action and the boundary quark and antiquark fields. Second there are $\mathrm{O}(a)$ effects from the bulk action which may be cancelled by the Sheikholeslami-Wohlert term, and third there are the $\mathrm{O}(a)$ effects associated with the composite operators in a given correlation function. It is interesting to note that $\mathrm{O}(a)$ cutoff effects from the bulk action are often quite large in SF correlation functions. This is illustrated in Figure 7 which shows the relative cutoff effects in the perturbative one-loop coefficient of the step-scaling function of the four-quark operator needed for B_K [52]. The operator here is unimproved, and the boundary effects remain uncancelled in order to mimick the non-perturbative procedure of Ref. 53. The most dramatic reduction of cutoff effects occurs when the Sheikholeslami-Wohlert term is included. Moreover, this has the side effect to reduce the ambiguity in the zero mass point, so that with the standard SF it makes sense to implement O(a) improvement even if it is not complete.

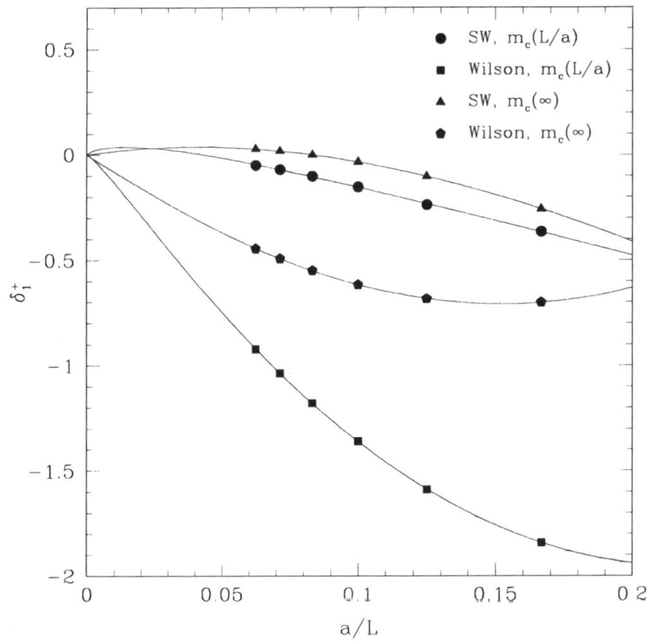

Figure 7. Relative cutoff effects in the one-loop coefficient of the step-scaling function of the B_K operator. Shown are two different regularisations ($c_{\rm sw} = 0, 1$) with two definitions of the zero mass point.

8.5. *The Schrödinger functional and $O(a)$ improvement*

The reason why automatic $O(a)$ improvement fails is that the $\gamma_5\tau^1$-transformation (54) changes the projectors of the quark boundary conditions,

$$P_{\pm}\gamma_5\tau^1 = \gamma_5\tau^1 P_{\mp}. \tag{79}$$

The boundary conditions, just like mass terms, define a direction in chiral flavour space. This means that the $\gamma_5\tau^1$-transformation yields inequivalent correlation functions even in the chiral limit. For a $\gamma_5\tau^1$-even operator O one finds

$$\langle O\rangle_{(m,\mu_q,P_+)} \;\rightarrow\; \langle O\rangle_{(-m,\mu_q,P_-)}. \tag{80}$$

It thus appears that the standard SF does not allow for the definition of $\gamma_5\tau^1$-even correlation functions, and bulk $\mathrm{O}(a)$ improvement is not automatic.

A possible solution is obtained by changing the projectors used to specify the Dirichlet components such that they commute with $\gamma_5\tau^1$. Allowing for an additional flavour structure one may think of $\frac{1}{2}(1 \pm \gamma_0\tau^3)$ or

$$Q_\pm = \tfrac{1}{2}(1 \pm i\gamma_0\gamma_5\tau^3). \tag{81}$$

Interestingly, the projectors $Q_\pm$ also appear in the chiral rotation of the SF by $\alpha = \pi/2$. Besides automatic $\mathrm{O}(a)$ improvement, the implementation of these boundary conditions may lead to some interesting checks of universality by comparing SF correlation functions in the standard framework and at maximal twist. Note that this direct comparison was not possible in Refs. 3, 54, where a twisted mass term was introduced whilst keeping the standard SF boundary conditions.

8.6. *The SF with chirally rotated boundary conditions*

The implementation of some given boundary conditions is not straightforward on the lattice, and some care has to be taken to ensure that one really ends up with the desired continuum theory. A successful implementation of the maximally twisted boundary conditions involving the projectors $Q_\pm$ has been described in Ref. 55, and relies on an orbifold construction to ensure the correct continuum limit.

8.6.1. *Symmetries and Counterterms*

Apart from the absence of a dimensionful parameter, the symmetries of the SF with maximally twisted boundary conditions are identical to those of tmQCD. One may then list the possible boundary counterterms of dimension 3 allowed by the symmetries:

$$K_1 = \bar{\psi} i\gamma_5\tau^3\psi, \qquad K_\pm = \bar{\psi}Q_\pm\psi. \tag{82}$$

As time reflection combined with a flavour permutation is a symmetry of the SF, it is enough to discuss the counterterms at $x_0 = 0$. K_1 corresponds to the logarithmically divergent boundary counterterm in the standard SF, which leads to a multiplicative renormalization of the quark boundary fields. The operator K_+ only involves Dirichlet components at $x_0 = 0$ and is therefore irrelevant for most correlation functions used in practice. The remaining operator K_- only contains non-Dirichlet boundary components. If rotated back to the primed basis it becomes proportional to $\bar{\psi}' i\gamma_5\tau^3 P_- \psi'$, which violates flavour symmetry and parity just like a twisted mass term. As these are symmetries which are restored in the continuum limit one concludes that this counterterm must be scale-independent. Its coefficient can be fixed by requiring that a parity violating SF correlation function vanishes at finite a.

This analysis can be extended to dimension 4 operators which appear as O(a) boundary counterterms [56]. It turns out that the situation is comparable to the standard SF, i.e. there are a couple of counterterms which one needs to tune in order to eliminate the O(a) boundary artefacts.

8.7. *An example from perturbation theory*

In perturbation theory, the values of all boundary counterterms are known, so that one may study both the equality of properly matched standard and twisted SF correlation functions, and confirm automatic bulk O(a) improvement. A first example is given by the SF coupling, which can be related perturbatively to the $\overline{\text{MS}}$-coupling,

$$\bar{g}^2(L) = g^2_{\overline{\text{MS}}}(\mu) + k_1(\mu L) g^4_{\overline{\text{MS}}}(\mu) + O(g^6). \tag{83}$$

The fermionic contribution to the one-loop coefficient, $k_1 = k_{1,0} + N_\mathrm{f} k_{1,1}$, has been computed in Ref. 49, yielding $k_{1,1} = -0.039863(2)/(4\pi)$. In practice, the perturbative data is obtained for a sequence of lattices, and one then expects the asymptotic large L/a behaviour:

$$f(L/a) \sim r_0 + (a/L)[r_1 + s_1 \ln(a/L)] + O(a^2). \tag{84}$$

Here $r_0 = k_{1,1}$ is the continuum limit value, and the O(a) effects lead to non-vanishing values of r_1 and s_1. In the standard Schrödinger functional set-up one expects that r_1 is eliminated by the boundary counterterm proportional to $\mathrm{tr}(F_{0k}F_{0k})$, whereas s_1 is due to bulk O(a) effects from the action, and thus proportional to $c_\mathrm{sw}^{(0)} - 1$. On the other hand, with twisted SF boundary conditions one expects that r_0 remains the same, due to universality, r_1 is cancelled again by a boundary counterterm, and s_1 should vanish independently of the value of $c_\mathrm{sw}^{(0)}$.This expectation is indeed confirmed numerically. A similar test can be performed with the tree level quark propagator in a non-vanishing gauge background field, induced by choosing non-vanishing gauge field boundary values C_k and C'_k. One then expects that, with the correct tree-level boundary counterterms, the bulk O(a) lattice artefacts will again be either proportional to $c_{\rm sw}^{(0)}-1$ (standard SF) or absent (twisted SF). Again this expectation is confirmed. However, in contrast to the SF coupling this test can not be extended beyond the tree level, unless one fixes the gauge.

9. Conclusions

Lattice QCD with Wilson type quarks remains an attractive regularisation of lattice QCD. Some of its problems can be alleviated by introducing a chirally twisted quark mass term. While the theories remain equivalent in the continuum limit, the twisted mass term supplies an infrared bound on the spectrum of the Wilson-Dirac operator which renders the quenched and partially quenched approximations well-defined. Some of the notorious lattice renormalisation problems of standard Wilson quarks can be by-passed, and tmQCD at maximal twist is automatically O(a) improved. These advantages are balanced by parity and flavour breaking and the fact that tmQCD comes naturally with an even number of quarks.

The Schrödinger functional has become an indispensable tool to tackle non-perturbative renormalisation problems in lattice QCD However, the standard set-up leads to a slow decoupling of heavy quarks, and is in conflict with automatic O(a) improvement of massless Wilson quarks. This motivates the application of a chiral twist to the SF boundary conditions. It is thus possible to extend equivalence between tmQCD and standard QCD to correlation functions derived from the Schrödinger functional. This allows for interesting tests of universality and the maximally twisted SF is compatible with automatic O(a) improvement, as I have illustrated with simple perturbative examples.

Acknowledgments

I would like to thank the organisers of the workshop, in particular Yoshinobu Kuramashi for the invitation to this very pleasant meeting in the beautiful city of Nara. Support by the Japanese government and by the Spanish government through a “Ramón y Cajal” fellowship is gratefully acknowledged.

References

1. R. Frezzotti, P. A. Grassi, S. Sint and P. Weisz, *Nucl. Phys. Proc. Suppl.* **83**, 941 (2000).
2. R. Frezzotti, P. A. Grassi, S. Sint and P. Weisz [Alpha collaboration], *JHEP* **0108**, 058 (2001).
3. R. Frezzotti, S. Sint and P. Weisz [ALPHA collaboration], *JHEP* **0107**, 048 (2001).
4. R. Frezzotti and G. C. Rossi, *JHEP* **0408**, 007 (2004).
5. A. Shindler, *PoS* **LAT2005**, 014 (2006).
6. M. Lüscher, S. Sint, R. Sommer and P. Weisz, *Nucl. Phys.* **B478**, 365 (1996).
7. M. Lüscher, *Commun. Math. Phys.* **54**, 283 (1977).
8. G. Schierholz et al., *Nucl. Phys. Proc. Suppl.* **73**, 889 (1999).
9. S. Duane, A. D. Kennedy, B. J. Pendleton and D. Roweth, *Phys. Lett.* **B195**, 216 (1987).
10. B. Joo et al. [UKQCD Collaboration], *Phys. Rev.* **D62**, 114501 (2000).
11. C. Bernard *et al.*, *Nucl. Phys. Proc. Suppl.* **106**, 199 (2002).
12. L. Giusti, *PoS* **LAT2006**, 009 (2006).
13. A. D. Kennedy, arXiv:hep-lat/0607038, these proceedings.
14. T. Reisz, *Nucl. Phys.* **B318**, 417 (1989).
15. M. Bochicchio, L. Maiani, G. Martinelli, G. C. Rossi and M. Testa, *Nucl. Phys.* **B262**, 331 (1985).
16. M. Lüscher, S. Sint, R. Sommer and H. Wittig, *Nucl. Phys.* **B491**, 344 (1997).
17. R. Frezzotti and S. Sint, *Nucl. Phys. Proc. Suppl.* **106**, 814 (2002).
18. K. Jansen, A. Shindler, C. Urbach and I. Wetzorke [XLF Collaboration], *Phys. Lett.* **B586**, 432 (2004).
19. K. Jansen, M. Papinutto, A. Shindler, C. Urbach and I. Wetzorke [XLF Collaboration], *JHEP* **0509**, 071 (2005).
20. C. W. Bernard, T. Draper, G. Hockney and A. Soni, *Nucl. Phys. Proc. Suppl.* **4**, 483 (1988).
21. S. R. Sharpe, "*Phenomenology from the lattice*", arXiv:hep-ph/9412243.
22. M. F. L. Golterman and K. C. Leung, *Phys. Rev.* **D66**, 010001 (2002).
23. K. Jansen *et al.*, *Phys. Lett.* **B372**, 275 (1996).
24. P. Dimopoulos et al. [ALPHA Collaboration], arXiv:hep-lat/0610077.
25. P. Dimopoulos et al. [ALPHA Collaboration], *Nucl. Phys.* **B749**, 69 (2006).
26. C. Pena, S. Sint and A. Vladikas, *JHEP* **0409**, 069 (2004).
27. R. Frezzotti and G. C. Rossi, *JHEP* **0410**, 070 (2004).
28. C. Pena, *PoS* **LAT2006**, 019 (2006), arXiv:hep-lat/0610109.
29. K. Symanzik, *Nucl. Phys.* **B226**, 187 (1983).
30. M. Lüscher and P. Weisz, *Commun. Math. Phys.* **97**, 59 (1985) [Erratum, *ibid.* **98**, 433 (1985)].
31. B. Sheikholeslami and R. Wohlert, *Nucl. Phys.* **B259**, 572 (1985).
32. T. Bhattacharya, R. Gupta, W. Lee, S. R. Sharpe and J. M. S. Wu, *Phys. Rev.* **D73**, 034504 (2006).
33. S. Sint, talk given at the International workshop "*Quantum Fields in the Era of Teraflops Computing*", November 22-25, 2004, Bielefeld, Germany.

34. R. Frezzotti, G. Martinelli, M. Papinutto and G. C. Rossi, *JHEP* **0604**, 038 (2006).
35. S. Aoki and O. Bär, *Phys. Rev.* **D74**, 034511 (2006).
36. S. R. Sharpe, arXiv:hep-lat/0607016, these proceedings.
37. S. Aoki and O. Bar, *Phys. Rev.* **D70**, 116011 (2004).
38. A. M. Abdel-Rehim, R. Lewis and R. M. Woloshyn, *Phys. Rev.* **D71**, 094505 (2005).
39. A. M. Abdel-Rehim, R. Lewis, R. M. Woloshyn and J. M. S. Wu, *Phys. Rev.* **74**, 014507 (2006).
40. K. Jansen *et al.* [XLF Collaboration], *Phys. Lett.* **B624**, 334 (2005).
41. R. Frezzotti and G. C. Rossi, *Nucl. Phys. Proc. Suppl.* **128**, 193 (2004).
42. T. Chiarappa *et al.*, arXiv:hep-lat/0606011.
43. K. Jansen and C. Urbach [ETM Collaboration], *PoS* **LAT2006**, 203 (2006), arXiv:hep-lat/0610015.
44. M. Lüscher, R. Narayanan, P. Weisz and U. Wolff, *Nucl. Phys.* **B384**, 168 (1992).
45. S. Sint, *Nucl. Phys.* **B421**, 135 (1994).
46. K. Symanzik, *Nucl. Phys.* **B190**, 1 (1981).
47. S. Sint, *Nucl. Phys.* **B451**, 416 (1995).
48. A. Bode, P. Weisz and U. Wolff [ALPHA collaboration], *Nucl. Phys.* **B576**, 517 (2000) [Erratum, *ibid.* **B600**, 453 (2001), Erratum, *ibid.* **B608**, 481 (2001)].
49. S. Sint and R. Sommer, *Nucl. Phys.* **B465**, 71 (1996).
50. P. Weisz, *Nucl. Phys. Proc. Suppl.* **47**, 71 (1996).
51. H. Georgi and H. D. Politzer, *Phys. Rev.* **D14**, 1829 (1976).
52. F. Palombi, C. Pena and S. Sint, *JHEP* **0603**, 089 (2006).
53. M. Guagnelli, J. Heitger, C. Pena, S. Sint and A. Vladikas [ALPHA Collaboration], *JHEP* **0603**, 088 (2006).
54. M. Della Morte, R. Frezzotti, J. Heitger and S. Sint [ALPHA collaboration], *JHEP* **0110**, 041 (2001).
55. S. Sint, *PoS* **LAT2005**, 235 (2006).
56. S. Sint, in preparation.

NON-PERTURBATIVE QCD: RENORMALIZATION, O(A)-IMPROVEMENT AND MATCHING TO HEAVY QUARK EFFECTIVE THEORY

RAINER SOMMER

DESY, Platanenallee 6,
15738 Zeuthen, Germany
E-mail: rainer.sommer@desy.de

We give an introduction to three topics in lattice gauge theory:

I. The Schrödinger functional and O(a) improvement.
O(a) improvement has been reviewed several times. Here we focus on explaining the basic ideas in detail and then proceed directly to an overview of the literature and our personal assessment of what has been achieved and what is missing.

II. The computation of the running coupling, running quark masses and the extraction of the renormalization group invariants.
We focus on the basic strategy and on the large effort that has been invested in understanding the continuum limit. We point out what remains to be done.

III. Non-perturbative Heavy Quark Effective Theory.
Since the literature on this subject is still rather sparse, we go beyond the basic ideas and discuss in some detail how the theory works in principle and in practice.

1. Introduction

Lattice QCD, the subject of this school, is our prime tool to make quantitative predictions in the low energy sector of QCD. Also connecting this sector to the perturbative high energy regime requires non-perturbative control over the theory, which can be achieved by lattice gauge theories. In these lectures the emphasis is on a non-perturbative treatment of the theory *including* its renormalization. Connecting the perturbative and the non-perturbative regions is one of the main subjects (II.).

Heavy quarks require special care on a lattice with spacing a, simply because their mass is of the order of the cutoff, a^{-1}, or higher. Effective theories can be used in this situation, in particular Heavy Quark Effective Theory (HQET) is appropriate for hadrons with a single heavy quark. It allows to compute the expansion of their properties in terms of the inverse quark mass (in practice the b-quark mass). The renormalization of this effective theory introduces couplings whose number grows with the order of the expansion. In order to preserve the predictivity of the theory, these couplings ought to be determined from the underlying theory, QCD. Again this step can be seen as the renormalization of the effective theory. As will be explained, non-perturbative precision is required if one wants to be able to take the continuum limit of the lattice effective field theory. Actually it is a general property of the expansion that a $1/m_{\rm b}$ correction is only defined once all parts including the matching are done non-perturbatively.

Renormalization is an ultraviolet phenomenon with relevant momentum scales of order a^{-1}. Since the QCD coupling becomes weak in the ultraviolet, one may expect to be able to perform renormalizations perturbatively, i.e. computed in a power series in the bare coupling g_0^2 as one approaches the continuum limit $a \to 0$.[a] However, one has to care about the following point. In order to keep the numerical effort of a simulation tractable, the number of degrees of freedom in the simulation may not be excessively large. This means that the lattice spacing a can not be taken very much smaller than the relevant physical length scales of the observable that is considered. Consequently the momentum scale a^{-1} that is relevant for the renormalization is not always large enough to justify the truncation of the perturbative series. In particular one has to remember that the bare coupling vanishes only logarithmically as $a \to 0$: $g_0^2 \sim 1/\log(a\Lambda_{\rm QCD})$. In order to obtain a truly non-perturbative answer, the renormalizations have to be performed non-perturbatively.

Depending on the observable, the necessary renormalizations are of different nature. I will use this introduction to point out the different types, and in particular explain the problem that occurs in a non-perturbative treatment of scale dependent renormalization.

1.1. *Basic renormalization: hadron spectrum*

The calculation of the hadron spectrum starts by choosing certain values for the bare coupling, g_0, and the bare masses of the quarks in units of the lattice spacing, $am_{0,i}$. The flavor index i assumes values $i = {\rm u, d, s, c, b}$ for the up, down, charm and bottom quarks that are sufficient to describe hadrons of up to a few GeV masses. We ignore the problem of simulating

[a] For simplicity we ignore here the cases of mixing of a given operator with operators of lower dimension where this statement does not hold.

the b-quark for the moment, neglect isospin breaking and take the light quarks to be degenerate, $m_{0,\mathrm{u}} = m_{0,\mathrm{d}} = m_{0,\mathrm{l}}$.

Next, from MC simulations of suitable correlation functions, one computes masses of five different hadrons H, e.g. $H = \mathrm{p}, \pi, \mathrm{K}, \mathrm{D}, \mathrm{B}$ for the proton, the pion and the K-,D- and B-mesons,

$$am_H = am_H(g_0, am_{0,\mathrm{l}}, am_{0,\mathrm{s}}, am_{0,\mathrm{c}}, am_{0,\mathrm{b}}) \ . \tag{1}$$

The theory is renormalized by first setting $m_\mathrm{p} = m_\mathrm{p}^\mathrm{exp}$, where $m_\mathrm{p}^\mathrm{exp}$ is the experimental value of the proton mass. This determines the lattice spacing via

$$a = (am_\mathrm{p})/m_\mathrm{p}^\mathrm{exp} \ . \tag{2}$$

Next one must choose the parameters $am_{0,i}$ such that (1) is indeed satisfied with the experimental values of the meson masses. Equivalently, one may say that at a given value of g_0 one fixes the bare quark masses from the condition

$$(am_H)/(am_\mathrm{p}) = m_H^\mathrm{exp}/m_\mathrm{p}^\mathrm{exp} \, , \quad H = \pi, \mathrm{K}, \mathrm{D}, \mathrm{B} \ . \tag{3}$$

and the bare coupling g_0 then determines the value of the lattice spacing through Eq. (2).

After this *renormalization*, namely *the elimination of the bare parameters in favor of physical observables*, the theory is completely defined and predictions can be made. E.g. the leptonic decay constant, F_π, of the pion can be determined,

$$F_\pi = a^{-1}[aF_\pi][1 + \mathrm{O}(a)] \ . \tag{4}$$

For the rest of this section, I assume that the bare parameters have been eliminated and consider the additional renormalizations of more complicated observables.

Note. Renormalization as described here is done without any reference to perturbation theory. One could in principle use the perturbative formula for $(a\Lambda_\mathrm{QCD})(g_0)$ for the renormalization of the bare coupling, where Λ_QCD denotes the Λ-parameter of the theory (in some scheme). Proceeding in this way, one obtains a further prediction namely $m_\mathrm{p}/\Lambda_\mathrm{QCD}$ but at the price of introducing $\mathrm{O}(g_0^2)$ errors in the prediction of the observables. As mentioned before, such errors decrease very slowly as one performs the continuum limit. A better method to compute the Λ-parameter will be discussed later.

1.2. *Scale dependent renormalization and fundamental parameters of QCD*

As we take the relevant length scales in correlation functions to be small or take the energy scale in scattering processes to be high, QCD is better and better approximated by weakly coupled quarks and gluons. The strength of the interaction may be measured for instance by the ratio of the production rate of three jets to the rate for two jets in high energy $e^+ \, e^-$ collisions[b]

$$\alpha(\mu) \propto \frac{\sigma(e^+ \, e^- \to q \, \bar{q} \, g)}{\sigma(e^+ \, e^- \to q \, \bar{q})}, \quad \mu^2 = q^2 = (p_{e^-} + p_{e^+})^2 \gg 10\text{GeV}^2 \ . \quad (5)$$

We observe the following points.

- The perturbative renormalization group tells us that $\alpha(\mu)$ decreases logarithmically with growing energy μ. In other words the renormalization from the bare coupling to a renormalized one is logarithmically scale dependent.
- Different definitions of α are possible; but with increasing energy, α depends less and less on the definition (or the process).
- In a similar way, one may define running quark masses $\overline{m}$ from combinations of observables at high energies.
- Using a suitable definition (scheme), the μ-dependence of α and $\overline{m}$ can be determined non-perturbatively and at high energies the short distance parameters α and $\overline{m}$ can be converted to the renormalization group invariants using perturbation theory in α. Being defined non-perturbatively, the latter are the natural fundamental parameters of QCD.

Explaining these points in detail is the main objective of the second lecture.

1.3. *Irrelevant operators*

Another category of renormalization is associated with the removal of lattice discretization errors such as the linear a-term in Eq. (4). Following Symanzik's improvement program, this can be achieved order by order in the lattice spacing by adding irrelevant operators, i.e. operators of dimension larger than four, to the lattice Lagrangian [1]. The coefficients of these operators are easily determined at tree level of perturbation theory, but in general they need to be renormalized. We will explain the general idea

---

[b]One should really use some rather inclusive process, e.g. one computable directly in the Euclidean theory. For explaining the principle we ignore this issue.

of the non-perturbative determination of the coefficients arising at order a and then briefly review the present status of $O(a)$ improvement.

Note also the alternative approach of removing lattice artifacts order by order in the coupling constant but non-perturbatively in the lattice spacing a described in the lectures by Peter Hasenfratz. Linear effects in a are automatically absent if the lattice regularization has enough chiral symmetry. Indeed chiral symmetry can be kept exactly in the discretized theory [2–4], but these theories are rather expensive to simulate. On the other hand also the "twisted mass" regularization [5,6] is automatically[c] $O(a)$-improved[7] (see the appendix of Ref. 8 and Stefan Sint's lectures at this school for a simple argument), but at the price of the violation of isospin symmetry.

1.4. *Heavy Quark Effective Theory*

This theory is very promising for B-physics. It approximates heavy-light bound state properties systematically in an expansion of $\Lambda_{\rm QCD}/m_{\rm b}$, a small expansion parameter. A non-trivial issue is the renormalization of the theory. Already at the lowest order in $1/m_{\rm b}$, the associated uncertainties are significant if renormalization is treated perturbatively. At that order renormalization can be carried out by the methods discussed in the second lecture[9–11], but when one includes $O(1/m_{\rm b})$ corrections one has to deal in addition with the mixing of operators of different dimensions.[d] The continuum limit of the effective theory then exists only if the *power divergent mixing coefficients* are computed non-perturbatively.

In the third lecture we will explain these issues in detail. We will formulate HQET non-perturbatively. The power divergent mixing coefficients can then be determined by matching the theory to QCD. A possible strategy will be explained. As an example we will show the computation of the b-quark mass including $1/m_{\rm b}$ corrections.

---

[c] "Automatically" still means that the standard mass term has to be tuned to zero, but that can be done by the use of the PCAC relation.

[d] Note that the computation of an order $\Lambda_{\rm QCD}$ term in the renormalized quark mass already constitutes a $O(1/m_{\rm b})$ correction to the leading term, although it is done in static approximation.

Lecture I.

The Schrödinger functional and O(a)-improvement of lattice QCD

I.1. The Schrödinger functional (SF)

For various applications, for instance scale dependent renormalization in QCD, O(a)-improvement and Heavy Quark Effective Theory, we need QCD in a finite volume with boundary conditions suitable for (easy) perturbative calculations and MC simulations. These are provided by the SF of QCD, which we introduce below. For a while we restrict the discussion to the pure gauge theory. In this part the presentation follows closely Ref. 12; we refer to this work for further details as well as proofs of the properties described below.

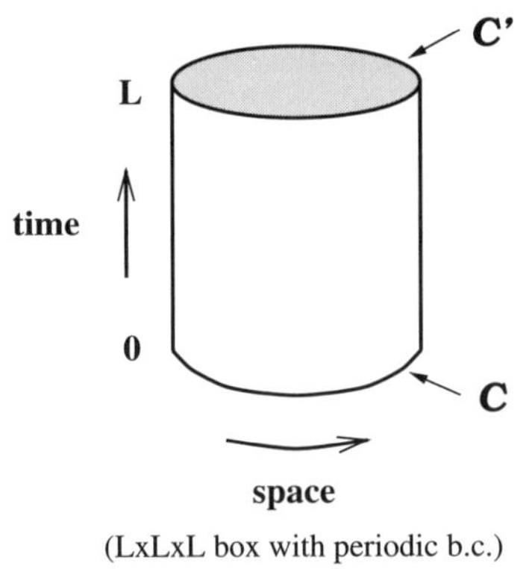

Fig. 1. Illustration of the Schrödinger functional.

I.1.1. *Definition*

Here, we give a formal definition of the SF in the Yang-Mills theory in continuum space-time, noting that a rigorous treatment is possible in the lattice regularized theory.

Space-time is taken to be a cylinder illustrated in Fig. 1. We impose

Dirichlet boundary conditions for the vector potentials[e] in time,

$$A_k(x) = \begin{cases} C_k^{\Lambda}(\mathbf{x}) \text{ at } & x_0 = 0 \\ C_k'(\mathbf{x}) \text{ at } & x_0 = L \end{cases} , \tag{6}$$

where C, C' are classical gauge potentials and A^{Λ} denotes the gauge transform of A,

$$A_k^{\Lambda}(\mathbf{x}) = \Lambda(\mathbf{x})A_k(\mathbf{x})\Lambda(\mathbf{x})^{-1} + \Lambda(\mathbf{x})\partial_k\Lambda(\mathbf{x})^{-1}, \qquad \Lambda \in \mathrm{SU}(N) \ . \tag{7}$$

In space, we impose periodic boundary conditions,

$$A_k(x + L\hat{k}) = A_k(x), \qquad \Lambda(\mathbf{x} + L\hat{k}) = \Lambda(\mathbf{x}) \ . \tag{8}$$

The (Euclidean) partition function with these boundary conditions defines the SF,

$$\begin{aligned} \mathcal{Z}[C', C] &\equiv \int \mathrm{D}[\Lambda] \int \mathrm{D}[A]\, \mathrm{e}^{-S_{\mathrm{G}}[A]} \ , \qquad (9) \\ S_{\mathrm{G}}[A] &= -\frac{1}{2g_0^2} \int \mathrm{d}^4 x \ \mathrm{tr}\, \{F_{\mu\nu} F_{\mu\nu}\} , \\ F_{\mu\nu} &= \partial_\mu A_\nu - \partial_\nu A_\mu + [A_\mu, A_\nu] \ , \\ \mathrm{D}[A] &= \prod_{\mathbf{x},\mu,a} \mathrm{d}A_\mu^a(x), \qquad \mathrm{D}[\Lambda] = \prod_{\mathbf{x}} \mathrm{d}\Lambda(\mathbf{x}) \ . \end{aligned}$$

Here $\mathrm{d}\Lambda(\mathbf{x})$ denotes the Haar measure of $\mathrm{SU}(N)$. It is easy to show that the SF is a gauge invariant functional of the boundary fields,

$$\mathcal{Z}[C'^{\Omega'}, C^{\Omega}] = \mathcal{Z}[C', C] \ , \tag{10}$$

where also large gauge transformations are permitted. The invariance under the latter is an automatic property of the SF defined on a lattice, while in the continuum formulation it is enforced by the integral over Λ in Eq. (9).

I.1.2. *Quantum mechanical interpretation*

The SF is the quantum mechanical transition amplitude from a state $|C\rangle$ to a state $|C'\rangle$ after a (Euclidean) time L. To explain the meaning of this statement, we introduce the Schrödinger representation. The Hilbert space consists of wave-functionals $\Psi[A]$ which are functionals of the spatial components of the vector potentials, $A_k^a(\mathbf{x})$. The canonically conjugate field

[e]We use anti-hermitian vector potentials.

variables are represented by functional derivatives, $E_k^a(\mathbf{x}) = \frac{1}{i}\frac{\delta}{\delta A_k^a(\mathbf{x})}$, and a scalar product is given by

$$\langle\Psi|\Psi'\rangle = \int \mathrm{D}[A]\,\Psi[A]^*\Psi'[A], \qquad \mathrm{D}[A] = \prod_{\mathbf{x},k,a} \mathrm{d}A_k^a(\mathbf{x}) \ . \tag{11}$$

The Hamilton operator,

$$\mathbb{H} = \int_0^L \mathrm{d}^3x \left\{ \frac{g_0^2}{2} E_k^a(\mathbf{x})E_k^a(\mathbf{x}) + \frac{1}{4g_0^2} F_{kl}^a(\mathbf{x})F_{kl}^a(\mathbf{x}) \right\} \ , \tag{12}$$

commutes with the projector, $\mathbb{P}$, onto the physical subspace of the Hilbert space (i.e. the space of gauge invariant states), where $\mathbb{P}$ acts as

$$\mathbb{P}\psi[A] = \int \mathrm{D}[\Lambda]\,\psi[A^\Lambda] \ . \tag{13}$$

Finally, each classical gauge field defines a state $|C\rangle$ through

$$\langle C|\Psi\rangle = \Psi[C] \ . \tag{14}$$

After these definitions, the quantum mechanical representation of the SF is given by

$$\begin{aligned} \mathcal{Z}[C',C] &= \langle C'|\mathrm{e}^{-\mathbb{H}T}\mathbb{P}|C\rangle \\ &= \sum_{n=0}^{\infty} \mathrm{e}^{-E_nT}\,\Psi_n[C']\Psi_n[C]^* \ . \end{aligned} \tag{15}$$

In Wilson's original lattice formulation, Eq. (15) can be derived rigorously and is valid with real energy eigenvalues E_n.

I.1.3. *Background field*

A complementary aspect of the SF is that it allows a treatment of QCD in a color background field in an unambiguous way. Let us assume that we have a solution B of the equations of motion, which satisfies also the boundary conditions Eq. (6). If, in addition,

$$S[A] > S[B] \tag{16}$$

for all gauge fields A that are not equal to a gauge transform B^Ω of B, then we call B the background field (induced by the boundary conditions). Here, $\Omega(x)$ is a gauge transformation defined for all x in the cylinder and its boundary and B^Ω is the corresponding generalization of Eq. (7). Background fields B, satisfying these conditions are known; we will describe a particular family of fields, later.

Due to Eq. (16), fields close to B dominate the path integral for weak coupling g_0 and the effective action,

$$\Gamma[B] \equiv -\ln \mathcal{Z}[C', C] \ , \tag{17}$$

has a regular perturbative expansion,

$$\Gamma[B] = \frac{1}{g_0^2}\Gamma_0[B] + \Gamma_1[B] + g_0^2\Gamma_2[B] + \dots \ , \tag{18}$$
$$\Gamma_0[B] \equiv g_0^2 S[B] \ .$$

Above we have used that due to our assumptions, the background field, B, and the boundary values C, C' are in one-to-one correspondence and have taken B as the argument of Γ.

I.1.4. *Perturbative expansion*

For the construction of the SF as a renormalization scheme, one needs to study the renormalization properties of the functional $\mathcal{Z}$. Lüscher, Narayanan, Weisz and Wolff have performed a one-loop calculation for arbitrary background field [12]. The calculation is done in dimensional regularization with $3-2\varepsilon$ space dimensions and one time dimension. One expands the field A in terms of the background field and a fluctuation field, q, as

$$A_\mu(x) = B_\mu(x) + g_0 q_\mu(x) \ . \tag{19}$$

Then one adds a gauge fixing term ("background field gauge") and the corresponding Fadeev-Popov term. Of course, care must be taken about the proper boundary conditions in all these expressions. Integration over the quantum field and the ghost fields then gives

$$\Gamma_1[B] = \tfrac{1}{2}\ln\det\hat{\Delta}_1 - \ln\det\hat{\Delta}_0 \ , \tag{20}$$

where $\hat{\Delta}_1$ is the fluctuation operator and $\hat{\Delta}_0$ the Fadeev-Popov operator defined in Ref. 12. The result can be cast in the form

$$\Gamma_1[B] \underset{\varepsilon\to 0}{=} -\frac{b_0}{\varepsilon}\Gamma_0[B] + \mathrm{O}(1) \ , \tag{21}$$

with the important result that the only (for $\varepsilon \to 0$) singular term is proportional to Γ_0.

After renormalization of the coupling, i.e. the replacement of the bare coupling by $\bar{g}_{\overline{\mathrm{MS}}}$ via

$$g_0^2 = \bar{\mu}^{2\varepsilon}\bar{g}_{\overline{\mathrm{MS}}}^2(\mu)[1 + z_1(\varepsilon)\bar{g}_{\overline{\mathrm{MS}}}^2(\mu)], \quad z_1(\varepsilon) = -\frac{b_0}{\varepsilon} \ , \tag{22}$$

the effective action is finite,

$$\Gamma[B]_{\varepsilon=0} = \left\{ \frac{1}{\bar{g}^2_{\overline{\rm MS}}} - b_0 \left[\ln \mu^2 - \tfrac{1}{16\pi^2} \right] \right\} \Gamma_0[B] \\ -\tfrac{1}{2}\zeta'(0|\Delta_1) + \zeta'(0|\Delta_0) + \mathrm{O}(\bar{g}^2_{\overline{\rm MS}}) \tag{23}$$

$$\zeta'(0|\Delta) = \frac{\mathrm{d}}{\mathrm{d}s}\zeta(s|\Delta)\bigg|_{s=0}, \qquad \zeta(s|\Delta) = \mathrm{Tr}\,\Delta^{-s} \ .$$

Here, $\zeta'(0|\Delta)$ is a complicated functional of B, which is not known analytically but can be evaluated numerically for specific choices of B.

The important result of this calculation is that (apart from field independent terms that have been dropped everywhere) the SF is finite after eliminating g_0 in favor of $\bar{g}_{\overline{\rm MS}}$. The presence of the boundaries does *not* introduce any extra divergences. In the following section we argue that this property is correct in general, not just in one-loop approximation.

I.1.5. *General renormalization properties*

The relevant question here is whether local quantum field theories formulated on space-time manifolds *with boundaries* develop divergences that are not present in the absence of boundaries (periodic boundary conditions or infinite space-time). In general the answer is "yes, such additional divergences exist". In particular, Symanzik studied the ϕ^4-theory with Schrödinger functional boundary conditions [13,14]. He presented arguments that to all orders of perturbation theory the Schrödinger functional is finite after

- renormalization of the self-coupling, λ, and the mass, m,
- *and* the addition of the boundary counter-terms

$$\int_{x^0=T} \mathrm{d}^3x \,\{Z_1\phi^2 + Z_2\phi\partial_0\phi\} + \int_{x^0=0} \mathrm{d}^3x \,\{Z_1\phi^2 - Z_2\phi\partial_0\phi\} \tag{24}$$

In addition to the standard renormalizations, one has to add counter-terms formed by local composite fields integrated over the boundaries. One expects that in general, all fields with dimension $d \leq 3$ have to be taken into account. Already Symanzik conjectured that counter-terms with this property are sufficient to renormalize the SF of any quantum field theory which is renormalizable when no boundaries are present.

Since this conjecture forms the basis for many applications of the SF, we note a few points concerning its status.

- As mentioned, a proof to all orders of perturbation theory does not exist. An application of power counting in momentum space in order to prove the conjecture is not possible due to the missing translation invariance.
- There is no gauge invariant local field with $d \leq 3$ in the Yang–Mills theory. Consequently no additional counter-term is necessary in accordance with the 1-loop result described in the previous subsection.
- In QCD it has been checked also by explicit 2–loop calculations [15,16]. MC simulations give further support for its validity beyond perturbation theory; we give examples in the second lecture.

Although a general proof is missing, there is little doubt that Symanzik's conjecture is valid in general. Concerning QCD, this puts us into the position to give an elegant definition of a renormalized coupling in finite volume.

I.1.6. *Renormalized coupling*

For the definition of a running coupling we need a quantity which depends only on one scale. We choose LB such that it depends only on one dimensionless variable η. In other words, the strength of the field is scaled as $1/L$. The background field is assumed to fulfill the requirements of Sect. I.1.3. Then, following the above discussion, the derivative

$$\Gamma'[B] = \frac{\partial}{\partial\eta}\Gamma[B] \ , \tag{25}$$

is finite when it is expressed in terms of a renormalized coupling like $\bar{g}_{\overline{\rm MS}}$ but Γ' is defined non-perturbatively. From Eq. (18) we read off immediately that a properly normalized coupling is given by

$$\bar{g}^2(L) = \Gamma'_0[B] \,/\, \Gamma'[B] \ . \tag{26}$$

Since there is only one length scale L, it is evident that $\bar{g}$ defined in this way runs with L.

A specific choice for the gauge group SU(3) is the abelian background field induced by the boundary values [17]

$$C_k = \tfrac{i}{L}\begin{pmatrix} \phi_1 & 0 & 0 \\ 0 & \phi_2 & 0 \\ 0 & 0 & \phi_3 \end{pmatrix}, \quad C'_k = \tfrac{i}{L}\begin{pmatrix} \phi'_1 & 0 & 0 \\ 0 & \phi'_2 & 0 \\ 0 & 0 & \phi'_3 \end{pmatrix}, \quad k = 1,2,3, \tag{27}$$

with

$$\begin{aligned} \phi_1 &= \eta - \tfrac{\pi}{3}, & \phi_1' &= -\phi_1 - \tfrac{4\pi}{3}, \\ \phi_2 &= -\tfrac{1}{2}\eta, & \phi_2' &= -\phi_3 + \tfrac{2\pi}{3}, \\ \phi_3 &= -\tfrac{1}{2}\eta + \tfrac{\pi}{3}, & \phi_3' &= -\phi_2 + \tfrac{2\pi}{3}. \end{aligned} \tag{28}$$

In this case, the derivatives with respect to η are to be evaluated at $\eta = 0$. The associated background field,

$$B_0 = 0, \qquad B_k = \left[x_0 C_k' + (L - x_0)C_k\right]/L, \quad k = 1,2,3 \ , \tag{29}$$

has a field tensor with non-vanishing components

$$G_{0k} = \partial_0 B_k = (C_k' - C_k)/L, \quad k = 1,2,3 \ . \tag{30}$$

It is a constant color-electric field.

I.1.7. *Quarks*

In the end, the real interest is in the renormalization of QCD and we need to consider the SF with quarks. We restrict our discussion to the original formulation of S. Sint [18].

Special care has to be taken in formulating the Dirichlet boundary conditions for the quark fields; since the Dirac operator is a first order differential operator, the Dirac equation has a unique solution when one half of the components of the fermion fields are specified on the boundaries. Indeed, a detailed investigation shows that the boundary condition

$$P_+\psi|_{x_0=0} = \rho, \quad P_-\psi|_{x_0=L} = \rho' , \qquad P_\pm = \tfrac{1}{2}(1 \pm \gamma_0) , \tag{31}$$

$$\overline{\psi}P_-|_{x_0=0} = \bar{\rho}, \quad \overline{\psi}P_+|_{x_0=L} = \bar{\rho}' , \tag{32}$$

lead to a quantum mechanical interpretation analogous to Eq. (15). The SF

$$\mathcal{Z}[C', \bar{\rho}', \rho'; C, \bar{\rho}, \rho] = \int \mathrm{D}[A]\mathrm{D}[\,\psi\,]\mathrm{D}[\,\overline{\psi}\,]\,\mathrm{e}^{-S[A,\overline{\psi},\psi\,]} \tag{33}$$

involves an integration over all fields with the specified boundary values. The full action may be written as

$$\begin{aligned} S[A, \overline{\psi}, \psi\,] &= S_{\mathrm{G}}[\overline{\psi}, \psi\,] + S_{\mathrm{F}}[A, \overline{\psi}, \psi\,] \\ S_{\mathrm{F}} &= \int \mathrm{d}^4x\, \overline{\psi}(x)[\gamma_\mu D_\mu + m]\psi(x) \\ &\quad - \int \mathrm{d}^3\mathbf{x}\, [\overline{\psi}(x)P_-\psi(x)]_{x_0=0} - \int \mathrm{d}^3\mathbf{x}\, [\overline{\psi}(x)P_+\psi(x)]_{x_0=L} \ , \end{aligned} \tag{34}$$

with S_G as given in Eq. (9). In Eq. (34) we use standard Euclidean γ-matrices. The covariant derivative, D_μ, acts as $D_\mu\psi(x) = \partial_\mu\psi(x) + A_\mu(x)\psi(x)$.

Let us now discuss the renormalization of the SF with quarks. In contrast to the pure Yang-Mills theory, gauge invariant composite fields of dimension three are present in QCD. Taking into account the boundary conditions one finds [18] that the counter-terms,

$$\overline{\psi}P_-\psi|_{x_0=0} \text{ and } \overline{\psi}P_+\psi|_{x_0=L} , \tag{35}$$

have to be added to the action with weight $1 - Z_b$ to obtain a finite renormalized functional. These counter-terms are equivalent to a multiplicative renormalization of the boundary values,

$$\rho_R = Z_b^{-1/2}\rho , \ \ldots \ , \bar{\rho}'_R = Z_b^{-1/2}\bar{\rho}' . \tag{36}$$

It follows that – apart from the renormalization of the coupling and the quark mass – no additional renormalization of the SF is necessary for *vanishing* boundary values $\rho, \ldots, \bar{\rho}'$. So, after imposing homogeneous boundary conditions for the fermion fields, a renormalized coupling may be defined as in the previous subsection.

As an important aside, we point out that the boundary conditions for the fermions introduce a gap into the spectrum of the Dirac operator (at least for weak couplings). One may hence simulate the lattice SF for vanishing physical quark masses. It is then convenient to supplement the definition of the renormalized coupling by the requirement $m = 0$. In this way, one defines a mass-independent renormalization scheme with simple renormalization group equations. In particular, the β-function remains independent of the quark mass.

I.1.7.1. *Correlation functions*

are given in terms of the expectation values of any product $\mathcal{O}$ of fields,

$$\langle\mathcal{O}\rangle = \left\{\frac{1}{\mathcal{Z}}\int \mathrm{D}[A]\mathrm{D}[\psi]\mathrm{D}[\overline{\psi}]\,\mathcal{O}\,\mathrm{e}^{-S[A,\overline{\psi},\psi]}\right\}_{\bar{\rho}'=\rho'=\bar{\rho}=\rho=0} , \tag{37}$$

evaluated for vanishing boundary values $\rho, \ldots, \bar{\rho}'$. Apart from the gauge field and the quark and anti-quark fields integrated over, $\mathcal{O}$ may involve the "boundary fields" [19]

$$\zeta(\mathbf{x}) = \frac{\delta}{\delta\bar{\rho}(\mathbf{x})}, \qquad \bar{\zeta}(\mathbf{x}) = -\frac{\delta}{\delta\rho(\mathbf{x})},$$
$$\zeta'(\mathbf{x}) = \frac{\delta}{\delta\bar{\rho}'(\mathbf{x})}, \qquad \bar{\zeta}'(\mathbf{x}) = -\frac{\delta}{\delta\rho'(\mathbf{x})} . \tag{38}$$

An application of fermionic correlation functions including the boundary fields is the definition of the renormalized quark mass in the SF scheme to be discussed next.

I.1.7.2. *Renormalized mass*

Just as in the case of the coupling constant, there is a great freedom in defining renormalized quark masses. A natural starting point is the PCAC relation which expresses the divergence of the axial current [f],

$$A_\mu^a(x) = \overline{\psi}(x)\gamma_5\gamma_\mu\tfrac{1}{2}\tau^a\psi(x) \ , \tag{39}$$

(for simplicity we have chosen just $N_\mathrm{f} = 2$ degenerate flavors and the Pauli matrix τ^a acts in this flavor space), in terms of the associated pseudo-scalar density,

$$P^a(x) = \overline{\psi}(x)\gamma_5\tfrac{1}{2}\tau^a\psi(x) \ , \tag{40}$$

via

$$\partial_\mu A_\mu^a(x) = 2mP^a(x) \ . \tag{41}$$

This operator identity is easily derived at the classical level (cf. Sect. I.2). After renormalizing the operators,

$$\begin{aligned} (A_\mathrm{R})_\mu^a &= Z_\mathrm{A} A_\mu^a \ , \\ P_\mathrm{R}^a &= Z_\mathrm{P} P^a \ , \end{aligned} \tag{42}$$

a renormalized current quark mass may be defined by

$$m_\mathrm{R} = \overline{m} = \frac{Z_\mathrm{A}}{Z_\mathrm{P}} m \ . \tag{43}$$

Here, m, is to be taken from Eq. (41) inserted into an arbitrary correlation function and Z_A can be determined from a proper chiral Ward identity [20–22]. Note that m does not depend on which correlation function is used because the PCAC relation is an operator identity. The definition of $\overline{m}$ is completed by supplementing Eq. (42) with a specific normalization condition for the pseudo-scalar density. The running mass $\overline{m}$ then inherits its scheme- and scale-dependence (μ) from the corresponding dependence of P_R. Such a normalization condition may be imposed through infinite volume correlation functions. Since we want to be able to compute $\overline{m}(\mu)$ for large energy scales μ, we do, however, need a finite volume definition (see

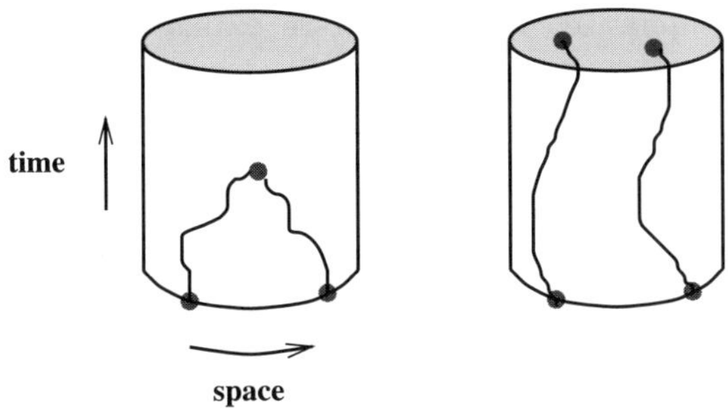

Fig. 2. $f_{\rm P}$ (left) and f_1 (right) in terms of quark propagators.

Sect. II.1.2). This is readily given in terms of correlation functions in the SF.

To start with, let us define (iso-vector) pseudo-scalar fields at the boundary of the SF,

$$\begin{aligned} \mathcal{O}^a &= \int \mathrm{d}^3\mathbf{u} \int \mathrm{d}^3\mathbf{v}\; \overline{\zeta}(\mathbf{u})\gamma_5 \tfrac{1}{2}\tau^a \zeta(\mathbf{v}), \\ \mathcal{O}'^a &= \int \mathrm{d}^3\mathbf{u} \int \mathrm{d}^3\mathbf{v}\; \overline{\zeta}'(\mathbf{u})\gamma_5 \tfrac{1}{2}\tau^a \zeta'(\mathbf{v}) \;, \end{aligned} \tag{44}$$

to build up the correlation functions

$$\begin{aligned} f_{\rm P}(x_0) &= -\tfrac{1}{3}\langle P^a(x)\mathcal{O}^a\rangle \;, \\ f_1 &= \langle \mathcal{O}'^a \mathcal{O}^a \rangle \;, \end{aligned} \tag{45}$$

which are illustrated in Fig. 2.

We then form the ratio

$$Z_{\rm P} = \text{const.}\sqrt{f_1}/f_{\rm P}(x)|_{x_0=L/2} \;, \tag{46}$$

such that the renormalization of the boundary quark fields, Eq. (36), cancels out. The proportionality constant is to be chosen such that $Z_{\rm P} = 1$ at tree level. To define the scheme completely one needs to further specify the boundary values C, C' and the boundary conditions for the quark fields in space. These details are of no importance, here.

We rather mention some more basic points about this renormalization scheme. Just like in the case of the running coupling, the only physical scale that exists in our definitions Eq. (43), Eq. (46) is the linear dimension of the SF, the length scale, L. So the mass $\overline{m}(\mu)$ runs with $\mu = 1/L$. We

[f] The reader is not to confuse $A_\mu^a(x)$, with the gauge vector potentials $A_\mu(x)$.

have already emphasized that $\bar{g}$ is to be evaluated at zero quark mass. It is advantageous to do the same for $Z_{\rm P}$. In this way we define a mass-independent renormalization scheme, with simple renormalization group equations.

By construction, the SF scheme is non-perturbative and independent of a specific regularization. For a concrete non-perturbative computation, we do, however, need to evaluate the expectation values by a MC-simulation of the corresponding lattice theory. We proceed to introduce the lattice formulation of the SF.

I.1.8. *Lattice formulation*

A detailed knowledge of the form of the lattice action is not required for an understanding of the following sections. Nevertheless, we give a definition of the SF in lattice regularization. This is done both for completeness and because it allows us to obtain a first impression about the size of discretization errors.

We choose a hyper-cubic Euclidean lattice with spacing a. A gauge field U on the lattice is an assignment of a matrix $U(x,\mu) \in \mathrm{SU}(N)$ to every lattice point x and direction $\mu = 0,1,2,3$. Quark and anti-quark fields, $\psi(x)$ and $\overline{\psi}(x)$, reside on the lattice sites and carry Dirac, color and flavor indices as in the continuum. To be able to write the quark action in an elegant form it is useful to extend the fields, initially defined only inside the SF manifold (cf. Fig. 1) to all times x_0 by "padding" with zeros. In the case of the quark field one sets

$$\psi(x) = 0 \quad \text{if } x_0 < 0 \text{ or } x_0 > L,$$

and

$$P_-\psi(x)|_{x_0=0} = P_+\psi(x)|_{x_0=L} = 0,$$

and similarly for the anti-quark field. Gauge field variables that reside outside the manifold are set to 1.

We may then write the fermionic action as a sum over all space-time points without restrictions for the time-coordinate,

$$S_{\rm F}[U,\bar{\psi},\psi] = a^4 \sum_x \bar{\psi}(D+m_0)\psi, \tag{47}$$

and with the standard Wilson-Dirac operator,

$$D = \tfrac{1}{2}\sum_{\mu=0}^{3}\{\gamma_\mu(\nabla_\mu^* + \nabla_\mu) - a\nabla_\mu^*\nabla_\mu\}\,. \tag{48}$$

Here, forward and backward covariant derivatives,

$$\nabla_\mu \psi(x) = \tfrac{1}{a}[U(x,\mu)\psi(x+a\hat{\mu}) - \psi(x)], \tag{49}$$

$$\nabla^*_\mu \psi(x) = \tfrac{1}{a}[\psi(x) - U(x-a\hat{\mu},\mu)^{-1}\psi(x-a\hat{\mu})] \ , \tag{50}$$

are used and m_0 is to be understood as a diagonal matrix in flavor space with elements m_0^f.

The gauge field action S_G is a sum over all oriented plaquettes p on the lattice, with the weight factors $w(p)$, and the parallel transporters $U(p)$ around p,

$$S_\mathrm{G}[U] = \frac{1}{g_0^2}\sum_p w(p)\ \mathrm{tr}\,\{1 - U(p)\} \ . \tag{51}$$

The weights $w(p)$ are 1 for plaquettes in the interior and

$$w(p) = \begin{cases} \tfrac{1}{2}c_s & \text{if } p \text{ is a spatial plaquette at } x_0 = 0 \text{ or } x_0 = L, \\ c_t & \text{if } p \text{ is time-like and attached to a boundary plane.} \end{cases} \tag{52}$$

The choice $c_s = c_t = 1$ corresponds to the standard Wilson action. However, these parameters can be tuned in order to reduce lattice artifacts, as will be briefly discussed below.

With these ingredients, the path integral representation of the Schrödinger functional reads [18],

$$\mathcal{Z} = \int \mathrm{D}[\psi]\mathrm{D}[\bar{\psi}]\mathrm{D}[U]\,\mathrm{e}^{-S}, \quad S = S_\mathrm{F} + S_\mathrm{G} \ , \tag{53}$$

$$\mathrm{D}[U] = \prod_{x,\mu} \mathrm{d}U(x,\mu) \ ,$$

with the Haar measure $\mathrm{d}U$.

I.1.8.1. *Boundary conditions and the background field*

The boundary conditions for the lattice gauge fields may be obtained from the continuum boundary values by forming the appropriate parallel transporters from $x + a\hat{k}$ to x at $x_0 = 0$ and $x_0 = L$. For the constant abelian boundary fields C and C' that we considered before, they are simply

$$U(x,k)|_{x_0=0} = \exp(aC_k), \qquad U(x,k)|_{x_0=L} = \exp(aC'_k), \tag{54}$$

for $k = 1,2,3$. All other boundary conditions are as in the continuum.

For the case of Eq. (27),Eq. (28), the boundary conditions (54) lead to a unique (up to gauge transformations) minimal action configuration V, the lattice background field. It can be expressed in terms of B (29),

$$V(x,\mu) = \exp\{aB_\mu(x)\} \ . \tag{55}$$

I.1.8.2. *Lattice artifacts*

Now we want to get a first impression about the dependence of the lattice SF on the value of the lattice spacing. In other words we study lattice artifacts. At lowest order in the bare coupling we have, just like in the continuum,

$$\Gamma = \frac{1}{g_0^2}\Gamma_0[V] + \mathrm{O}((g_0)^0)\,, \quad \Gamma_0[V] \equiv g_0^2 S_\mathrm{G}[V]\ . \tag{56}$$

Furthermore one easily finds the action for small lattice spacings,

$$\begin{aligned} S_\mathrm{G}[V] &= \left[1+(1-c_\mathrm{t})\tfrac{2a}{L}\right] \frac{3L^4}{g_0^2} \sum_{\alpha=1}^{N} \left\{ \frac{2}{a^2} \sin\left[\frac{a^2}{2L^2}(\phi'_\alpha - \phi_\alpha)\right]\right\}^2 \\ &= \frac{3}{g_0^2} \sum_{\alpha=1}^{N} (\phi'_\alpha - \phi_\alpha)^2 \left[1+(1-c_\mathrm{t})\tfrac{2a}{L} + \mathrm{O}(a^4)\right]\ . \end{aligned} \tag{57}$$

We observe: at tree-level of perturbation theory, all linear lattice artifacts are removed when one sets $c_\mathrm{t} = 1$. Beyond tree-level (and in the theory without quarks), one has to tune the coefficient c_t as a function of the bare coupling. We will show the effect, when this is done to first order in g_0^2, below. Note that the existence of linear $\mathrm{O}(a)$ errors in the Yang-Mills theory is special to the SF; they originate from dimension four operators $F_{0k}F_{0k}$ and $F_{kl}F_{kl}$ which are irrelevant terms (i.e. they carry an explicit factor of the lattice spacing) when they are integrated over the surfaces. c_s, which can be tuned to cancel the effects of $F_{kl}F_{kl}$, does not appear for the electric field that we discussed above.

Once quark fields are present, there are more irrelevant operators that can generate $\mathrm{O}(a)$ effects as discussed in detail in Ref. 19. Here we emphasize a different feature of Eq. (57): once the $\mathrm{O}(a)$-terms are canceled, the remaining a-effects are tiny. This special feature of the abelian background field is most welcome for the numerical computation of the running coupling; it allows for reliable extrapolations to the continuum limit.

I.1.8.3. *Explicit expression for* Γ'

Let us finally explain that Γ' is an observable that can easily be calculated in a MC simulation. From its definition we find immediately

$$\Gamma' = -\frac{\partial}{\partial\eta} \ln\left\{\int \mathrm{D}[\psi]\mathrm{D}[\bar\psi]\mathrm{D}[U]\,\mathrm{e}^{-S}\right\} = \left\langle \frac{\partial S_\mathrm{G}}{\partial\eta}\right\rangle + \left\langle \frac{\partial S_\mathrm{F}}{\partial\eta}\right\rangle\ . \tag{58}$$

The derivative $\frac{\partial S_G}{\partial \eta}$ evaluates to the (color 8 component of the) electric field at the boundary,

$$\frac{\partial S_{\rm G}}{\partial \eta} = -\frac{2}{g_0^2 L} a^3 \sum_{\mathbf{x}} \{E_k^8(\mathbf{x}) - (E_k^8)'(\mathbf{x})\} \ , \tag{59}$$

$$E_k^8(\mathbf{x}) = \frac{1}{a^2} \mathrm{Re\ tr} \left\{ i\lambda_8 U(x,k) U(x+a\hat{k},0) U(x+a\hat{0},k)^{-1} U(x,0)^{-1} \right\}_{x_0=0} \ ,$$

where $\lambda_8 = \mathrm{diag}(1,-1/2,-1/2)$. (A similar expression holds for $(E_k^8)'(\mathbf{x})$). The second term $\frac{\partial S_F}{\partial \eta}$, which is only non-zero in the $\mathrm{O}(a)$-improved formulation is numerically less relevant. An explicit expression is given in Ref. 23.

The renormalized coupling is related to the expectation value of a local operator; no correlation function is involved. This means that it is easy and fast in computer time to evaluate it. It further turns out that a good statistical precision is reached with a moderate size statistical ensemble.

I.1.9. *More literature*

We here give some guide for further reading on the SF. Independently of the work of Symanzik, G. C. Rossi and M. Testa discussed different boundary conditions imposed at fixed time[24,25]. The renormalization properties of that functional have not yet been discussed.

There are also rather recent developments. Different formulations of the lattice Schrödinger functional with overlap fermions satisfying the Ginsparg Wilson relation have been found by Y. Taniguchi using an orbifold construction [26], and by M. Lüscher using a general universality argument concerning QFT's with boundaries [14]; see also Ref. 27. As the Schrödinger functional breaks chiral symmetry by the boundary conditions, it is relevant into which direction in flavor space the mass term is introduced. The Schrödinger functional with a twisted mass term [5,6] and the boundary conditions specified above differs from the SF with a standard mass term (at finite quark mass). S. Sint found a modification of the boundary conditions, which yields the standard SF as the continuum limit of the lattice theory with a twisted mass [28]. An even number of flavors is required in this formulation. It offers also advantages in the massless limit, where "automatic bulk $\mathrm{O}(a)$-improvement" is achieved *after the tuning of one counter-term*. It is discussed in detail in S. Sint's lectures at this school. Another Schrödinger functional with automatic bulk $\mathrm{O}(a)$-improvement is proposed in Ref. 29.

I.2. Chiral symmetry and O(a)-improvement

The main focus of this section is on the O(a) improvement of Wilson's lattice QCD. However, we also mention the finite normalization of isovector currents. Both of these problems have the same origin, namely that chiral symmetry is broken in Wilson's regularization and then also the same solution: chiral Ward identities. The possibility to use these to normalize the currents has first been discussed by Refs. 20, 30. Here, we describe their application in the computation of the O(a)-improved action and currents. A difference to the aforementioned work is that we insist that only on-shell improvable correlation functions are used in the normalization conditions in order to be compatible with on-shell improvement.

Before going into more details, we would like to convey the general idea of the application of chiral Ward identities. For simplicity we assume an isospin doublet of mass-degenerate quarks. Consider first a regularization of QCD which preserves the full $\mathrm{SU}(2)_{\mathrm{V}} \times \mathrm{SU}(2)_{\mathrm{A}}$ flavor symmetry as it is present in the continuum Lagrangian of massless QCD. Indeed, such regularizations exist [2–4], see Peter Hasenfratz' lectures.

We can derive chiral Ward identities in the Euclidean formulation of the theory. These then provide exact relations between different correlation functions. Immediate consequences of these relations are that there are currents V_μ^a, A_μ^a which do not get renormalized ($Z_{\mathrm{A}} = Z_{\mathrm{V}} = 1$) and the quark mass does not have an additive renormalization.

In a general discretization, such as the Wilson formulation, lattice QCD does *not* have the $\mathrm{SU}(2)_{\mathrm{A}}$ flavor symmetry for finite values of the lattice spacing. Then, the Ward identities are not satisfied exactly. From universality, we do, however, expect that the correlation functions may be renormalized such that they obey the same Ward identities as before – up to O(a) corrections. Therefore we may impose those Ward identities for the renormalized currents, to fix their normalizations.

Furthermore, following Symanzik [1], it suffices to a add a few local irrelevant terms to the action and to the currents in order to obtain an improved lattice theory, where the continuum limit is approached with corrections of order a^2. The coefficients of these terms can be determined by imposing improvement conditions. For example one may require certain chiral Ward identities to be valid at finite lattice spacing a.

I.2.1. *Chiral Ward identities*

For the moment we do not pay attention to a regularization of the theory and derive the Ward identities in a formal way. As mentioned above these identities are exact in a regularization that preserves chiral symmetry. To derive the Ward identities, one starts from the path integral representation of a correlation function and performs the change of integration variables

$$\begin{aligned}\psi(x) &\to \mathrm{e}^{i\frac{\tau^a}{2}[\epsilon^a_{\mathrm{A}}(x)\gamma_5+\epsilon^a_{\mathrm{V}}(x)]}\psi(x)\\ &= \psi(x)+i\epsilon^a_{\mathrm{A}}(x)\delta^a_{\mathrm{A}}\psi(x)+i\epsilon^a_{\mathrm{V}}(x)\delta^a_{\mathrm{V}}\psi(x)\ ,\\ \overline{\psi}(x) &\to \overline{\psi}(x)\mathrm{e}^{i\frac{\tau^a}{2}[\epsilon^a_{\mathrm{A}}(x)\gamma_5-\epsilon^a_{\mathrm{V}}(x)]}\\ &= \overline{\psi}(x)+i\epsilon^a_{\mathrm{A}}(x)\delta^a_{\mathrm{A}}\overline{\psi}(x)+i\epsilon^a_{\mathrm{V}}(x)\delta^a_{\mathrm{V}}\overline{\psi}(x)\ ,\end{aligned} \tag{60}$$

where we have taken $\epsilon^a_{\mathrm{A}}(x), \epsilon^a_{\mathrm{V}}(x)$ infinitesimal and introduced the variations

$$\begin{aligned}\delta^a_{\mathrm{V}}\psi(x) &= \tfrac{1}{2}\tau^a\psi(x), & \delta^a_{\mathrm{V}}\overline{\psi}(x) &= -\overline{\psi}(x)\tfrac{1}{2}\tau^a\ ,\\ \delta^a_{\mathrm{A}}\psi(x) &= \tfrac{1}{2}\tau^a\gamma_5\psi(x), & \delta^a_{\mathrm{A}}\overline{\psi}(x) &= \overline{\psi}(x)\gamma_5\tfrac{1}{2}\tau^a\ .\end{aligned} \tag{61}$$

The Ward identities then follow from the invariance of the path integral representation of correlation functions with respect to such changes of integration variables. They obtain contributions from the variation of the action and the variations of the fields in the correlation functions. The variations of the currents

$$V^a_\mu(x) = \overline{\psi}(x)\gamma_\mu\tfrac{1}{2}\tau^a\psi(x) \tag{62}$$

and A^a_μ, Eq. (39), is given by

$$\begin{aligned}\delta^a_{\mathrm{V}}V^b_\mu(x) &= -i\epsilon^{abc}V^c_\mu(x), & \delta^a_{\mathrm{A}}V^b_\mu(x) &= -i\epsilon^{abc}A^c_\mu(x),\\ \delta^a_{\mathrm{V}}A^b_\mu(x) &= -i\epsilon^{abc}A^c_\mu(x), & \delta^a_{\mathrm{A}}A^b_\mu(x) &= -i\epsilon^{abc}V^c_\mu(x)\ .\end{aligned} \tag{63}$$

They form a closed algebra under these variations.

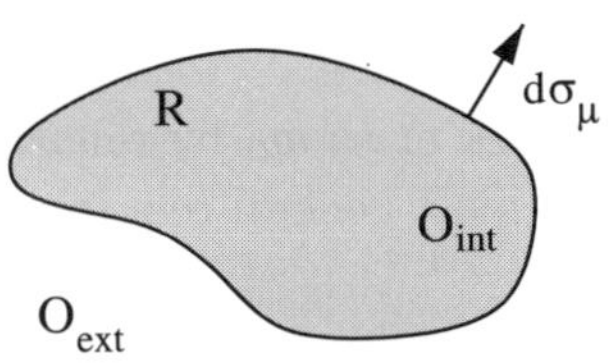

Since this is convenient for our applications, we write the Ward identities in an integrated form. Let R be a space-time region with smooth boundary ∂R. Suppose $\mathcal{O}_{\mathrm{int}}$ and $\mathcal{O}_{\mathrm{ext}}$ are polynomials in the basic fields localized in the interior and exterior of R respectively (see left). The general vector current Ward identity then reads

$$\int_{\partial R} \mathrm{d}\sigma_\mu(x)\ \left\langle V^a_\mu(x)\mathcal{O}_{\mathrm{int}}\mathcal{O}_{\mathrm{ext}}\right\rangle = -\left\langle \left(\delta^a_{\mathrm{V}}\mathcal{O}_{\mathrm{int}}\right)\mathcal{O}_{\mathrm{ext}}\right\rangle, \tag{64}$$

while for the axial current one obtains

$$\int_{\partial R} d\sigma_\mu(x)\,\langle A^a_\mu(x)\mathcal{O}_{\rm int}\mathcal{O}_{\rm ext}\rangle = -\,\langle(\delta^a_{\rm A}\mathcal{O}_{\rm int})\,\mathcal{O}_{\rm ext}\rangle \\ +2m\int_R d^4x\,\langle P^a(x)\mathcal{O}_{\rm int}\mathcal{O}_{\rm ext}\rangle\ . \tag{65}$$

Here volume integrals over for example $\partial_\mu A^a_\mu(x)$ have been changed to surface integrals. The integration measure $d\sigma_\mu(x)$ points along the outward normal to the surface ∂R and $P^a(x)$ was defined in Eq. (40).

We may also write down the precise meaning of the PCAC-relation Eq. (41). It is Eq. (65) in a differential form,

$$\langle\left[\partial_\mu A^a_\mu(x) - 2mP^a(x)\right]\mathcal{O}_{\rm ext}\rangle = 0\ , \tag{66}$$

where now $\mathcal{O}_{\rm ext}$ may have support everywhere except for at the point x.

Going through the same derivation in the lattice regularization, one finds equations of essentially the same form as the ones given above, but with additional terms [20]. At the classical level these terms are of order a. More precisely, in Eq. (66) the important additional term originates from the variation of the Wilson term, $a\,\delta^a_{\rm A}(\overline{\psi}\nabla^*_\mu\nabla_\mu\psi)$, and is a local field of dimension 5. Such $\mathrm{O}(a)$-corrections are present in any observable computed on the lattice and are no reason for concern. However, as is well known in field theory, such operators mix with the ones of lower and equal dimensions when one goes beyond the classical approximation. In the present case, the dimension five operator mixes among others also with $\partial_\mu A^a_\mu(x)$ and $P^a(x)$. This means that part of the classical $\mathrm{O}(a)$-terms turn into $\mathrm{O}(g_0^2)$ in the quantum theory. The essential observation is now that this mixing can simply be written in the form of a renormalization of the terms that are already present in the Ward identities, since all dimension three and four operators with the right quantum number are already there.

We conclude that the identities, which we derived above in a formal manner, are valid in any proper lattice regularization after

- replacing the bare fields A, V, P and quark mass m_0 by renormalized ones, where one must allow for the most general renormalizations,

$$(A_{\rm R})^a_\mu = Z_{\rm A}A^a_\mu\ ,\quad (V_{\rm R})^a_\mu = Z_{\rm V}V^a_\mu\ ,$$
$$(P_{\rm R})^a = Z_{\rm P}P^a\ ,\quad m_{\rm R} = Z_{\rm m}m_{\rm q}\,,\quad m_{\rm q} = m_0 - m_c\ ,$$

- allowing for the usual $\mathrm{O}(a)$ lattice artifacts.

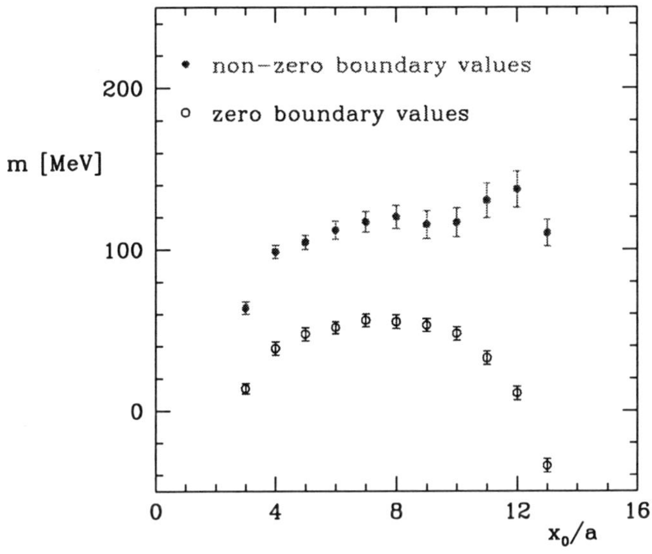

Fig. 3. Dependence of current quark mass m on the boundary condition and the time coordinate [32]. The calculation is done in the quenched approximation on a 16×8^3 lattice at $\beta = 6.4$, which corresponds to a lattice spacing of $a \approx 0.05\,\text{fm}$. "Boundary values" refer to the gauge field boundary conditions in the SF. Their values are given in Ref. 32.

Note that the additive quark mass renormalization m_c diverges like $\mathrm{O}(g_0^2/a)$ for dimensional reasons.

As a result of this discussion, the formal Ward identities may be used to determine the normalizations of the currents. We refer the reader to Refs. 31, 21 for details and explain here the general idea how one can use the Ward identities to determine improvement coefficients.

I.2.2. *On-shell* O(a)*-improvement*

I.2.2.1. *Motivation*

Let us first recall why one wants to remove lattice spacing effects linear in a. The prime reason is as follows. If linear effects are present, one has to vary a in the numerical simulations over a large range in order to be able to get a reasonable estimate of their magnitude.[g] In contrast if the

[g]Obviously it does not really help to have a large range by considering large values of a. Then one enters the regime where either the higher order terms are significant or – more likely – the whole asymptotic expansion in a breaks down.

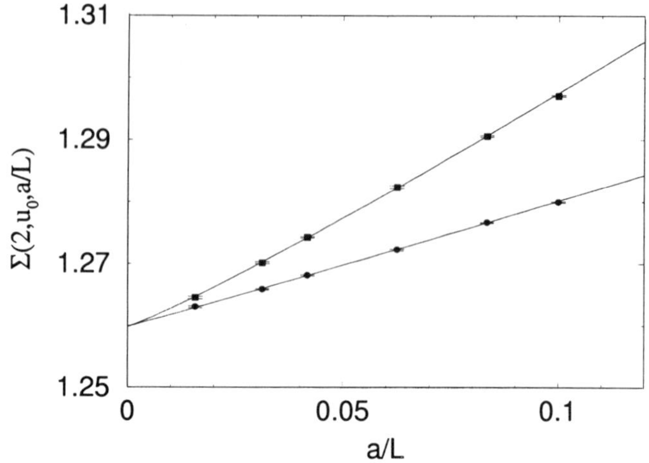

Fig. 4. Lattice spacing dependence of the step scaling function of the LWW coupling in the 2-d O(3) sigma model[36] for coupling $u_0 = 1.0595$. The data points with the smaller cutoff effects are for the standard nearest neighbor action.

cutoff effects are $O(a^2)$, a range of $0.05\,\mathrm{fm} \leq a \leq 0.1\,\mathrm{fm}$ typically allows to check well whether they contribute significantly. In fact a reasonably well controlled extrapolation to the continuum can then be done allowing for a term proportional to a^2 and also a smaller range in a may be sufficient. Examples can be found e.g. in Ref. 33. In addition, it does turn out that linear a effects can be quite large. Let us give here just two examples.

The first one is the current quark mass m defined by the PCAC relation. As detailed below, its value is independent of kinematical variables such as the boundary conditions. Dependences on such variables are pure lattice artifacts. We examined the current quark mass in the valence approximation by numerical Monte Carlo simulations and found large lattice artifacts even for quite small lattice spacings [32] (cf. Fig. 3).

The second example is the mass of the vector meson, made dimensionless by multiplying with r_0. This quantity has large cutoff effects in the quenched approximation [34]. Depending on the quark mass, a-effects of around 20% and more are seen at $a \approx 0.15\,\mathrm{fm}$; see for example Fig. 1 of Ref. 34, Fig. 1 of Ref. 35.

I.2.2.2. *A warning from two dimensions*

Before entering the discussion of the O(a) improvement programme, we mention some unexpected results from thorough examinations of 2-d O(N) sigma models. The theoretical basis for the discussion and removal of a-effects is Symanzik's effective theory, see Sect. I.2.2.3. O(N) sigma models were the second class of models investigated by Symanzik in order to establish this theory. For these models the basic statement is that (up to logarithmic modifications) the cutoff effects are quadratic in a, when a is small enough.

It therefore came as a surprise that Hasenfratz and Niedermayer found in a numerical study of the Lüscher-Weisz-Wolff (LWW) renormalized coupling[37] of the 2-d O(3) sigma model that its step scaling function shows an a-dependence which is roughly linear in a for quite small a (large correlation length)[38]. With a further improved algorithm, a Bern-Berlin collaboration confirmed this behavior with even higher precision and smaller lattice spacing[36]. We show their result in Fig. 4. On the other hand it was known that the cutoff effects of the step scaling function are O(a^2) in the large N limit of the O(N) models[37]. Subsequent numerical studies for $N = 4, 8$ showed no conclusive results: just like in the $N = 3$ case, the cutoff effects look linear when judged by eye, but they can also be fitted with O(a^2) functions, in particular when the expected logarithmic modifications are taken into account[39].

Later the $1/N$ correction was worked out at finite lattice spacing[39]. Recall that at order $(1/N)^0$ one has O(a^2) effects. The cutoff effect proportional to $1/N$ is shown in Fig. 5. Over a large range in a it is almost a linear function of a, but close to $a = 0$ it is dominated by an O(a^2) term. Thus our personal conclusion is that there is no conflict with Symanzik's effective theory in the O(N) models. One should also note that all the a-effects discussed here are rather small.

However, there is a clear warning that, depending on model and observable, the lattice spacing may have to be rather small for the leading correction term in the effective theory to dominate. On the more practical side, long continuum extrapolations with significant slopes may be dangerous, since in QCD we do not have much information where the asymptotic expansion in a is accurate[40]. This is one of the reasons why we will spend much time on understanding the cutoff effects in the QCD step scaling function of the coupling in Sect. II.2.2 and Sect. II.2.3.

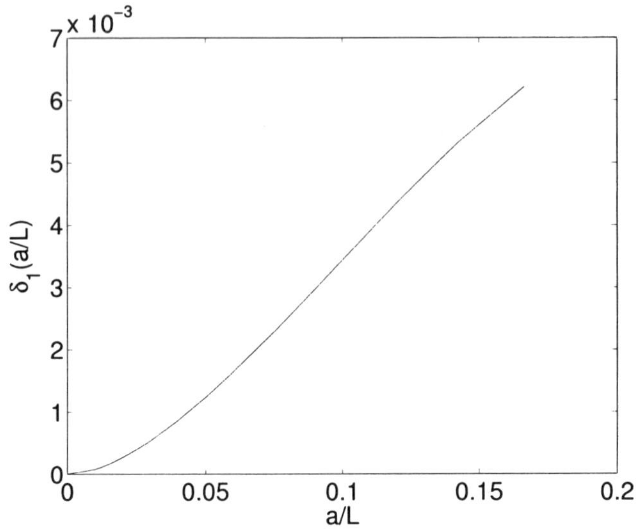

Fig. 5. Coefficient of $1/N$ in the $1/N$-expansion of the cutoff effects of the step scaling function of the LWW coupling of 2-d O(N) sigma models. Graph prepare by U. Wolff based on Ref. 39.

I.2.2.3. *Symanzik's local effective theory (SET)*

In the following explanation of the theory we follow quite closely Ref. 19. We consider QCD on an infinitely extended lattice with two degenerate light Wilson quarks of bare mass m_0 [41]. The action is then given as in Sect. I.1.8 except that no boundary conditions or boundary terms are necessary.

Quite some time ago, Symanzik provided arguments that a lattice theory can be described in terms of a local effective theory, when the lattice spacing is small enough [1]. The effective action,

$$S_{\text{eff}} = S_0 + aS_1 + a^2 S_2 + \dots , \tag{67}$$

has as a leading order, S_0, the action of the continuum theory [h]. The terms S_k, $k = 1, 2, \dots$, are space-time integrals of Lagrangians $\mathcal{L}_k(x)$. These are given as general linear combinations of local gauge-invariant composite fields which respect the exact symmetries of the lattice theory and have canonical dimension $4 + k$. We use the convention that explicit (non-negative) powers of the quark mass m are included in the dimension

[h] If more rigor is desired one may define it on a lattice with spacing $\varepsilon \ll a$.

counting. A possible basis of fields for the Lagrangian $\mathcal{L}_1(x)$ reads

$$\begin{aligned}
\mathcal{O}_1 &= \overline{\psi}\,\sigma_{\mu\nu}F_{\mu\nu}\psi, \\
\mathcal{O}_2 &= \overline{\psi}\,D_\mu D_\mu\psi + \overline{\psi}\,\overleftarrow{D}_\mu\overleftarrow{D}_\mu\psi, \\
\mathcal{O}_3 &= m\,\mathrm{tr}\left\{F_{\mu\nu}F_{\mu\nu}\right\}, \\
\mathcal{O}_4 &= m\left\{\overline{\psi}\,\gamma_\mu D_\mu\psi - \overline{\psi}\,\overleftarrow{D}_\mu\gamma_\mu\psi\right\}, \\
\mathcal{O}_5 &= m^2\overline{\psi}\psi,
\end{aligned} \tag{68}$$

where $F_{\mu\nu}$ is the gluon field tensor and $\sigma_{\mu\nu} = \frac{i}{2}[\gamma_\mu, \gamma_\nu]$.

When considering correlation functions of local gauge invariant fields the action is not the only source of cutoff effects. If $\phi(x)$ denotes such a lattice field (e.g. the axial density or the isospin currents), one expects the connected n-point function

$$G_n(x_1, \ldots, x_n) = (Z_\phi)^n \left\langle\phi(x_1)\ldots\phi(x_n)\right\rangle_{\mathrm{con}} \tag{69}$$

to have a well-defined continuum limit, provided the renormalization constant Z_ϕ is correctly tuned and the space-time arguments $x_1, \ldots, x_n$ are kept at a physical distance from each other.

In the effective theory the renormalized lattice field $Z_\phi\phi(x)$ is represented by an effective field,

$$\phi_{\mathrm{eff}}(x) = \phi_0(x) + a\phi_1(x) + a^2\phi_2(x) + \ldots, \tag{70}$$

where the $\phi_k(x)$ are linear combinations of composite, local fields with the appropriate dimension and symmetries. For example, in the case of the axial current (39), ϕ_1 is given as a linear combination of the terms

$$\begin{aligned}
(\mathcal{O}_6)^a_\mu &= \overline{\psi}\,\gamma_5\tfrac{1}{2}\tau^a\sigma_{\mu\nu}\left[D_\nu - \overleftarrow{D}_\nu\right]\psi, \\
(\mathcal{O}_7)^a_\mu &= \overline{\psi}\,\tfrac{1}{2}\tau^a\gamma_5\left[D_\mu + \overleftarrow{D}_\mu\right]\psi, \\
(\mathcal{O}_8)^a_\mu &= m\overline{\psi}\,\gamma_\mu\gamma_5\tfrac{1}{2}\tau^a\psi.
\end{aligned} \tag{71}$$

The convergence of $G_n(x_1, \ldots, x_n)$ to its continuum limit can now be studied in the effective theory,

$$\begin{aligned}
G_n(x_1, \ldots, x_n) = {} & \left\langle\phi_0(x_1)\ldots\phi_0(x_n)\right\rangle_{\mathrm{con}} \\
& -a\int \mathrm{d}^4y\,\left\langle\phi_0(x_1)\ldots\phi_0(x_n)\mathcal{L}_1(y)\right\rangle_{\mathrm{con}} \\
& +a\sum_{k=1}^{n}\left\langle\phi_0(x_1)\ldots\phi_1(x_k)\ldots\phi_0(x_n)\right\rangle_{\mathrm{con}} + \mathrm{O}(a^2),
\end{aligned} \tag{72}$$

where the expectation values on the right-hand side are to be taken in the continuum theory with action S_0.

Using the field equations

For most applications, it is sufficient to compute on-shell quantities such as particle masses, S-matrix elements and correlation functions at space-time arguments, which are separated by a physical distance. It is then possible to make use of the field equations to reduce first the number of basis fields in the effective Lagrangian $\mathcal{L}_1$ and, in a second step, also in the O(a) counter-term ϕ_1 of the effective composite fields.

Of course, in the quantum theory, the field equations have to be used with care. Performing changes in the integration variables of the path integral shows immediately that the field equations are only valid up to additional terms. In particular, if one uses the field equations in the Lagrangian $\mathcal{L}_1$ under the space-time integral in Eq. (72), the errors made are contact terms that arise when y comes close to one of the arguments $x_1, \ldots, x_n$. Using the operator product expansion and the dimensions and symmetries, one easily verifies that these contact terms must have the same structure as the insertions of ϕ_1 in the last term of Eq. (72). The use of the field equations in $\mathcal{L}_1$ therefore just means a redefinition of the coefficients in the counter-term ϕ_1 (apart from contributions that are absorbed in the renormalization factor Z_ϕ, which originate from the mixing of ϕ_1 with ϕ).

It turns out that one may eliminate two of the terms in Eq. (68) by using the field equations. A possible choice is to stay with the terms $\mathcal{O}_1$, $\mathcal{O}_3$ and $\mathcal{O}_5$, which yields the effective continuum action for on-shell quantities to order a. Having made this choice one may apply the field equations once again to simplify the term ϕ_1 in the effective field as well. In the example of the axial current it is then possible to eliminate the term $\mathcal{O}_6$ in Eq. (71).

I.2.2.4. *Improved lattice action and fields*

The on-shell O(a) improved lattice action is obtained by adding a counter-term to the unimproved lattice action such that the effects of the action S_1 in the effective theory are canceled in on-shell amplitudes. This can be achieved by adding lattice representatives of the terms $\mathcal{O}_1$, $\mathcal{O}_3$ and $\mathcal{O}_5$ to the unimproved lattice Lagrangian, with coefficients that are functions of the bare coupling g_0 only. Leaving the discussion of suitable improvement conditions for later, we here note that the fields $\mathcal{O}_3$ and $\mathcal{O}_5$ already appear in the unimproved theory and thus merely lead to a re-parametrization of the bare parameters g_0 and m_0. In the following, we will not consider these terms any further. Their relevance in connection with massless renormalization schemes is discussed in detail in Ref. 19.

We choose the standard discretization $\widehat{F}_{\mu\nu}$ of the field tensor [19] and add the improvement term to the Wilson-Dirac operator, Eq. (48),

$$D_{\text{impr}} = D + c_{\text{sw}} \frac{ia}{4} \sigma_{\mu\nu} \widehat{F}_{\mu\nu}. \tag{73}$$

With a properly chosen coefficient $c_{\text{sw}}(g_0)$, this yields the on-shell $\mathrm{O}(a)$ improved lattice action which has first been proposed by Sheikholeslami and Wohlert [42].

The perturbative expansion of c_{sw} reads $c_{\text{sw}} = 1 + c_{\text{sw}}^{(1)} g_0^2 + \mathrm{O}(g_0^4)$, with [43,44] $c_{\text{sw}}^{(1)} = 0.26590(7)$ where the Wilson plaquette gauge action is assumed.

The $\mathrm{O}(a)$ improved isospin currents and the axial density can be parametrized as follows,

$$\begin{aligned}
(A_{\mathrm{R}})_\mu^a &= Z_{\mathrm{A}}(1 + b_{\mathrm{A}} a m_{\mathrm{q}}) \{ A_\mu^a + a c_{\mathrm{A}} \tilde{\partial}_\mu P^a \}, \\
(V_{\mathrm{R}})_\mu^a &= Z_{\mathrm{V}}(1 + b_{\mathrm{V}} a m_{\mathrm{q}}) \{ V_\mu^a + a c_{\mathrm{V}} \tilde{\partial}_\nu T_{\mu\nu}^a \}, \\
(P_{\mathrm{R}})^a &= Z_{\mathrm{P}}(1 + b_{\mathrm{P}} a m_{\mathrm{q}}) P^a ,
\end{aligned} \tag{74}$$

where

$$T_{\mu\nu}^a = i \overline{\psi} \sigma_{\mu\nu} \tfrac{1}{2} \tau^a \psi \,.$$

We have included the normalization constants $Z_{\mathrm{A,V,P}}$, which have to be fixed by appropriate normalization conditions. Again, the improvement coefficients $b_{\mathrm{A,V,P}}$ and $c_{\mathrm{A,V}}$ are functions of g_0 only. At tree level of perturbation theory, they are given by $b_{\mathrm{A}} = b_{\mathrm{P}} = b_{\mathrm{V}} = 1$ and $c_{\mathrm{A}} = c_{\mathrm{V}} = 0$ [45,19] and to 1-loop accuracy and with the plaquette gauge action we have [43,46]

$$\begin{aligned}
c_{\mathrm{A}} &= -0.005680(2) \times g_0^2 C_{\mathrm{F}} + \mathrm{O}(g_0^4), \\
c_{\mathrm{V}} &= -0.01225(1) \times g_0^2 C_{\mathrm{F}} + \mathrm{O}(g_0^4), \\
b_{\mathrm{A}} &= 1 + 0.11414(4) \times g_0^2 C_{\mathrm{F}} + \mathrm{O}(g_0^4), \\
b_{\mathrm{V}} &= 1 + 0.11492(4) \times g_0^2 C_{\mathrm{F}} + \mathrm{O}(g_0^4), \\
b_{\mathrm{P}} &= 1 + 0.11484(2) \times g_0^2 C_{\mathrm{F}} + \mathrm{O}(g_0^4).
\end{aligned} \tag{75}$$

Here $C_{\mathrm{F}} = 4/3$. Non-perturbative determinations will be mentioned below.

I.2.3. *The PCAC relation*

We assume for the moment that on-shell $\mathrm{O}(a)$ improvement has been fully implemented, i.e. the improvement coefficients are assigned their correct values. If $\mathcal{O}$ denotes a renormalized on-shell $\mathrm{O}(a)$ improved field localized in a region not containing x, we thus expect that the PCAC relation,

$$\tilde{\partial}_\mu \langle (A_{\mathrm{R}})^a_\mu(x)\, \mathcal{O}\rangle = 2m_{\mathrm{R}} \langle (P_{\mathrm{R}})^a(x)\, \mathcal{O}\rangle \tag{76}$$

$$\tilde{\partial}_\mu = \tfrac{1}{2}(\partial^*_\mu + \partial_\mu) \tag{77}$$

holds up to corrections of order a^2. At this point we note that the field $\mathcal{O}$ need not be improved for this statement to be true. To see this we use again Symanzik's local effective theory and denote the $\mathrm{O}(a)$ correction term in $\mathcal{O}_{\mathrm{eff}}$ by ϕ_1. Equation (76) then receives an order a contribution

$$a\big\langle \big\{\partial_\mu (A_{\mathrm{R}})^a_\mu(x) - 2m_{\mathrm{R}}(P_{\mathrm{R}})^a(x)\big\}\, \phi_1\big\rangle, \tag{78}$$

which is to be evaluated in the continuum theory. The PCAC relation holds exactly in the continuum and the extra term (78) thus vanishes, a conclusion that holds also in the Schrödinger functional.

I.2.4. *Non-perturbative improvement*

The coefficients of the different improvement terms need to be fixed by suitable improvement conditions. One considers pure lattice artifacts, i.e. combinations of observables that are known to vanish in the continuum limit of the theory. Improvement conditions require these lattice artifacts to vanish, thus defining the values of the improvement coefficients as a function of the lattice spacing.

In perturbation theory, lattice artifacts can be obtained from any (renormalized) quantity by subtracting its value in the continuum limit. The improvement coefficients are unique and some of them have been cited above.

Beyond perturbation theory, one wants to determine the improvement coefficients by MC calculations and it requires significant effort to take the continuum limit. It is therefore advantageous to use lattice artifacts that derive from a symmetry of the continuum field theory that is not respected by the lattice regularization. One may require rotational invariance of the static potential $V(\mathbf{r})$, e.g.

$$V(\mathbf{r} = (2,2,1)r/3) - V(\mathbf{r} = (r,0,0)) = 0\,,$$

or Lorentz invariance,

$$[E(\mathbf{p})]^2 - [E(\mathbf{0})]^2 - \mathbf{p}^2 = 0\,,$$

for the momentum dependence of a one-particle energy E.

For $\mathrm{O}(a)$ improvement of QCD it is advantageous to use violations of the PCAC relation (66), instead. PCAC can be used in the context of the Schrödinger functional (SF), where one has a large flexibility to choose appropriate improvement conditions and can compute the improvement coefficients also for small values of the bare coupling g_0, making contact with their perturbative expansions. A further – and maybe the most significant – advantage of the SF in this context is the following. As we have seen earlier, in the SF we may choose boundary conditions such that the induced background field, has non-vanishing components $F_{\mu\nu}$. Remembering Eq. (73), we observe that correlation functions are then sensitive to the improvement coefficient $c_{\rm sw}$ already at tree level of perturbation theory. With a vanishing background field this would be the case at higher orders only. Also in the non-perturbative regime this is the basis for a good sensitivity of the improvement conditions to $c_{\rm sw}$.

As a consequence of the freedom to choose improvement conditions, the resulting values of improvement coefficients such as $c_{\rm sw}$, $c_{\rm A}$ depend on the exact choices made. The corresponding variation of $c_{\rm sw}$, $c_{\rm A}$ is of order a. It changes the effects of order a^2 in physical observables computed after improvement. We will come back to this point, but first we proceed to sketch the non-perturbative calculation of the coefficient $c_{\rm sw}$ [19,43].

We define a bare current quark mass, m, viz.

$$m \equiv \frac{\langle [\partial_\mu (A_{\rm I})^a_\mu(x)]\, \mathcal{O}^a \rangle}{2\,\langle P^a(x)\mathcal{O}^a\rangle}\,, \quad (A_{\rm I})^a_\mu = A^a_\mu + a c_{\rm A}\tfrac{1}{2}(\partial_\mu + \partial^*_\mu)P^a \tag{79}$$

with the pseudo scalar boundary operator $\mathcal{O}^a$ from Eq. (44). When all improvement coefficients have their proper values, the renormalized quark mass, defined by the renormalized PCAC-relation, is related to m by

$$m_{\rm R} = \frac{Z_{\rm A}(1 + b_{\rm A} a m_{\rm q})}{Z_{\rm P}(1 + b_{\rm P} a m_{\rm q})} m + {\rm O}(a^2)\ . \tag{80}$$

It now suffices to choose 3 different versions of Eq. (79) by different choices of the time coordinate x_0 and/or boundary conditions and obtain 3 different values of m, denoted by m_1, m_2, m_3. Since the prefactor in front of m in Eq. (80) is just a numerical factor, we may conclude that all m_i have to be equal in the improved theory up to errors of order a^2. $c_{\rm sw}$ and $c_{\rm A}$ may therefore be computed by requiring [i]

$$m_1 = m_2 = m_3\ . \tag{81}$$

This simple idea has first been used to compute $c_{\rm sw}$ and $c_{\rm A}$ as a function of g_0 in the quenched approximation [19]. A good accuracy has been reached

[i] In the practical calculations actually four different masses m_i were used to compute $c_{\rm sw}$ and two more to extract $c_{\rm A}$.

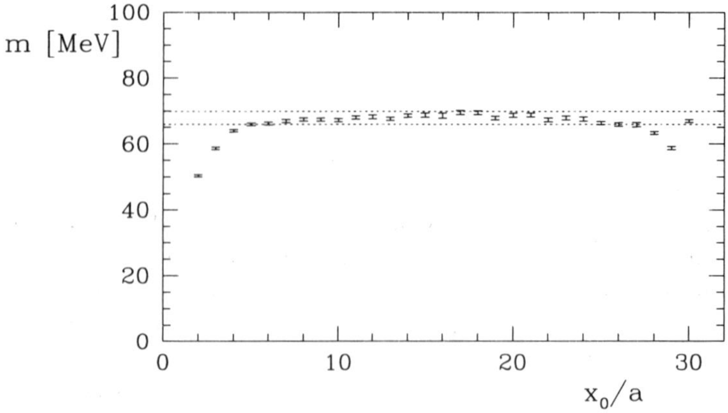

Fig. 6. Unrenormalized current quark mass m in the improved theory, with non-perturbatively determined c_{sw} and c_A, as a function of the time x_0 on a 32×16^3 lattice at $\beta = 6.2$ ($a \approx 0.07\,\text{fm}$)and $\kappa = 0.1350$. The width of the corridor bounded by the dotted horizontal lines is 4 MeV.

in these determinations for $a \leq 0.1\,\text{fm}$. We comment on more recent determinations below, after showing some tests of the effectiveness of O(a)-improvement, namely the test whether cutoff effects are indeed significantly reduced in practical calculations.

Verification of O(a)*-improvement*

The first test of the size of residual O(a^2) effects is provided again by the PCAC relation. To set the scale, remember that the cutoff effects in the PCAC mass m were as large as several tens of MeV before improvement (cf. Fig. 3). The situation after improvement, and for a somewhat larger value of the lattice spacing, is illustrated in Fig. 6. Away from the boundaries, where the effect of states with energies of the order of the cutoff induces noticeable effects, m is independent of time to within $\pm 2\,\text{MeV}$. Compared to the situation before O(a) improvement, this is a change by more than an order of magnitude.

The second test is in the scaling of the vector meson mass, which is improved dramatically in the quenched approximation. In the range $0.01 \leq a^2/r_0^2 \leq 0.035$ ($a \leq 0.1\,\text{fm}$)[j] and with statistical errors of 1-3%, no a effect was found (see Fig. 3, Ref. 33). This is a significant test because i) the a-effects were large before improvement, ii) the statistical precision is good

[j]We use the reference scale r_0 defined in terms of the static quark potential.[47] It has a phenomenological value of $r_0 \approx 0.5\,\text{fm}$.

and iii) a^2 varies by more than a factor 3.

In the quenched approximation several other tests of cutoff effects in the improved theory have been carried out [48,33] revealing that $\mathrm{O}(a)$-improvement works. More precisely we mean by this that for $a \leq 0.1$ fm cut-off effects found after improvement are generally reasonably small ($\mathrm{O}(5\,\%)$) and well compatible with an $\mathrm{O}(a^2)$ behavior. Still, no miracles have been achieved: at $a \approx 0.1$ fm also cutoff effects of about 15% have been found [33].

We now proceed to discuss a rather relevant detail in the non-perturbative determination of improvement coefficients.

I.2.4.1. *The constant physics condition*

We have mentioned before that beyond perturbation theory the improvement coefficients are not unique; they depend on the improvement condition. This ambiguity is of $\mathrm{O}(a)$ in the coefficients and then of $\mathrm{O}(a^2)$ in the physical observables after improvement. So everything is correct, but if these ambiguities are large, one has to take extra care.

To give an explicit example, consider the quenched approximation but with non-degenerate quark masses, $m_{\mathrm{R},i} \neq m_{\mathrm{R},j}$. In this theory the renormalized, improved currents and axial densities are given by a straight forward generalization of what we wrote down earlier[49],

$$
\begin{aligned}
(A_\mathrm{R})_\mu^{ij} &= Z_\mathrm{A}(1 + b_\mathrm{A}\tfrac{1}{2}(am_{\mathrm{q},i} + am_{\mathrm{q},j}))\{A_\mu^{ij} + ac_\mathrm{A}\tilde{\partial}_\mu P^{ij}\},\\
(P_\mathrm{R})^{ij} &= Z_\mathrm{P}(1 + b_\mathrm{P}\tfrac{1}{2}(am_{\mathrm{q},i} + am_{\mathrm{q},j}))P^{ij}\,,\\
A_\mu^{ij} &= \overline{\psi}_i\gamma_\mu\gamma_5\psi_j\,, \quad P^{ij} = \overline{\psi}_i\gamma_5\psi_j\,,
\end{aligned} \tag{82}
$$

and the bare quark masses, $m_{0,i}$, of the Lagrangian are related to the renormalized ones via [k]

$$
m_{\mathrm{R},i} = \frac{Z_\mathrm{A}}{Z_\mathrm{P}}\, Z\, m_{\mathrm{q},i}(1 + a\, b_\mathrm{m} m_{\mathrm{q},i})\,, \quad m_{\mathrm{q},i} = m_{0,i} - m_\mathrm{c}\,. \tag{83}
$$

One then derives the (unrenormalized) PCAC relation [49]

$$
\begin{aligned}
&\frac{\partial_\mu A_\mu^{ij} + a\partial_\mu^*\partial_\mu P^{ij}}{P^{ij}}[1 + a\tfrac{1}{2}(b_\mathrm{A} - b_\mathrm{P})\,(m_{\mathrm{q},i} + m_{\mathrm{q},j})]\\
&= Z\,[m_{\mathrm{q},i} + m_{\mathrm{q},j} + a\, b_\mathrm{m}(m_{\mathrm{q},i}^2 + m_{\mathrm{q},j}^2)] + \mathrm{O}(a^2)\,.
\end{aligned} \tag{84}
$$

This operator identity has to be understood as in Eq. (76). Applying it with a few external operators/kinematical conditions, one can extract $b_\mathrm{A} - b_\mathrm{P}$,

---

[k]In the improved action, b_m is the coefficient of the field $\mathcal{O}_5$.

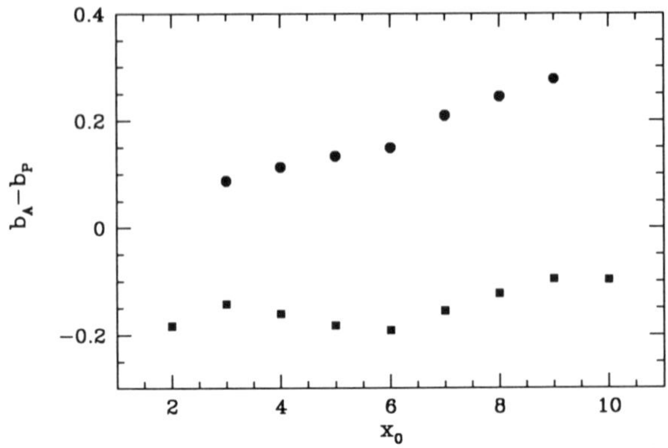

Fig. 7. The improvement coefficient $b_A - b_P$ extracted on a 12×8^3 lattice and a lattice spacing of $a \approx 0.1\,\mathrm{fm}$. Axial current and density are inserted at distance x_0. Squares are obtained with standard lattice representatives of the derivatives ∂_μ and $\partial^*_\mu \partial_\mu$, while circles come from improved (and therefore less local) derivatives [51].

b_m and Z [49,50]. Suitable correlation functions have been implemented in the Schrödinger functional . As shown in Fig. 7, the result for $b_A - b_P$ at $a \approx 0.1\,\mathrm{fm}$ does depend a lot on the details, namely here x_0 and the exact lattice representation of derivatives. This is an extreme example of an improvement coefficient that is difficult to determine. The reasons are surely that the O(a)-effects are not dominating over the O(a^2) effects in the considered correlation functions. Presumably the O(a)-effects are just not very large. We proceed to discuss what one should do in these cases.

When the improvement conditions are formulated through Schrödinger functional correlation functions, we can also study them in perturbation theory. One chooses kinematical variables such as $T/L\,, x_0/L$ exactly as in the non-perturbative set up. One then computes the expansion of an improvement coefficient, here denoted generically by b, as a series

$$b(g_0, a/L) \sim b^{(0)}(a/L) + g_0^2\, b^{(1)}(a/L) + \mathrm{O}(g_0^4)\,. \tag{85}$$

Taking the case of $b_A - b_P$ and $L/a = 8, T/a = 12$ as in Fig. 7, the perturbative coefficients $b^{(0)}$, $b^{(1)}$ show a similar dependence on the kinematics as the non-perturbative results [51].

While generically the functions $b^{(0)}(a/L)$, $b^{(1)}(a/L)$ depend on the kinematical choices made in the improvement coefficients, the values $b^{(0)}(0)$, $b^{(1)}(0)$ are unique. This is the precise meaning of our earlier statement that improvement coefficients are unique in perturbation theory.

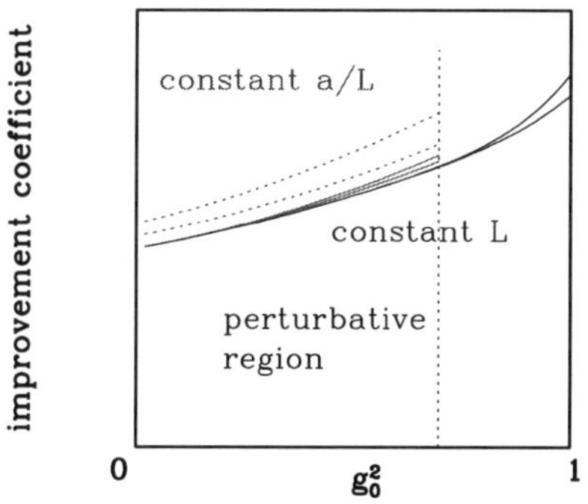

Generic behavior of improvement coefficients. Dotted lines correspond to two different improvement conditions at fixed a/L, while the full lines are for fixed L in physical units.

On the left we illustrates what happens generically. The continuous lines correspond to two different improvement conditions at *constant physics*. By this we mean that all length scales in the correlation functions which define the improvement condition are kept fixed in physical units, for example in units of r_0. When g_0 is varied, only the lattice spacing a changes and this is the situation to which the SET can be applied. The two different improvement coefficients then very rapidly go to an almost unique function of g_0 as $g_0 \to 0$.

On the other hand, one may want to set improvement conditions at fixed L/a for practical reasons[1]. In this setting, the O(a) ambiguity in the improvement coefficients does not go to zero as $g_0 \to 0$ (and $a \to 0$) (dotted lines).

If the dependence of $b^{(i)}(a/L)$ on a/L is rather weak, say

$$|b^{(i)}(a/L) - b^{(i)}(0)| \ll 10^{-1}\,, \tag{86}$$

for the relevant a/L, one may also choose fixed a/L. Still it is clear from our discussion that fixed L/r_0 is to be preferred whenever possible. In cases where it is advantageous to work at finite quark masses, also the combination $r_0\, m_{\mathrm{R},i}$ should be kept constant. Note that these *constant physics* conditions do not have to be satisfied very precisely since we are talking about a correction to an O(a) term.

Such conditions have first been imposed in Ref. 51 – exactly because of Fig. 7. We show the result for $b_{\mathrm{A}} - b_{\mathrm{P}}$, with the constant physics condition implemented, in Fig. 8. In this extreme case the ambiguity is of the same order as the improvement coefficient itself. In such a situation it is rather tempting to just put the coefficient to zero. However, it should be

[1]In fact this was done originally for c_{sw}, c_{A} [52]. Note, however, that the small tree-level a-effect was subtracted from the non-perturbative ones to insure the improvement coefficients go to their tree-level values exactly and further the conditions were chosen such that also $|b^{(1)}(a/L) - b^{(1)}(0)|$ is very small.

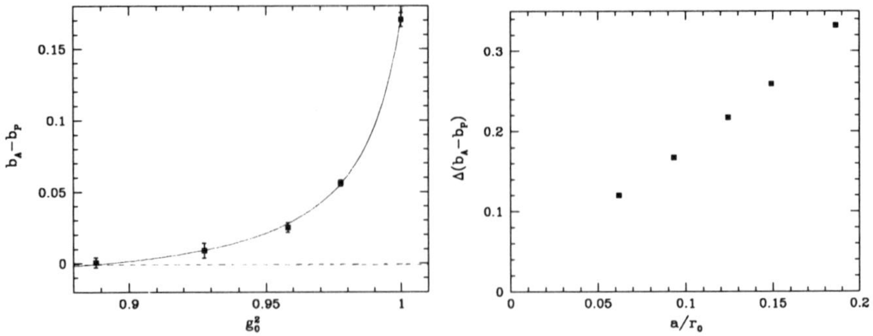

Fig. 8. Non-perturbative $b_A - b_P$ for $N_f = 0$ and plaquette gauge action (left) as well as its ambiguity $\Delta(b_A - b_P)$. The latter is the difference of $b_A - b_P$ for one improvement condition and $b_A - b_P$ for another one.

obvious that it is then not guaranteed that linear a-effects are absent after improvement. An extrapolation to the continuum using an $O(a^2)$ model function for the cutoff effects might then give significantly wrong results. We repeat this relevant fact in different words. While an unfortunately chosen improvement condition, but implemented at constant physics, may even enlarge the cutoff effects for intermediate a, it guarantees[m] that linear a-effects are absent. Putting the improvement coefficient to zero or some perturbative approximation does not guarantee the latter and the linear effects should at least be estimated in some way.

Before we review what is known in $O(a)$-improvement, we return once more to the improvement coefficients c_{sw}, c_A. With plaquette gauge action, $a/L = 1/8$ and the chosen improvement condition, Eq. (86) is very well satisfied for $i = 0, 1$ [43]. It appears justified to subtract these effects perturbatively as it was done in Refs. 52, 53. For c_A there is evidence for a significant $O(a)$ ambiguity [50,54,55]. As a consequence, the computation with $N_f = 2$ has been carried out with an improvement condition at constant physics [56].

I.2.4.2. *Summary of results*

A summary of the available results for $O(a)$ improvement coefficients is listed in Table 1. Many investigations have been carried out accumulating a quite advanced knowledge. It is impossible for me to review it all, rather

---

[m]Here we use a strong wording. Remember of course that everything is based on the SET, which has not been proved. Remember also Sect. I.2.2.2.

Table 1. Literature on O(a) improvement coefficients. NP stands for non-perturbative, CP for constant physics condition, Wilson for the Wilson plaquette action, "1×1 & 1×2" for the gauge action including a 1×2 loop and Iwasaki for the action where the coefficient of that term is set to the value proposed by Y. Iwasaki.

improvement coefficient or improved fields	order in PT	N_f	gauge action	Reference
c_{sw}, c_A	1-loop	–	Wilson	43, 44
b_g	1-loop	–	Wilson	19, 57
b_m, b_A, b_P, b_V, c_V V^a_μ, A^a_μ	1-loop	–	Wilson	46, 58
c_{sw}	1-loop	–	various	59
c_{sw}, c_A $b_m, b_A - b_P, b_V, c_A, c_V$	NP	0	Wilson	52
V^a_μ, A^a_μ	NP	0	Wilson	50
c_A	NP	0	Wilson	54
$b_A - b_P, b_V - b_S, b_m$	NP	0	Wilson	49
$b_A - b_P, b_m$	NP at CP	0	Wilson	51
$b_A - b_P, b_m$	NP at CP	0	Wilson	60
c_{sw}	NP	2	Wilson	53
c_{sw}	NP	3	Wilson	61
c_A	NP at CP	2	Wilson	56
$b_m, b_A, b_P, b_V, c_V, b_T$ $V^a_\mu, A^a_\mu, T^a_{\mu\nu}$	1-loop	–	1×1 & 1×2	58
c_{sw}	1-loop	–	1×1 & 1×2	62
c_{sw}	NP at CP	0,2,3	Iwasaki	63
c_A	NP at CP	3	Iwasaki	CP-PACS & JLQCD

I add only a few comments.

- In the quenched approximation there is a rather complete knowledge for the Wilson gauge action, but c_A has not been determined with a constant physics condition. Conservative continuum extrapolations in the improved theory usually start only with values of a which are a bit below 0.1 fm.
The problem with "exceptional configurations"[64], which is specific to the quenched approximation, is enhanced by $c_{sw} > 0$[65]. Thus small quark masses can only be reached near the continuum limit.
- For $N_f = 2$ and Wilson gauge action, the determination of improvement coefficients is quite advanced. Interesting computations of physical observables can be done in the improved theory.
- For $N_f = 3$ and Wilson gauge action, evidence for the existence of a first order phase transition in the (β, κ)-plane has been reported[66]. The value of c_{sw} was fixed to typical values found for $N_f = 0, 2$. The

authors conclude that one has to remain with lattice spacings significantly below $a = 0.1\,\mathrm{fm}$ for this action. Recent investigations[67] have shown that simulations with Wilson fermions at small quark masses are only algorithmically stable when the physical volume is sufficiently large. [n] It is possible that the simulations that lead to the evidence for a phase transition[66] suffer from this problem. We therefore consider the existence (and even more its position) of this phase transition *not* as settled. Unfortunately it appears that only large volume simulations with good statistics can tell. Note that this statement may be controversial.

- With Iwasaki gauge action, c_{sw} is now known[63] non-perturbatively for $N_{\mathrm{f}} = 0, 2, 3$. Non-perturbative computations of c_{A} are in progress and the relevant other improvement coefficients are known to 1-loop order[58]. A small grain of salt is the following. The perturbative expansion seems less trustworthy for this action, since the bare coupling is significantly larger than for the Wilson gauge action at the same lattice spacing, while the perturbative coefficients are of a similar size.
- In Sect. I.2.4.4 we will come to the question whether more information is needed to do interesting computations in QCD.

I.2.4.3. $\mathrm{O}(a^2)$ *effects after improvement for* $N_{\mathrm{f}} = 2$

The ALPHA collaboration has performed a large series of simulations with the Schrödinger functional in small volumes of a linear extent of at most $L \approx 0.5\,\mathrm{fm}$ [69,70,23]. The purpose of these simulations was the non-perturbative determination of the running of scale dependent quantities (see Sect. II.1). Extracting the continuum limit of the step scaling functions would have been impossible without $\mathrm{O}(a)$-improvement. It is fair to say that $\mathrm{O}(a)$-improvement worked marvelously well in this situation. The interested reader may in particular study the discussion in Ref. 69. The emerging picture for QCD in the Wilson formulation is that the Symanzik expansion and improvement works very well for lattice spacings around $0.05\,\mathrm{fm}$ and below.

At larger lattice spacings, the situation is unclear at present. In particular indirect evidence has been shown that $\mathrm{O}(a^2)$ effects are rather large at $a \approx 0.1\,\mathrm{fm}$[71,22]. Large scale simulations in large volume will tell, whether

---

[n] The lattice spacing should also be sufficiently small, but this is less relevant here. See also Ref. 68.

the remaining $O(a^2)$ effects are a problem in practice. We emphasize that there is no evidence anywhere that the Symanzik theory of a-effects is invalid. The remaining issue is how small a has to be made for it to be precise when truncated to the first one or two correction terms. It is also expected that smaller lattice spacings are needed for smaller quark masses. This issue is discussed in Steve Sharpe's lectures.[72]

I.2.4.4. *Do we need more?*

The question arises, whether the knowledge described above is sufficient to do QCD computations.

First, we recall that we assumed that one is interested in on-shell quantities. In the popular MOM-scheme[73] used for the renormalization of composite operators, one does, however, consider off-shell correlation functions. While improvement in that situation has been investigated, too[74–76], we continue under the assumption that off-shell correlation functions are avoided. So far we have restricted our discussion to the case of QCD with mass-degenerate quarks. This does of course not correspond to real QCD. General quark masses are no complication for quenched quarks; we already referred to that situation in Sect. I.2.4.1. However, dynamical quarks of different masses allow for a whole series of new $O(a)$ counter-terms which are proportional to the difference of quark masses [77]. Determining them non-perturbatively would be a considerable challenge. However, let us look at their influence in practice[o]. For $N_\mathrm{f} = 3$, the biggest mass difference is of order m_s. For $a \leq 0.1\,\mathrm{fm}$, we have

$$am_{\mathrm{q,s}} \leq 1/30\,. \tag{87}$$

Thus we are talking about very small effects even at the largest typical lattice spacing. It is then sufficient to know the coefficient with, an accuracy of 0.25 or so which is surely possible by perturbation theory. The same argument applies to b_g[19]. Note, however, that for non-degenerate masses, Eq. (83) is modified already at the level a^0 and this has to be taken into account[77], or better it is avoided by relying on the PCAC masses only.

Obviously, for a dynamical charm quark, the question should be reconsidered carefully.

Another issue is the improvement of more complicated composite operators, such as 4-fermion operators. Many improvement terms may be necessary. For these difficult cases one may consider a mixed action approach,

[o]We thank M. Lüscher for emphasizing this point.

where (some part) of chiral symmetry forbids the $\mathrm{O}(a)$ terms[78]. Also twisted mass lattice QCD offers an interesting strategy for these problems. We refer to Stefan Sint's lectures.

In conclusion, a very large class of interesting problems do not seem to need more beyond what is presently known or is being determined. One may therefore take advantage of the simplicity of the formulation and work on the algorithms [79–84] to be able to work at small lattice spacings and quark masses; see also the lectures of Tony Kennedy[85].

Lecture II.
Fundamental parameters of QCD from the lattice

II.1. The problem of scale dependent renormalization

Let us investigate the extraction of short distance parameters (Sect. 1.2a) in some detail. After some brief comments on the conventional way of obtaining α from experiments we explain how one can compute it at large energy scales using lattice QCD. This paves the road for a computation of the Λ-parameter and then also for the renormalization group invariant quark masses. We will finally comment briefly on other scale-dependent renormalization.

II.1.1. *The extraction of α from experiments*

One considers experimental observables O_i depending on an overall energy scale q and possibly some additional kinematical variables denoted by y. The observables can be computed in a perturbative series which is usually written in terms of the $\overline{\rm MS}$ coupling $\alpha_{\overline{\rm MS}}$, [p]

$$O_i(q,y) = \alpha_{\overline{\rm MS}}(\mu) + A_i(y)\alpha^2_{\overline{\rm MS}}(\mu) + \ldots , \quad \mu = q \ . \tag{88}$$

For example O_i may be constructed from jet cross sections and y may be related to the details of the definition of a jet.

The renormalization group describes the energy dependence of α in a general scheme ($\alpha \equiv \bar{g}^2/(4\pi)$),

$$\mu \frac{\partial \bar{g}}{\partial \mu} = \beta(\bar{g}) \ , \tag{89}$$

where the β-function has an asymptotic expansion

$$\begin{aligned} \beta(\bar{g}) &\overset{\bar{g}\to 0}{\sim} -\bar{g}^3\left\{b_0 + \bar{g}^2 b_1 + \ldots\right\} \ , \\ b_0 &= \tfrac{1}{(4\pi)^2}\left(11 - \tfrac{2}{3}N_{\rm f}\right) \ , \quad b_1 = \tfrac{1}{(4\pi)^4}\left(102 - \tfrac{38}{3}N_{\rm f}\right) \ , \end{aligned} \tag{90}$$

with higher order coefficients b_i, $i > 1$ that depend on the scheme. Eq. (90) entails asymptotic freedom: at energies that are high enough for Eq. (90) to be applicable and for a number of quark flavors, $N_{\rm f}$, that is not too large,

[p] We can always arrange the definition of the observables such that they start with a term α. For simplicity we neglect all quark mass dependences; they are irrelevant for the main points of the present discussion.

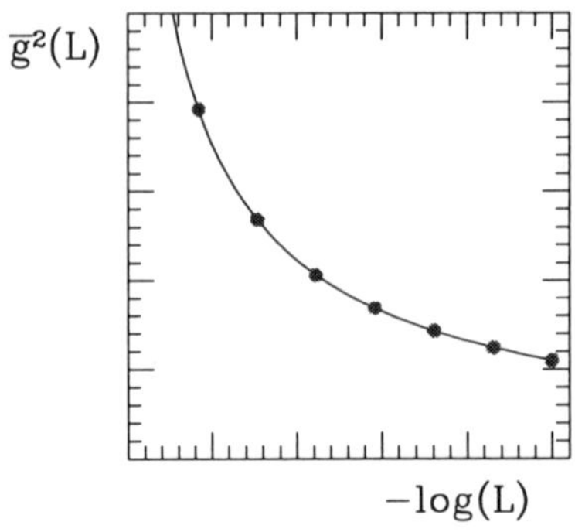

Fig. 9. Running of $\bar{g}^2$ and its reconstruction from the step scaling function (points).

α decreases with increasing energy as indicated in Fig. 9. The solution of Eq. (89) contains an integration constant, the renormalization group invariant parameter Λ. It is (exactly) given by

$$\Lambda = \mu \left(b_0 \bar{g}^2\right)^{-b1/(2b_0^2)} \mathrm{e}^{-1/(2b_0\bar{g}^2)} \exp\left\{-\int_0^{\bar{g}} \mathrm{d}x \left[\frac{1}{\beta(x)} + \frac{1}{b_0 x^3} - \frac{b_1}{b_0^2 x}\right]\right\} , \tag{91}$$

where $\bar{g} \equiv \bar{g}(\mu)$ Note that Λ is different in each scheme. If a coupling α_X is related to another one α_Y at the same energy scale via

$$\alpha_Y(\mu) = \alpha_X(\mu) + c_1^{XY} [\alpha_X(\mu)]^2 + c_2^{XY} [\alpha_X(\mu)]^3 + \dots , \tag{92}$$

their Λ-parameters are converted via

$$\Lambda_X/\Lambda_Y = \exp\{-c_1^{XY}/(8\pi b_0)\} . \tag{93}$$

From the above equations it is easy to show that Eq. (93) is exact. For large μ one reads off the asymptotics

$$\bar{g}^2 \overset{\mu\to\infty}{\sim} \frac{1}{b_0 \ln(\mu^2/\Lambda^2)} - \frac{b_1 \ln[\ln(\mu^2/\Lambda^2)]}{b_0^3[\ln(\mu^2/\Lambda^2)]^2} + \mathrm{O}\left(\frac{\{\ln[\ln(\mu^2/\Lambda^2)]\}^2}{[\ln(\mu^2/\Lambda^2)]^3}\right) . \tag{94}$$

We note that – neglecting experimental uncertainties – $\alpha_{\overline{\rm MS}}$ extracted in this way is obtained with a precision given by the terms that are left out in Eq. (88). In addition to α^3-terms, there are non-perturbative contributions which may originate from "renormalons", "condensates" (the two possibly being related), "instantons" or – most importantly – may have an origin that no physicist has yet uncovered. Empirically, one observes that values of $\alpha_{\overline{\rm MS}}$ determined at different energies and evolved to a common reference point using the renormalization group equation Eq. (89) including b_2 agree

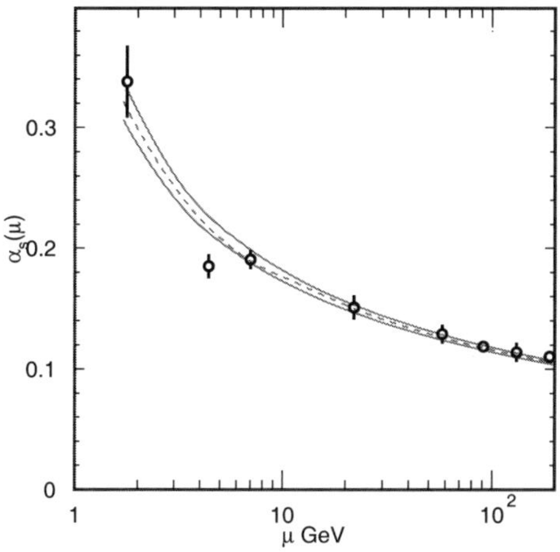

Fig. 10. The running coupling in the $\overline{\rm MS}$ scheme extracted from various scattering experiments compared to the perturbative scale dependence. Graph of the particle data group [86].

rather well with each other; the aforementioned uncertainties are apparently not very large (Fig. 10). Nevertheless, determinations of α are limited in precision because of these uncertainties and in particular if there was a significant discrepancy between α determined at different energies it would be difficult to say whether this was due to the terms left out in Eq. (88) or was due to terms missing in the Standard Model Lagrangian.

It is an obvious possibility and at the same time a challenge for lattice QCD to achieve a determination of α in one (non-perturbatively) well defined scheme and evolve this coupling to high energies. There one may use Eq. (91) with a perturbative approximation for $\beta(\bar{g})$. For a good precision b_2 should be known.

The Λ-parameter can then serve as an input for perturbative predictions of jet cross sections or the hadronic width of the Z-boson and compare to high energy experiments to test the agreement between theory and experiment. Since in the lattice regularization, QCD is naturally renormalized through the hadron spectrum, such a calculation provides the connection between low energies and high energies, verifying that one and the same theory describes both the hadron spectrum and the properties of jets.

Note. A dis-satisfying property of $\alpha_{\overline{\rm MS}}$ is that it is *only* defined in a perturbative framework; strictly speaking there is no meaning of phrases like "non-perturbative corrections" in the extraction of $\alpha_{\overline{\rm MS}}$ from experiments. The way that I have written Eq. (88) suggests immediately what should be done instead. An observable O_i itself may be taken as a definition of α – of course with due care. Such schemes called *physical schemes* are defined without ambiguities. This is what will be done below for observables that are easily handled in MC-simulations of QCD. For an additional example see Refs. 87, 88.

II.1.2. *Reaching large scales in lattice QCD*

Let us simplify the discussion and restrict ourselves to the pure Yang-Mills theory without matter fields in this section. A natural candidate for a non-perturbative definition of α is the following. Consider a quark and an anti-quark separated by a distance r and in the limit of infinite mass. They feel a force $F(r)$, the derivative of the static potential $V(r)$, which can be computed from Wilson loops (see e.g. Ref. 89). A physical coupling is defined as

$$\alpha_{\rm q\bar{q}}(\mu) \equiv \tfrac{1}{C_F} r^2 F(r) , \quad \mu = 1/r, \quad C_F = 4/3 \ . \tag{95}$$

It is related to the $\overline{\rm MS}$ coupling by Eq. (92) with a certain constant $c_1^{\overline{\rm MS}\,{\rm q\bar{q}}}$, which also determines the ratio of the Λ-parameters (Eq. (93)). Note that $\alpha_{\rm q\bar{q}}$ is a renormalized coupling defined in continuum QCD.

Problem. If we want to achieve what was proposed in the previous subsection, the following criteria must be met.

- Compute $\alpha_{\rm q\bar{q}}(\mu)$ at energy scales of $\mu \sim 10\,{\rm GeV}$ or higher in order to be able to make the connection to other schemes with controlled perturbative errors.
- Keep the energy scale μ removed from the cutoff a^{-1} to avoid large discretization effects and to be able to extrapolate to the continuum limit.
- Of course, only a finite system can be simulated by MC. To avoid finite size effects one must keep the box size L large compared to both the mass of the lightest physical state (the pion) as well as to a typical QCD scale, say the potential scale r_0 [47].

These conditions are summarized by

$$L \gg r_0, \frac{1}{m_\pi} \sim \frac{1}{0.14\mathrm{GeV}} \gg \frac{1}{\mu} \sim \frac{1}{10\mathrm{GeV}} \gg a \ , \tag{96}$$

which means that one must perform a MC-computation of an N^4 lattice with $N \equiv L/a \gg 70$. In the near future it is impossible to perform such a computation. The origin of this problem is simply that the extraction of short distance parameters requires that one has control over physical scales that are quite disparate. To cover these scales in one simulation requires a very fine resolution, which is too demanding for a MC-calculation.

Of course, one may attempt to compromise in various ways. E.g. one may perform phenomenological corrections for lattice artifacts, keep $1/\mu \sim a$ and at the same time reduce the value of μ compared to what I quoted in Eq. (96). Calculations of $\alpha_{\mathrm{q\bar{q}}}$ along these lines have been performed in the Yang-Mills theory [90–92]. It is difficult to estimate the uncertainties due to the approximations that are necessary in this approach. More recently, results in the continuum limit could be obtained up to $\mu = 4\,\mathrm{GeV}$ by simulating very large lattices[93,94], and still it is not obvious that one has reached the perturbative region. We will come back to this.

Solution. Fortunately these compromises can be avoided altogether [37]. The solution is to identify the two physical scales, above,

$$\mu = 1/L \ . \tag{97}$$

In other words, one takes a finite size effect as the physical observable. The evolution of the coupling with μ can then be computed in several steps, changing μ by factors of order 2 in each step. In this way, no large scale ratios appear and discretization errors are small for $L/a \gg 1$.

The complete strategy to compute short distance parameters is summarized in Fig. 11. One first renormalizes QCD replacing the bare parameters by hadronic observables. This defines the hadronic scheme (HS) as explained in Sect. 1.1. At a low energy scale $\mu = 1/L_{\mathrm{max}}$ this scheme can be related to the finite volume scheme denoted by SF in the graph. Within this scheme one then computes the scale evolution up to a desired energy $\mu = 2^n/L_{\mathrm{max}}$. As we will see it is no problem to choose the number of steps n large enough to be sure that one is in the perturbative regime. There perturbation theory (PT) is used to evolve further to infinite energy and compute the Λ-parameter and the renormalization group invariant quark masses. Inserted into perturbative expressions these provide predictions for jet cross sections or other high energy observables. In the graph all arrows

$$L_{\rm max} = {\rm O}(\tfrac{1}{2}{\rm fm}): \qquad {\rm HS} \longrightarrow {\rm SF}(\mu = 1/L_{\rm max})$$

$$\downarrow$$

$${\rm SF}(\mu = 2/L_{\rm max})$$

$$\downarrow$$

$$\bullet$$

$$\bullet$$

$$\bullet$$

$$\downarrow$$

$${\rm SF}(\mu = 2^n/L_{\rm max})$$

$${\rm PT:} \ \downarrow$$

$${\rm DIS,\ jet\text{-}physics,}\ R_{\rm had}\ {\rm at}\ s = M_Z^2 \quad \overset{\rm PT}{\longleftarrow} \quad \Lambda_{\rm QCD}, M$$

Fig. 11. The strategy for a non-perturbative computation of short distance parameters. SF refers to the Schrödinger functional scheme introduced in Sect. I.1.6

correspond to relations in the continuum; the whole strategy is designed such that lattice calculations for these relations can be extrapolated to the continuum limit.

For the practical success of the approach, the finite volume coupling (as well as the corresponding quark mass) must satisfy a number of criteria.

- They should have an easy perturbative expansion, such that the β-function (and τ-function, which describes the evolution of the running masses) can be computed to sufficient order.
- They should be easy to calculate in MC (small variance!).
- Discretization errors must be small to allow for safe extrapolations to the continuum limit.

Careful consideration of the above points led to the introduction of renormalized coupling and quark mass through the Schrödinger functional (SF) of QCD [12,95,17,18,32], introduced in Sect. I.1. In the Yang-Mills theory, an alternative finite volume coupling was studied in detail in Refs. 96, 97.

The criteria Eq. (96) apply quite generally to any scale dependent renormalization, e.g. the one of 4-fermion operators of the effective weak Hamiltonian at scales $\mu \ll M_{\rm W}$. Indeed, details have been worked out for several cases[98,9,10,99,100,11,101].

A frequently applied alternative is to search for a "window" where μ is high enough to apply PT but not too close to a^{-1}, the cutoff [102]. An essential advantage of the details of the approach of Ref. 102 as applied to the renormalization of composite quark operators is its simplicity: formulating the renormalization conditions in a *MOM*-scheme [q], one may use results from perturbation theory in infinite volume in the perturbative part of the matching. Since, however, it is difficult to reach high energies μ in this approach, we will not discuss it further and refer to Refs. 102, 103, 104, 105 for an idea of the present status and further references, instead.

II.2. The computation of $\alpha(\mu)$ and Λ

We are now in the position to explain the details of Fig. 11 [37,98]. The problem has been solved "completely" in the SU(N) Yang-Mills theories for $N = 2, 3$ [95,97,17,98,106,93]. In QCD with only two dynamical quarks the strategy has been carried out well [69,70], except for the last line in the graph, which needs more work. In the present context, the Yang-Mills theory is of course equivalent to the quenched approximation of QCD or the limit of zero flavors. We will therefore also refer to results in quenched QCD.

Our central observable is the step scaling function that describes the scale-evolution of the coupling, i.e. moving vertically in Fig. 11. The analogous function for the running quark mass will be discussed in the following section.

II.2.1. *The step scaling function*

We start from a given value of the coupling, $u = \bar{g}^2(L)$. When we change the length scale by a factor s, the coupling has a value $\bar{g}^2(sL) = u'$.[r] The step scaling function, σ is then defined as

$$\sigma(s, u) = u' \ . \tag{98}$$

The interpretation is obvious. $\sigma(s, u)$ is a discrete β-function. Its knowledge allows for the recursive construction of the running coupling at discrete values of the length scale,

$$u_k = \bar{g}^2(s^{-k}L) \ , \tag{99}$$

once a starting value $u_0 = \bar{g}^2(L)$ is specified (cf. the points in Fig. 9). The step scaling function, σ, which is readily expressed as an integral of the

[q] We do not use the prefix *RI* (regularization independent), since also the SF-scheme is regularization independent.
[r] Figure 11 is for the most frequently used case $s = 2$.

β-function, has a perturbative expansion

$$\sigma(s,u) = u + 2b_0 \ln(s)u^2 + \ldots \quad . \tag{100}$$

On a lattice with finite spacing, a, the step scaling function will have an additional dependence on the resolution a/L. We define

$$\Sigma(s,u,a/L) = u' \ , \tag{101}$$

with

$$\bar{g}^2(L) = u, \quad \bar{g}^2(sL) = u' \, , \quad g_0 \text{ fixed}, \ L/a \text{ fixed} \ . \tag{102}$$

The continuum limit $\sigma(s,u) = \Sigma(s,u,0)$ is then reached by performing calculations for several different resolutions and extrapolation $a/L \to 0$. The computation of $\sigma(2,u)$ is illustrated in Fig. 12. In detail, one performs the following steps:

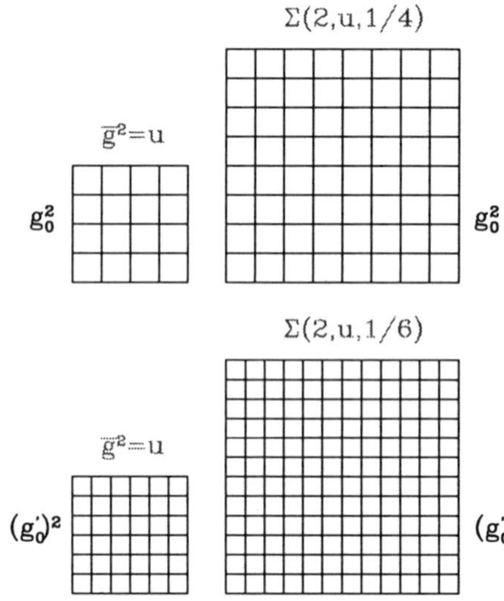

Fig. 12. The computation of a lattice step scaling function.

1. Choose a lattice with L/a points in each direction.
2. Tune the bare coupling g_0 such that the renormalized coupling $\bar{g}^2(L)$ has the value u and tune the bare quark mass, m_0, such that the PCAC-mass, defined at fixed physical kinematics [69], vanishes.[s]

[s]The details of the definition of the kinematics affects only the cutoff effects, see Sect. I.2.2.

3. At the same value of g_0, simulate a lattice with twice the linear size; compute $u' = \bar{g}^2(2L)$. This determines the lattice step scaling function $\Sigma(2, u, a/L)$.
4. Repeat steps 1.–3. with different resolutions L/a and extrapolate $a/L \to 0$.

Note that step 2. takes care of the renormalization and 3. determines the evolution of the *renormalized* coupling.

The presently most advanced numerical results are displayed in Fig. 13. The coupling used is exactly the one defined in Sect. I.1.6 and the calculation is done in the theory with $N_\mathrm{f} = 2$ flavors of O(a)-improved fermions. One observes that the dependence on the resolution is very weak, in fact it is not observable within the precision of the data in Fig. 13. We now investigate in more detail how the continuum limit of Σ is reached. As a first step, we turn to perturbation theory.

II.2.2. *Lattice spacing effects in perturbation theory*

Symanzik has investigated the cutoff dependence of field theories in perturbation theory [107]. Generalizing his discussion to the present case, one concludes that the lattice spacing effects have the expansion

$$\frac{\Sigma(2,u,a/L) - \sigma(2,u)}{\sigma(2,u)} = \delta_1(a/L)\,u + \delta_2(a/L)\,u^2 + \ldots \tag{103}$$

$$\delta_n(a/L) \overset{a/L\to 0}{\sim} \sum_{k=0}^{n} e_{k,n}[\ln(\tfrac{a}{L})]^k (\tfrac{a}{L}) + d_{k,n}[\ln(\tfrac{a}{L})]^k (\tfrac{a}{L})^2 + \ldots\ .$$

We expect that the continuum limit is reached with corrections O(a/L) also beyond perturbation theory. In this context O(a/L) summarizes terms that contain at least one power of a/L and may be modified by logarithmic corrections as it is the case in Eq. (103). To motivate this expectation recall Sect. 1.3, where we explained that lattice artifacts correspond to irrelevant operators which carry explicit factors of the lattice spacing. Of course, an additional a-dependence comes from their anomalous dimension, but in an asymptotically free theory such as QCD, this just is a logarithmic (in a) modification.

As mentioned in Sect. I.1.8, the lattice artifacts may be reduced to O($(a/L)^2$) by canceling the leading irrelevant operators. In the case at hand, this is achieved by a proper choice of $c_\mathrm{t}(g_0)$. It is interesting to note, that by using the perturbative approximation

$$c_\mathrm{t}(g_0) = 1 + c_\mathrm{t}^{(1)} g_0^2 \tag{104}$$

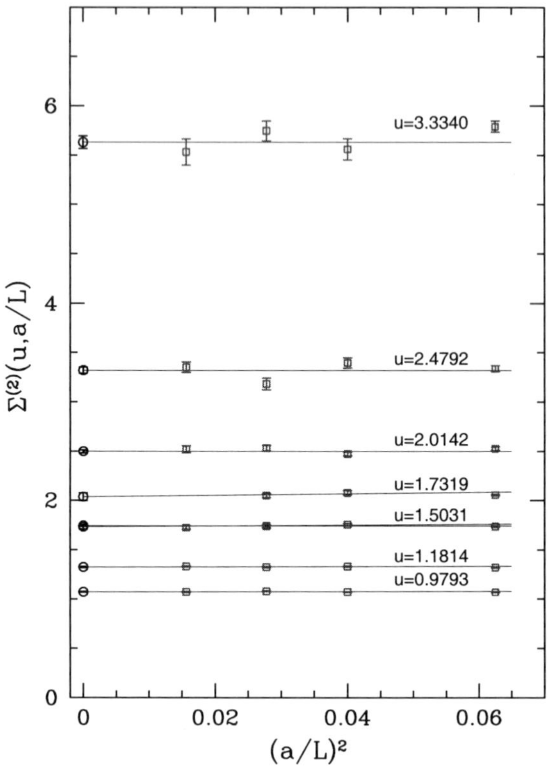

Fig. 13. The lattice step scaling function after 2-loop observable improvement for $N_\mathrm{f} = 2$.

one does not only eliminate $e_{k,n}$ for $n = 0, 1$ but also the logarithmic terms generated at higher orders $(n > 1)$are reduced,

$$e_{n,n} = 0, \quad e_{n-1,n} = 0 \ . \tag{105}$$

For tree-level improvement, $c_\mathrm{t}(g_0) = 1$, the corresponding statement is $e_{n,n} = 0$. Heuristically, the latter is easy to understand. Tree-level improvement means that the propagators and vertexes agree with the continuum ones up to corrections of order $\mathrm{O}(a^2)$. Terms proportional to a can then arise only through a linear divergence of the loop integrals. Once this happens, one cannot have the maximum number of logarithmic divergences any more; consequently $e_{n,n}$ vanishes.

To demonstrate further that the abelian field introduced in the previous section induces small lattice artifacts, we show $\delta_1(a/L)$ for the one loop

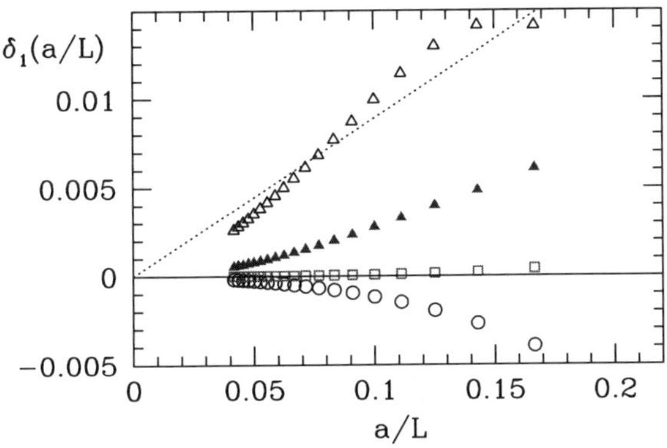

Fig. 14. Lattice artifacts at 1-loop order. Points show $\delta_1(a/L)$ for the SU(3) Yang-Mills theory with 1-loop improvement. Wilson (circles), Symanzik (squares), Iwasaki (filled triangles) and DBW2 (empty triangles) are shown. The dotted line corresponds to the linear piece in a, when only tree-level improvement is used, instead (only for Wilson).

improved case in Fig. 14 (circles). The term that is canceled by the proper choice $c_t^{(1)} = -0.089$ is shown as a dashed line. The left over $O((a/L)^2)$-terms are below the 1% level for couplings $u \leq 2$ and lattice sizes $L/a \geq 6$. For not too large coefficient of the 1×2 loop in the action, they are close to quadratic in a/L in the range of interest. Also the fermion contribution and the 2-loop cutoff effects are known [57,16]. When the $O(a)$-improved theory is used (and the coupling has reasonable values), they turn out to be smaller than the 1-loop terms discussed here.

Of course the cutoff effects also depend on the gauge action. In one class of such actions, one adds a 1×2 rectangular loop. Its coefficient has originally been determined to have tree-level Symanzik $O(a^2)$ improvement in the pure gauge theory [108]. Later other choices have been proposed by Y. Iwasaki [109] and the QCD-TARO collaboration [110], based on renormalization group considerations. We refer to these as "Symanzik,Iwasaki,DBW2". The choice of the action close to the Schrödinger functional boundaries does of course involve an additional freedom discussed in Refs. 111, 112. For the choice favored in Ref. 112, we included the corresponding $\delta_1(a/L)$ in Fig. 14. For the tree-level Symanzik improved theory the cutoff effects are remarkably small. One might think that this is so by construction, but note that here we are discussing 1-loop effects and in addition also boundary operators contribute to the SET expansion of the cutoff effects.

Altogether, we now understand better why the a/L-dependence is so

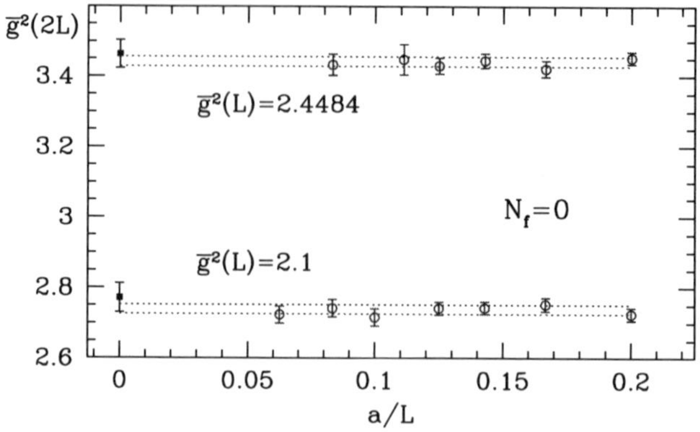

Fig. 15. The lattice step scaling function after 2-loop observable improvement for $N_{\rm f} = 0$. Data from Refs. 98, 113. The point at $a = 0$ is the continuum limit of Ref. 98 taken with a subset of the data.

small in Fig. 13. Still, because the continuum limit is so important, and we have seen that unexpected difficulties may be present (Sect. I.2.2.2), we continue its discussion in the pure gauge theory. There, numerical simulations with better resolutions have been carried out.

II.2.3. *The continuum limit – universality*

In fact, it is not only of interest to investigate how exactly the continuum limit is approached. Its very existence and its universality, i.e. the independence of the renormalized continuum observable from the lattice action, is worth testing by precise MC computation.

Above we have discussed how one can add boundary terms to the *action* to systematically reduce the cutoff effects linear in a. Given the knowledge of the observables in perturbation theory at finite lattice spacing, we can go further and remove the cutoff-effects for arbitrary a/L, i.e. for instance including $\mathrm{O}((a/L)^2)$, *but up to a fixed order in* u. This is achieved by the *improvement of the observable* [97,69,70]

$$\Sigma^{(k)}(2,u,a/L) = \frac{\Sigma(2,u,a/L)}{1+\sum_{i=1}^{k}\delta_i(a/L)\,u^i}\,. \tag{106}$$

The approach has been carried out with $k = 2$ for the $N_{\rm f} = 2$ data (Fig. 13) and is tested thoroughly for $N_{\rm f} = 0$ in Fig. 15.

Recently, a nice universality test has been carried out by the CP-PACS collaboration [114]. Obviously the results in Fig. 16 are well compatible with

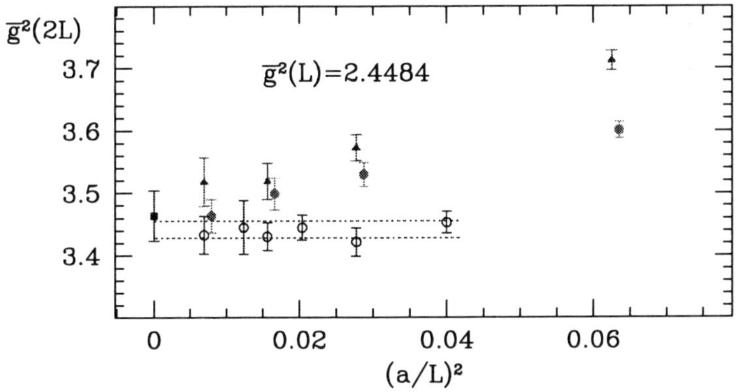

Fig. 16. Universality test in the SU(3) Yang Mills theory. The data from top (triangles) to bottom (open circles) are for the Iwasaki, Symanzik and Wilson gauge action. Both the boundary improvement of the action and the improvement of the observables are carried out. At present this is possible at the 2-loop level for the Wilson gauge action (same as Fig. 15) and at the 1-loop level otherwise. Data are from Refs. 114, 69.

an $a \to 0$ limit which is independent of the action. Similar results exist for the SU(2) theory [97]. These results leave little doubt that the continuum limit of the Schrödinger functional exists and is independent of the lattice action. In turn this also supports the statement that the Schrödinger functional is renormalized after the renormalization of the coupling constant.

We return to the extraction of σ in $N_\mathrm{f} = 2$ QCD. Even though a-effects are not statistically significant, one has to be careful how one extracts the continuum limit. The worry does not so much concern the central values, but the correct estimate of their uncertainties. For example just averaging data at all values of a/L produces an unrealistically small statistical error, because one has then assumed that a-effects are entirely absent, although the data tell only that they are smaller than statistical uncertainties. One possible strategy for the continuum extrapolation is thus a fit to a constant that uses the lattices with $L/a = 6, 8$ only. As a check of this procedure, different variants of a combined continuum extrapolation of all the data sets, but excluding $L/a = 4$ were carried out. For example the ansatz

$$\Sigma^{(2)}(2, u, a/L) = \sigma(2, u) + \rho\, u^4\, (a/L)^2$$

with a constant ρ was fitted to the data. The final conclusion was that the simple fit to a constant (for $L/a = 6, 8$) yields realistic error estimates for the existing data set [70]. A further check of this procedure is Fig. 15 where the dotted lines represent the error band obtained in this way and the four points at smaller lattice spacing are in perfect agreement with this band.

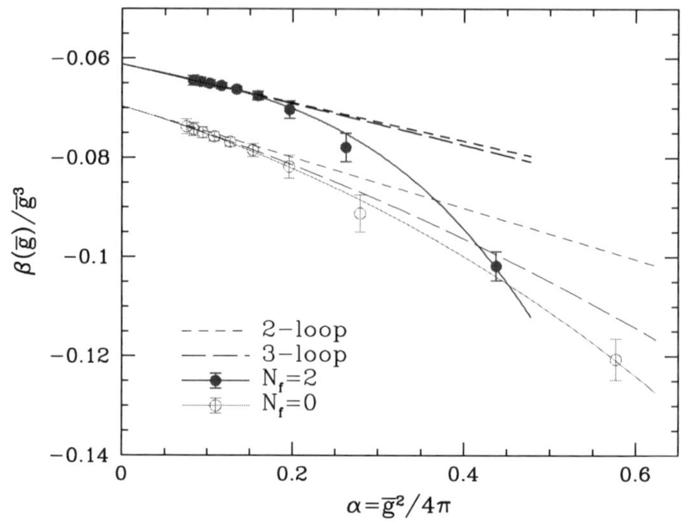

Fig. 17. The QCD β–function in the SF scheme.

After this long - but important - discussion of cutoff effects, we are convinced that we have continuum results for the step scaling function with realistic uncertainties. They are ready to be used to construct the running coupling and the Λ-parameter.

II.2.4. *The running of the coupling*

The numerical values of $\sigma(u)$ are next represented by a smooth interpolating function (a polynomial in u). With this function the running coupling $\bar{g}^2(2^{-i}L_{\rm max}) \equiv u_i$ can be constructed from the recursion

$$u_{\rm max} \equiv u_0 = \bar{g}^2(L_{\rm max})\,, \quad \sigma(u_{i+1}) = u_i\,,\; i = 0 \ldots n\,; \tag{107}$$

the result is shown in Fig. 18 for the (arbitrary) choice $u_{\rm max} = 5.5$. One can also set up a recursion for the β-function itself [70],

$$\beta(\sqrt{u_i}) = \sqrt{u_{i+1}/u_i}\,\sigma'(u_{i+1})\,\beta(\sqrt{u_{i+1}})\,. \tag{108}$$

Together with a start value for the β-function taken from perturbation theory (3-loop) at the weakest coupling ($\alpha \approx 0.08$) this yields the numerical results Fig. 17. Their agreement with perturbation theory is excellent at weak couplings $\alpha < 0.2$, while at the largest couplings significant deviations from perturbation theory are present for $N_f = 2$. Indeed the difference between non-perturbative points and 3-loop can't be described by an effective 4-loop term *with a reasonable coefficient*. At the same time the perturbative series just by itself does not show signs of its failure at, say, $\alpha \approx 0.3$: instead successive orders yield smaller and smaller corrections.

We return to the running couplings shown in Fig. 18. In the zero flavor case, also the region of μ of around 250 MeV was investigated with a specifically adapted strategy [113]. In this region, the SF coupling shows the rapid growth expected from a strong coupling expansion.

Initially, the graphs Fig. 18 are obtained for μ in units of $\mu_{\rm min} = 1/L_{\rm max}$. One chooses $u_{\rm max}$ relatively large, but within the range covered by the non-perturbative computation of $\sigma(u)$. The artificial scale $L_{\rm max}$ has been replaced by the Λ parameter by use of Eq. (91). We proceed to explain this step.

II.2.5. *The Λ parameter*

We may evaluate Eq. (91) for the last few data points in Fig. 18 using the 3-loop approximation to the β-function in the SF-scheme. The resulting Λ-values are essentially independent of the starting point, since the data follow the perturbative running very accurately at large μ. This excludes a sizable contribution to the β-function beyond 3-loops in this region. Indeed, a typical estimate of a 4-loop term in the β-function would change the value of Λ by a tiny amount. The corresponding uncertainty can be neglected compared to the statistical errors.

Changing Λ from the SF scheme to the $\overline{\rm MS}$ scheme we then have

$$N_f = 0\,,\;\; u_{\rm max} = 3.48\;:\;\ln(\Lambda_{\overline{\rm MS}} L_{\rm max}) = -0.84(8)\,, \tag{109}$$

$$N_f = 2\,,\;\; u_{\rm max} = 4.61\;:\;\ln(\Lambda_{\overline{\rm MS}} L_{\rm max}) = -0.40(7)\,. \tag{110}$$

It remains to connect the artificially defined length scale $L_{\rm max}$ to an experimentally measurable low energy scale of QCD such as the proton mass or the kaon decay constant, F_K.

So far it has been convenient to first evaluate $L_{\rm max}$ in units of the low energy scale r_0. This reference scale is precisely defined through the QCD static quark potential [47] but related to experiments only through potential models: $r_0 \approx 0.5$ fm. For $N_f = 0$ a detailed investigation resulted in the continuum limit [106,93]

$$N_f = 0\,,\;\; u_{\rm max} = 3.48\;:\;L_{\rm max}/r_0 = 0.738(16) \tag{111}$$

and thus

$$\Lambda^{(0)}_{\overline{\rm MS}} r_0 = 0.60(5)\,. \tag{112}$$

In the $N_{\rm f} = 2$ theory, the situation is illustrated in Table 2 which relies on results for r_0/a from Refs. 115, 116. On the one hand, all the numbers in italic are consistent, indicating that lattice spacing effects are small, on the other hand the first column shows that a is not yet varied very much. At the moment

$$\Lambda^{(2)}_{\overline{\rm MS}} r_0 = 0.62(4)(4)$$

is quoted, where the second error generously covers the range of numbers in italic in Table 2 and the first one comes from Eq. (110).

II.2.6. *Discussion*

The scale dependence of the SF coupling is close to perturbative below $\alpha_{\rm SF} = 0.2$ and becomes non-perturbative above $\alpha_{\rm SF} = 0.25$. In fact a strong coupling expansion suggests that *this* coupling grows exponentially for large L. In the $N_{\rm f} = 0$ theory it was possible to verify this behavior for L close to 1 fm [113] (Fig. 18). Apart from the determination of Λ, an achievement

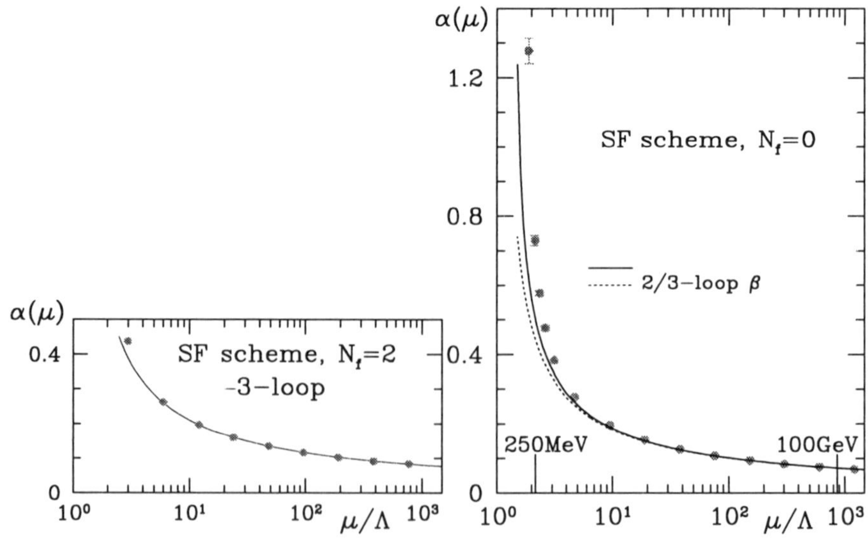

Fig. 18. Running coupling for $N_{\rm f} = 2$ (left) and $N_{\rm f} = 0$ (right).

Table 2. Λ–parameter in units of r_0 for $N_f = 2$ for different resolutions in the low energy part of the calculation. Two different values of u_{max} are considered, but $\Lambda_{\overline{MS}} r_0$ should be independent of these when cutoff effects are small.

		$u_{max} = 3.65$		$u_{max} = 4.61$	
$\beta = 6/g_0^2$	r_0/a	L_{max}/a	$\Lambda_{\overline{MS}} r_0$	L_{max}/a	$\Lambda_{\overline{MS}} r_0$
5.00	5.45(5)(20)	4.00(6)	0.655(27)	6.00(8)	*0.610(25)*
5.29	6.01(4)(22)	4.67(6)	*0.619(25)*	6.57(6)	*0.614(24)*
5.40	7.01(5)(15)	5.43(9)	*0.621(17)*	7.73(10)	*0.609(16)*

of this investigation is the confirmation that the transition between this non-perturbative region and the perturbative one is very smooth.

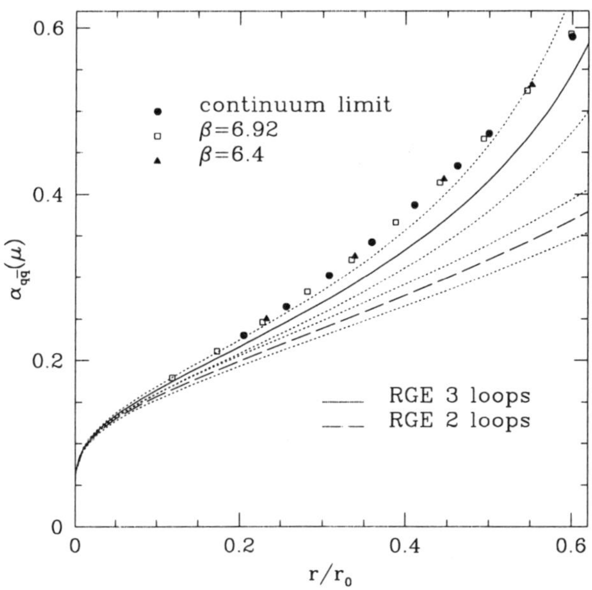

Fig. 19. Running coupling in the q$\bar{q}$-scheme (Eq. (95)). The perturbative prediction is given by the relation Eq. (91) between the running coupling and $\Lambda^{(0)}_{q\bar{q}}$. The dotted lines show the uncertainty of Λ Eq. (112).

With some effort this transition region was also bridged for $\alpha_{q\bar{q}}$ in the SU(3) pure gauge theory. As shown in Fig. 19, the non-perturbative continuum results agree reasonably well with the perturbative prediction at the largest μ (small r). The prediction is parameter free, since the Λ-parameter,

Eq. (112), can be changed to the q$\bar{q}$-scheme, Eq. (95), where also the 3-loop β-function is known [117,118,94]. The figure also illustrates that a reliable determination of Λ is not possible from $\alpha_{q\bar{q}}$ because for the accessible μ the 3-loop contribution to the β-function is still very significant.

Table 3. Results for $\Lambda_{\overline{MS}} r_0$ for different number of flavors.

source	$\Lambda_{\overline{MS}}\, r_0$			
	$N_f = 0$	$N_f = 2$	$N_f = 4$	$N_f = 5$
ALPHA[70,17]	0.60(5)	0.62(6)		
world average[119]			0.74(10)	0.54(8)
DISNNLO[120]			0.57(8)	

In Table 3 we compare our results for $\Lambda_{\overline{MS}} r_0$ to selected phenomenological ones, where we set $r_0 = 0.5\,\text{fm}$. There appears to be an irregular N_f-dependence, but we note that

1. the errors are not very small yet,
2. the 4-flavor Λ is obtained from the 5-flavor one by perturbation theory [121]. For this to be accurate, perturbation theory has to be accurate for $\mu \ll m_{\text{beauty}}$, which is not completely obvious.

II.2.7. *Improvements are necessary*

The most urgently needed improvement of the present results is to eliminate the model dependence which is intrinsic in the use of r_0. So one should replace $L_{\max}/r_0$ by $F_K \times L_{\max}$ (computed at small enough light quark masses and small a) and an effort is presently being made. Also the strange quark sea has to be included ("2+1") and one should estimate the effects of the charm quark. Such 2+1 simulations are for example being carried out by JLQCD and CP-PACS, who have also studied the computation of the SF coupling with the gauge actions they are using [114].

These improvements will come and I am convinced that lattice results will yield the best controlled and most precise results for Λ in the long run. The reason is simple: I essentially described *all* the sources of systematic errors. The kind of assumptions one has to make are minimal.

Let me also mention that there is a large number of other results for α from lattice gauge theory, where $\alpha(\mu \sim a^{-1})$ is extracted from quantities related to the cutoff. Some of them cite a very small error [122]. As discussed in Ref. 31, it is not easy to estimate the systematic errors in these computations, mainly because one cannot separately discuss higher order perturbative corrections and discretization errors. We thus think it is very desirable to carry out the program of Fig. 11 with good precision and the relevant number of flavors.

II.3. Renormalization group invariant quark masses

The computation of running quark masses and the renormalization group invariant (RGI) quark mass proceeds in analogy to the computation of $\alpha(\mu)$. Since we are using a mass-independent renormalization scheme (cf. Sect. I.1.7.2), the renormalization (and thus the scale dependence) is independent of the flavor of the quark. When we consider "the" running mass below, any one flavor can be envisaged; the scale dependence is the same for all of them.

The renormalization group equation for the coupling Eq. (89) is now accompanied by one describing the scale dependence of the mass,

$$\mu \frac{\partial \overline{m}}{\partial \mu} = \tau(\bar{g}) \ , \tag{113}$$

where τ has an asymptotic expansion

$$\tau(\bar{g}) \overset{\bar{g}\to 0}{\sim} -\bar{g}^2 \left\{ d_0 + \bar{g}^2 d_1 + \ldots \right\} , \qquad d_0 = 8/(4\pi)^2 \ , \tag{114}$$

with higher order coefficients d_i, $i > 0$ which depend on the scheme.

Similarly to the Λ-parameter, we may define a renormalization group invariant quark mass, M, by the asymptotic behavior of $\overline{m}$,

$$M = \lim_{\mu\to\infty} \overline{m}(\mu)[2b_0\bar{g}(\mu)^2]^{-d_0/2b_0} \ . \tag{115}$$

It is an easy exercise to show that M does not depend on the renormalization scheme. It can be computed in the SF-scheme and used afterward to obtain the running mass in any other scheme by inserting the proper β- and τ-functions in the renormalization group equations.

To compute the scale evolution of the mass non-perturbatively, we introduce a new step scaling function,

$$\Sigma_{\mathrm{P}}(u, a/L) = \left. \frac{Z_{\mathrm{P}}(2L)}{Z_{\mathrm{P}}(L)} \right|_{\bar{g}^2(L)=u} \ , \tag{116}$$

with Z_{P} of Eq. (46). Results for Σ_{P} at finite lattice spacing and the extracted continuum limit are displayed in Fig. 20. Details may be found in Refs. 98, 23.

Applying σ_{P} and σ recursively one then obtains the series,

$$\overline{m}(2^{-k}L_{\max})/\overline{m}(2L_{\max}) \, , \ k = 0, 1, \ldots \ , \tag{117}$$

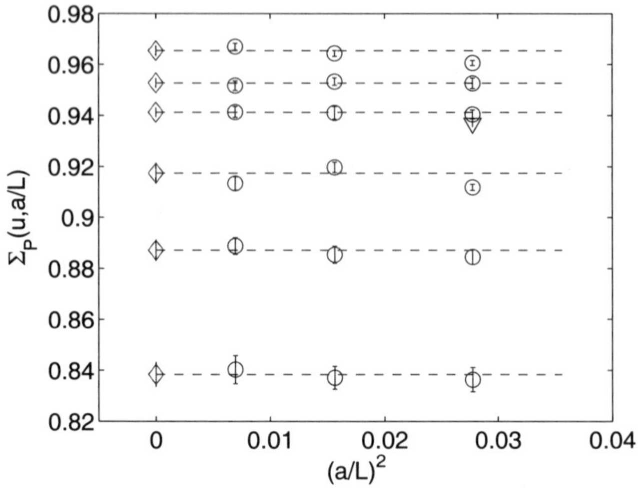

Fig. 20. Lattice spacing dependence of the step scaling function for the quark mass for $N_f = 2$. The coupling u ranges from $u = 0.979$ to $u = 3.33$.

up to a largest value of k, which corresponds to the smallest $\bar{g}$ that was considered in Fig. 20. From there on, the perturbative 2-loop approximation to the τ-function and 3-loop approximation to the β-function (in the SF-scheme) may be used to integrate the renormalization group equations to infinite energy, or equivalently to $\bar{g} = 0$. The result is the renormalization group invariant mass,

$$M = \overline{m}\,(2b_0\bar{g}^2)^{-d_0/2b_0} \exp\left\{-\int_0^{\bar{g}} \mathrm{d}g \left[\frac{\tau(g)}{\beta(g)} - \frac{d_0}{b_0 g}\right]\right\} . \tag{118}$$

In this way, one is finally able to express the running mass $\overline{m}$ in units of the renormalization group invariant mass, M, as shown in Fig. 21. Remember that M has the same value in all renormalization schemes, in contrast to the running mass $\overline{m}$.

The *perturbative* evolution of the quark masses follows very accurately the non-perturbative results down to rather low energy scales. Of course, this result may not be generalized to running masses in other schemes.

The point at lowest scale μ in Fig. 21 corresponds to

$$M/\overline{m} = 1.296(16) \quad \text{at} \quad L = 2L_{\max} . \tag{119}$$

Remembering the very definition of the renormalized mass Eq. (43), one can use this result to relate the renormalization group invariant mass and

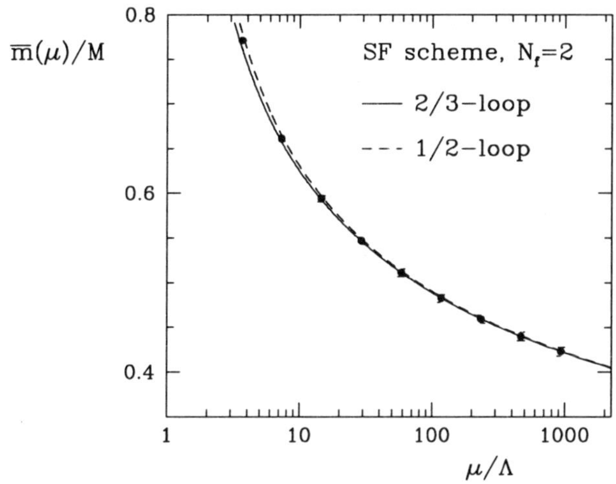

Fig. 21. The running quark mass as a function of $\mu \equiv 1/L$ for $N_\mathrm{f} = 2$.

the bare current quark mass m on the lattice through

$$M = m \times 1.296(16) \times Z_\mathrm{A}(g_0)/Z_\mathrm{P}(g_0, 2L_\mathrm{max}/a) = Z_\mathrm{M}(g_0)\, m \ . \quad (120)$$

In this last step, one then inserts the bare current quark mass, e.g. of the strange quark, and extrapolates the result to the continuum limit. The bare current quark masses themselves are the ones for which the appropriate pseudo scalar masses are fixed to their experimental values. The presently available results from this strategy are listed in Table 4. Note that the computation of the b-quark mass required a detour through Heavy Quark Effective Theory, the subject of the following lecture.

For the charm and the beauty quark masses determinations with $N_\mathrm{f} > 0$ and NP renormalization are still missing. However, in our opinion it is even quite early concerning the determinations of the light quark masses. Although some N_f-dependence of the strange quark mass has been reported

Table 4. Quark masses determined with full NP renormalization and continuum limit. We use $r_0 = 0.5\,\mathrm{fm}$.

i	N_f	input	M_i/GeV	Ref.
strange	0	m_K, r_0	0.137(05)	33
strange	2	m_K, r_0	0.137(27)	23
charm	0	m_D, r_0	1.654(45)	123
beauty	0	$m_{\mathrm{B_s}}, m_{\mathrm{B_s^*}}, r_0$	6.771(99)	124

in the literature, one can presently not exclude that this is due to perturbative uncertainties or discretization errors. Note that the quark masses computed in the quenched approximation were in a similar stage in 1996 but very soon afterward the uncertainties shrunk by an order of magnitude due to NP renormalization and continuum extrapolations. This remains to be achieved for the real theory with $N_{\rm f} > 0$!

II.4. Renormalization scale dependence of other composite operators

Due to our definition of the renormalized quark mass, its scale dependence is given by the one of the composite operator $P^a(x)$. Other composite operators can be considered and indeed the strategy described here has been applied to 4-fermion operators in the weak effective Hamiltonian [100,99,11,101], the HQET axial current [9,10] and in the operator which yields $\langle x \rangle$ of the non-singlet structure functions [125–127]. It is worth pointing out that the non-perturbative scale dependence disagrees more (and significantly) from the perturbative one in some of these cases. They provide examples which emphasize that a fully non-perturbative renormalization is necessary to control the associated uncertainties. We show two examples in Fig. 22. For details we have to refer to the cited papers.

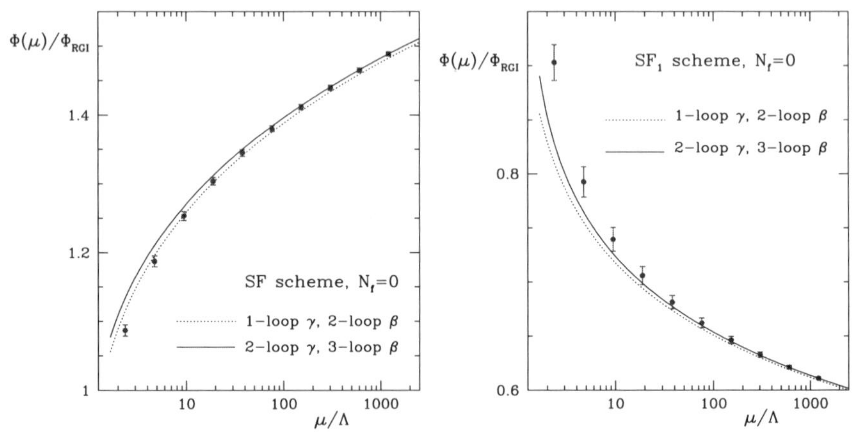

Fig. 22. The running of the static-light axial current $\mu \equiv 1/L$ for $N_{\rm f} = 0$ (left)[10] and a left-left 4-fermion operator (right)[100].

However, we want to emphasize again one general feature, which was noted already for the $N_{\rm f} = 2$ running coupling. Significant differences between the perturbative and the non-perturbative result are present in cases, where the subsequent order of perturbation theory are very close to each other. The difference of 2/3-loop order running to the non-perturbative results can't be parameterized by the next order perturbative term with a reasonable coefficient.

Lecture III.

Non-perturbative Heavy Quark Effective Theory

III.1. Introduction

Similarly to the scale dependent renormalization covered in the previous lecture, the inclusion of heavy quarks in a lattice gauge theory simulation is a multi-scale problem. Such problems are always difficult and require the development of new techniques. By "heavy quark" we mean a quark whose mass is large compared to the intrinsic scale of QCD, $\Lambda_{\rm QCD}$. If we take $\Lambda_{\rm QCD} \approx 500\,{\rm MeV}$, the charm quark mass is about a factor 2 higher, but the bottom quark mass has

$$m_{\rm b} \approx 5\,{\rm GeV} \sim 10\Lambda_{\rm QCD}\,. \tag{121}$$

Here $m_{\rm b}$ is a quark mass defined at the scale $\mu = m_{\rm b}$, but the scheme is irrelevant at this point; we will continue to use the symbol $m_{\rm b}$ when this is the case. Since practical large volume simulations do not (yet) reach lattice spacings as small as $a = 1/(5\,{\rm GeV})$, we are faced with

$$am_{\rm b} > 1\,, \tag{122}$$

and a b-quark does not propagate properly on the lattice, at least when it is discretized with a standard relativistic QCD Lagrangian.

For the charm quark, values $am_{\rm c} \ll 1$ are achievable, but still care has to be taken. In Fig. 23 we show the RGI mass of the charm quark in the quenched approximation, computed for three definitions, which differ at finite lattice spacing. While an extrapolation to a common continuum value is convincing, it is already clear from this figure, that the four times heavier bottom quark can't be treated this way. In fact, the main point is that the cutoff-effects become entirely non-linear in a^2, when am is too large. This breakdown of the Symanzik expansion can be seen explicitly in perturbation theory [128]. In a 1-loop calculation, it has been estimated to happen around [128] $am_{\rm b} \approx 1/2$ or $aM_{\rm b} \approx 0.7$.

Various ways of coping with this problem have been proposed and investigated. Referring the reader to reviews for other approaches[129–131], we directly turn to HQET. Already in 1987 Estia Eichten suggested that to describe the non-perturbative dynamics of hadrons with a single heavy quark, it is a good approximation to consider this quark to be static, i.e. it propagates only in time (in the rest frame of the hadron) [132]. This static approximation describes the correct asymptotics of bound state properties

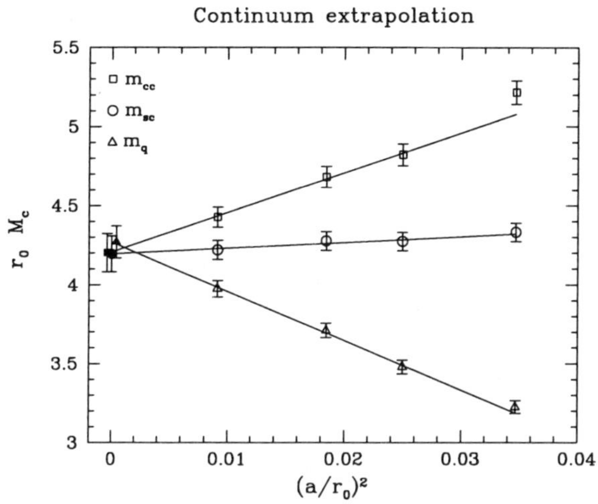

Fig. 23. The RGI mass of the charm quark in units of r_0, evaluated with the O(a)-improved formulation, as a function of the lattice spacing [123]. The physical input is the mass of the D_s meson (and m_K, r_0). The RGI-mass at finite lattice spacing is defined through the PCAC-mass, Eq. (80), of a charm-charm correlator (m_{cc}) or a charm-strange correlator (m_{sc}) and through the bare quark mass, m_q, and Eq. (83).

as $m_b \to \infty$ and corrections of order $1/m_b$ can be included systematically. The expansion in $1/m_b$ is then given by an effective field theory [133–135]. It has been extended to transition form factors, e.g. between B- and D-mesons, assuming that also the charm quark can be described by the effective field theory[136,137,134]. This is done by considering heavy quark fields with finite velocities in the limit of large mass. We will here ignore this phenomenologically very interesting possibility and restrict ourselves to HQET at zero velocity (The formulation of HQET in Euclidean space and at non-zero velocity is more subtle [138]).

In this theory considerable progress has been made recently, which we want to explain. In particular, in order to compute $1/m_b$ corrections, one has to know the leading order (including its renormalization) non-perturbatively (Sect. III.3.1.2). This necessitates a so-called non-perturbative matching of effective theory and QCD. A strategy for this [139,140] will be described in Sect. III.5 including results for $1/m_b$ corrections in a test case. The strategy solves at the same time the problem of a proper definition and computation of $1/m_b$-corrections, which is present in any regularization of the theory, and the problem of power divergences ($\sim a^{-n}$) in a theory with a hard cutoff such as the lattice regularization. After a more general introduction we will concentrate on the inclusion of

$1/m_{\rm b}$ corrections. Because of limitations of space we mention the also very relevant developments in the static approximation only rather briefly.

III.1.1. *Derivation of the classical theory*

We here go through some steps to derive the effective theory at the classical level. The main point is to see what assumptions have to be made and to present the explicit form of the Lagrangian. We follow the idea of Ref. 141, but work in Euclidean space since we are ultimately interested in the lattice theory.

In this section, we keep the dependence of the fields on the space-time coordinates implicit and also drop the label b on the quark field and its mass. We start from the Dirac-Lagrangian of a b-quark with a large mass, m, in the continuum,

$$\mathcal{L} = \overline{\psi}(D_\mu\gamma_\mu + m)\psi \tag{123}$$

$$= \psi^\dagger \mathcal{D}\psi\,, \quad \mathcal{D} = m\gamma_0 + D_0 + \gamma_0 D_k\gamma_k\,. \tag{124}$$

The light quark fields and gauge fields are not touched by our considerations. We write $\psi^\dagger$, but it is just another independent Grassmann integration variable in the path integral. Since we are considering the classical theory, we can assume that the fields are smooth. We can therefore perform an expansion in D_μ. More precisely, we have to refer to a special kinematical situation. We want to describe the dynamics of a hadron containing one heavy quark, where the hadron is at rest. For infinite mass, the heavy quark propagates only in time. Denoting the expansion parameter by ε, the dynamics thus dictates

$$D_0/m = \mathrm{O}(1)\,, \quad D_k/m = \mathrm{O}(\varepsilon)\,, \tag{125}$$

when these derivatives act on the heavy quark fields. This is often called a power counting scheme. In the quantum theory we will have $\varepsilon = \Lambda_{\rm QCD}/m$. Obviously quantities such as $F_{\mu\nu} = \mathrm{O}(1)$ are not touched by this consideration. At the lowest order in this expansion the ("large components") quark field ($P_\pm = \frac{1\pm\gamma_0}{2}$)

$$\psi_{\rm h} = P_+\psi\,, \quad \overline{\psi}_{\rm h} = \overline{\psi}P_+, \tag{126}$$

propagates forward in time, while the anti-quark field,

$$\psi_{\bar{\rm h}} = P_-\psi\,, \quad \overline{\psi}_{\bar{\rm h}} = \overline{\psi}P_-\,, \tag{127}$$

propagates backward. In a somewhat sloppy notation we will often write $\mathrm{O}(1/m)$ instead of $\mathrm{O}(\varepsilon)$. The $\mathrm{O}(1/m)$ terms in the Lagrangian

$$\mathcal{L} = \mathcal{L}_{\mathrm{h}}^{\mathrm{stat}} + \mathcal{L}_{\bar{\mathrm{h}}}^{\mathrm{stat}} + \mathrm{O}(\tfrac{1}{m}) \tag{128}$$

$$\mathcal{L}_{\mathrm{h}}^{\mathrm{stat}} = \overline{\psi}_{\mathrm{h}}(D_0 + m)\psi_{\mathrm{h}}\,, \quad \mathcal{L}_{\bar{\mathrm{h}}}^{\mathrm{stat}} = \overline{\psi}_{\bar{\mathrm{h}}}(-D_0 + m)\psi_{\bar{\mathrm{h}}}\,, \tag{129}$$

connect quark and anti-quark fields. They can be decoupled through a Foldy-Wouthuysen rotation,

$$\mathcal{L} = \phi^{\dagger}\mathcal{D}'\phi\,, \quad \phi = \mathrm{e}^{S}\psi\,,\ \phi^{\dagger} = \psi^{\dagger}\mathrm{e}^{-S} \tag{130}$$

$$\mathcal{D}' = \mathrm{e}^{S}\mathcal{D}\mathrm{e}^{-S}\,, \quad S = \tfrac{1}{2m}D_k\gamma_k = -S^{\dagger} = \mathrm{O}(\tfrac{1}{m})\,, \tag{131}$$

which yields explicitly

$$\begin{aligned}\mathcal{D}' &= \mathcal{D} + \tfrac{1}{2m}[D_k\gamma_k, \mathcal{D}] + \tfrac{1}{8m^2}[D_l\gamma_l, [D_k\gamma_k, \mathcal{D}]] + \mathrm{O}(\tfrac{1}{m^2}) \\ &= \mathcal{D} + \tfrac{1}{2m}[D_k\gamma_k, \mathcal{D}] - \tfrac{1}{4m}[D_l\gamma_l, \gamma_0 D_k\gamma_k] + \mathrm{O}(\tfrac{1}{m^2}) \\ &= \gamma_0\left\{\gamma_0 D_0 + m + \tfrac{1}{2m}(-D_kD_k - \frac{1}{2i}F_{kl}\sigma_{kl}) + \tfrac{1}{2m}F_{k0}\gamma_0\gamma_k\right\} + \mathrm{O}(\tfrac{1}{m^2})\,.\end{aligned} \tag{132}$$

The Lagrangian then reads

$$\mathcal{L} = \mathcal{L}_{\mathrm{h}}^{\mathrm{stat}} + \mathcal{L}_{\bar{\mathrm{h}}}^{\mathrm{stat}} + \left\{\mathcal{L}_{\mathrm{h}}^{(1)} + \mathcal{L}_{\bar{\mathrm{h}}}^{(1)} + \mathcal{L}_{\mathrm{h}\bar{\mathrm{h}}}^{(1)}\right\} + \mathrm{O}(\tfrac{1}{m^2}) \tag{133}$$

$$\mathcal{L}_{\mathrm{h}}^{(1)} = \tfrac{1}{2m}\overline{\psi}_{\mathrm{h}}(-D_kD_k - \frac{1}{2i}F_{kl}\,\sigma_{kl})\psi_{\mathrm{h}}\,, \tag{134}$$

$$\sigma_{\mu\nu} = \frac{i}{2}[\gamma_\mu, \gamma_\nu]\,, \quad F_{kl} = [D_k, D_l]\,. \tag{135}$$

For hadrons (or correlation functions) with a single b-quark (or anti-b-quark) only double insertions of $\mathcal{L}_{\mathrm{h}\bar{\mathrm{h}}}^{(1)}$ contribute. These are of order $1/m^2$ and may be dropped at the order written explicitly.

For later convenience we introduce the short hand

$$\mathcal{L}_{\mathrm{h}}^{(1)} = -\tfrac{1}{2m}(\mathcal{O}_{\mathrm{kin}} + \mathcal{O}_{\mathrm{spin}})\,, \tag{136}$$

$$\mathcal{O}_{\mathrm{kin}} = \overline{\psi}_{\mathrm{h}}\, D_kD_k\, \psi_{\mathrm{h}} = \overline{\psi}_{\mathrm{h}}\, \mathbf{D}^2\, \psi_{\mathrm{h}}\,, \tag{137}$$

$$\mathcal{O}_{\mathrm{spin}} = \overline{\psi}_{\mathrm{h}}\, \frac{1}{2i}F_{kl}\,\sigma_{kl}\,\psi_{\mathrm{h}} = \overline{\psi}_{\mathrm{h}}\,\boldsymbol{\sigma}\cdot\mathbf{B}\,\psi_{\mathrm{h}}\,. \tag{138}$$

We note that $\mathcal{L}$, Eq. (133) is a low energy effective Lagrangian [142–144]. It describes the long wave length modes of the fields accurately and makes truncation errors, which are of increasing relevance for shorter wave lengths. This becomes particularly apparent when we remove the mass terms from the static Lagrangian and define

$$\mathcal{L}_{\mathrm{h}}^{\mathrm{stat}} = \overline{\psi}_{\mathrm{h}}(D_0 + \epsilon)\psi_{\mathrm{h}}\,, \quad \mathcal{L}_{\bar{\mathrm{h}}}^{\mathrm{stat}} = \overline{\psi}_{\bar{\mathrm{h}}}(-D_0 + \epsilon)\psi_{\bar{\mathrm{h}}}\,, \tag{139}$$

where the limit $\epsilon \to 0_+$ is to be understood in order to select the proper propagation in time. Replacing Eq. (129) by Eq. (139) corresponds *exactly* to an energy shift by an amount m of all states containing a single heavy quark or anti-quark. For Euclidean correlation functions it just leads to an additional factor of $\exp(-m\,(y_0 - x_0))$ for correlation functions where a quark propagates from x_0 to $y_0 \geq x_0$. (For the anti-quark there is a factor $\exp(+m\,(y_0 - x_0))$ with $y_0 \leq x_0$).

We note again that the essential assumption is Eq. (125), namely the spatial covariant derivatives are counted as small compared to the mass term and the time derivative. This is the correct physical situation in a frame where the hadron is at rest and therefore at lowest order also the quark is at rest.

Instead of carrying out the expansion of the action, one could also expand the heavy quark propagator in terms of $1/m$.

Quantum fluctuations are not smooth and invalidate the above "derivation". However, one expects that they do not modify the structure of the effective Lagrangian, but rather only modify the coefficients of the various terms by non-trivial renormalizations due to these *short distance* fluctuations. After all, arguing heuristically, long wavelength terms have been identified correctly and are described by local interaction terms. In local quantum field theory, also effective local quantum field theory, such terms are renormalized by a renormalization of the coefficients of the local fields. Below, we will discuss this in some detail.

III.2. The effective quantum field theory

III.2.1. *The static approximation and its symmetries*

We start with the lowest order and just consider the heavy quark; the antiquark action and propagator is completely analogous. In continuum Euclidean space, the classic Lagrangian density $\mathcal{L}_{\rm h}^{\rm stat}$, Eq. (139), contains local fields of a mass dimension $D \leq 4$. It is power counting renormalizable. The static effective field theory is thus expected to be renormalizable in the usual sense, i.e. by a finite number of counter-terms. The possible counter-terms are restricted by the symmetries. Apart from the usual ones (parity, gauge symmetry ...), there is a well known invariance under spin rotations [148,136,137,134]. The infinitesimal variations of these transformations can be written as

$$\delta_\sigma^k \psi_{\rm h} = \sigma_k \psi_{\rm h}\,, \qquad \delta_\sigma^k \overline{\psi}_{\rm h} = -\overline{\psi}_{\rm h}\sigma_k\,, \qquad \sigma_k \equiv -\tfrac{1}{2}\epsilon_{ijk}\sigma_{ij}\,, \tag{140}$$

with $[\sigma_k, \sigma_l] = i2\epsilon_{klm}\sigma_m$. In addition, the action is invariant under phase transformations,

$$\psi_\mathrm{h} \longrightarrow \mathrm{e}^{i\eta(\mathbf{x})}\,\psi_\mathrm{h}, \qquad \overline{\psi}_\mathrm{h} \longrightarrow \overline{\psi}_\mathrm{h}\mathrm{e}^{-i\eta(\mathbf{x})}, \tag{141}$$

with an arbitrary space- (but not time-) dependent parameter $\eta(\mathbf{x})$. This invariance corresponds to the local conservation of b-quark number, ensuring that the quark propagates only in time. The only counter-term of dimension $D \leq 4$, which involves $\psi_\mathrm{h}, \overline{\psi}_\mathrm{h}$ and is invariant under these symmetries is $\overline{\psi}_\mathrm{h}\psi_\mathrm{h}$. Denoting its coefficient by δm, the formal continuum quantum Lagrangian is thus simply

$$\mathcal{L}_\mathrm{h}^\mathrm{stat} = \overline{\psi}_\mathrm{h}(D_0 + \delta m)\psi_\mathrm{h}\,. \tag{142}$$

In order to discuss the equivalence of the effective theory and QCD for correlation functions, we introduce also the time component of the axial current in the effective theory,

$$A_0^\mathrm{stat}(x) = \overline{\psi}_\mathrm{l}(x)\gamma_0\gamma_5\psi_\mathrm{h}(x)\,, \tag{143}$$

as a prototype for a composite field. There is no other operator with its quantum numbers and with dimension $D \leq 3$. Hence $A_0^\mathrm{stat}(x)$ renormalizes multiplicatively.

III.2.1.1. *Lattice formulation*

Just like the continuum theory we formulate the lattice theory in terms of formally 4-component fields, satisfying

$$P_+\psi_\mathrm{h} = \psi_\mathrm{h}\,, \quad \overline{\psi}_\mathrm{h}P_+ = \overline{\psi}_\mathrm{h}\,. \tag{144}$$

The time-doubler is removed by choosing the lattice backward derivative

$$D_0^\mathrm{W}\psi_\mathrm{h}(x) = \frac{1}{a}\left[\psi_\mathrm{h}(x) - W^\dagger(x - a\hat{0}, 0)\psi_\mathrm{h}(x - a\hat{0})\right]\,, \tag{145}$$

in the action

$$S_\mathrm{h}^\mathrm{W} = a^4 \frac{1}{1 + a\,\delta m_\mathrm{W}} \sum_x \overline{\psi}_\mathrm{h}(x)(D_0^\mathrm{W} + \delta m_\mathrm{W})\psi_\mathrm{h}(x)\,. \tag{146}$$

With $W(x,0) = U(x,0)$, the standard time-like links, this is the Eichten-Hill action[133]. For the Monte Carlo evaluation it is however of a considerable advantage to define the theory with more general parallel transporters $W(x,0)$, equivalent to $U(x,0)$ up to $\mathrm{O}(a^2)$. In this way statistical errors of correlation functions at large separation x_0 can be reduced exponentially (in x_0) and at the same time discretization errors have been found to be somewhat smaller [145,146]. Concentrating on conceptual issues, we refer to the cited papers for details on this more practical issue.

It is an easy exercise to show that the static propagator in the presence of gauge fields is ($a\widehat{\delta m} = \ln(1 + a\delta m)$)

$$G_{\rm h}^{\rm W}(x,y) = \theta(x_0 - y_0)\,\delta(\mathbf{x} - \mathbf{y})\,\exp\big(-\widehat{\delta m}_{\rm W}\,(x_0 - y_0)\big)\,\mathcal{P}^{\rm W}(y,x)^\dagger\,P_+\,,$$
$$\mathcal{P}^{\rm W}(x,x) = 1\,,\quad \mathcal{P}^{\rm W}(x,y+a\hat{\mu}) = \mathcal{P}^{\rm W}(x,y)W(y,\mu)\,, \tag{147}$$

when the fields are normalized as in Eq. (146). Here $\theta(x_0), \delta(\mathbf{x})$ are straight forward lattice transcriptions of the continuum θ- and δ-functions. While in the continuum case, an ϵ-prescription is necessary to select the forward propagation $\theta(x_0 - y_0)$, the propagator of the lattice action with backward derivative, Eq. (145), has this property automatically. The symmetries Eq. (140) and Eq. (141) are preserved by the lattice regularization Eq. (146).

Improvement

With these symmetries one easily goes through the steps introduced in Sect. I.2.2 to find the structure of Symanzik's effective action. It turns out that the only allowed O(a) terms are proportional to the light quark masses[9]. They are therefore numerically not very relevant and in addition their coefficients vanish for $N_{\rm f} = 0$. It is interesting to note that through the static effective theory one can also give a convincing argument that the force between static quarks is free of linear a-effects, if the light quark action is O(a)-improved [93].

For ease of notation we drop the sub- and superscript W from now on.

III.2.1.2. *Renormalization*

The only term needed for the renormalization of the action, δm, has been included above. The explicit form of the propagator shows that δm enters in a purely kinematical way, just as an energy shift by an amount $\widehat{\delta m}$ compared to the unrenormalized, $\delta m = 0$ case. Thus all masses of hadrons (we consider only those with a single heavy quark) have the same shift and their splittings can be predicted in static approximation up to $\Lambda_{\rm QCD}^2/m_{\rm b}$ corrections without adjusting any parameter except for those in the "light part of QCD". Obviously, the splittings are also the same if the b-quark is replaced by the charm quark, but now up to $\Lambda_{\rm QCD}^2/m_{\rm c}$, if $m_{\rm c}$ is large enough for the $1/m_{\rm c}$ expansion to make sense.

But also in the static approximation non-trivial renormalizations occur as soon as hadronic matrix elements are considered. The most prominent example is the computation of the B-meson decay constant, $F_{\rm B}$. In QCD it can be obtained from the correlation function [t]

$$C_{\rm AA}(x_0) = Z_{\rm A}^2 a^3 \sum_{\mathbf{x}} \left\langle A_0(x)(A_0)^\dagger(0) \right\rangle \tag{148}$$

with the heavy-light axial current in QCD, $A_\mu = \overline{\psi}_{\rm l}\gamma_\mu\gamma_5\psi_{\rm b}$, and $A_\mu^\dagger = \overline{\psi}_{\rm b}\gamma_\mu\gamma_5\psi_{\rm l}$. In QCD, $Z_{\rm A}(g_0)$ is fixed by chiral Ward identities [20,21], which shows that it does not depend on a renormalization scale. The decay constant may e.g. be obtained from (we use the finite volume normalization $\langle B|B\rangle = 1$ for the zero momentum state $|B\rangle$)

$$\begin{aligned}\left[\Phi^{\rm QCD}\right]^2 &\equiv F_{\rm B}^2\, m_{\rm B} \\ &= 2L^3 \left[\langle B|Z_{\rm A} A_0|0\rangle\right]^2 = 2 \lim_{x_0\to\infty} \exp(x_0\,\Gamma_{\rm AA}(x_0)) C_{\rm AA}(x_0)\,,\end{aligned} \tag{149}$$

where the effective mass ($\widetilde{\partial}_0 f(x_0) = \frac{1}{2a}[f(x_0+a) - f(x_0-a)]$)

$$\Gamma_{\rm AA}(x_0) = -\widetilde{\partial}_0 \ln(C_{\rm AA}(x_0)) \tag{150}$$

appears. In the static approximation, the QCD current is represented by

$$(A_{\rm R}^{\rm stat})_0(x) = Z_{\rm A}^{\rm stat}(g_0, \mu a) A_0^{\rm stat}(x)\,, \tag{151}$$

where we have allowed for a μ-dependence of the renormalization. This is expected because chiral symmetry is not just broken softly, but in the effective theory it is not present at all. We will come back to it in Sect. III.3.1.

III.2.2. *Including $1/m_{\rm b}$ corrections*

We work directly in lattice regularization. The continuum formulae are completely analogous. The expressions for $\mathcal{O}_{\rm kin}, \mathcal{O}_{\rm spin}$, Eq. (137), Eq. (138), are discretized in a straight forward way, $D_k D_k \to \nabla_k^* \nabla_k$ and $F_{kl} \to \widehat{F}_{kl}$ with the latter defined in Ref. 19. Of course other discretizations of these composite fields are possible.

Apart from the terms in the classical Lagrangian, renormalization can in principle introduce new local fields compatible with the symmetries (but not Eq. (140) and Eq. (141), which are broken by $\mathcal{O}_{\rm spin}, \mathcal{O}_{\rm kin}$) and with dimension $D \leq 5$. Also the field equations can be used to eliminate terms.

---

[t] It is technically of advantage to consider so-called smeared-smeared and local-smeared correlation functions, but this is irrelevant in the present discussion.

With these rules one easily finds that no new terms are needed and it suffices to treat the coefficients of $\mathcal{O}_{\rm spin}, \mathcal{O}_{\rm kin}$ as free parameters which depend on the bare coupling of the theory and on $m_{\rm b}$.

The $1/m_{\rm b}$ Lagrangian then reads

$$\mathcal{L}_{\rm h}^{(1)}(x) = -(\omega_{\rm kin}\,\mathcal{O}_{\rm kin}(x) + \omega_{\rm spin}\,\mathcal{O}_{\rm spin}(x))\,. \tag{152}$$

Since these terms are fields of dimension five, the theory defined with a path integral weight $P \propto \exp(-a^4\sum_x[\mathcal{L}_{\rm light}(x) + \mathcal{L}_{\rm h}^{\rm stat}(x) + \mathcal{L}_{\rm h}^{(1)}(x)])$ is *not* renormalizable. In perturbation theory, new divergences will occur at each order in the loop expansion, which necessitate to introduce new counter-terms. The continuum limit of the lattice theory will then not exist. However, that effective theory is NRQCD not HQET. Since the effective theory is "only" supposed to reproduce the $1/m_{\rm b}$ expansion of the observables order by order in $1/m_{\rm b}$, we expand the weight P in $1/m_{\rm b}$, counting $\omega_{\rm kin} = {\rm O}(1/m_{\rm b}) = \omega_{\rm spin}$. This defines HQET. The same step has already been used in Symanzik's effective theory.

Up to and including ${\rm O}(1/m_{\rm b})$, expectation values in HQET are defined as

$$\begin{aligned}\langle\mathcal{O}\rangle &= \langle\mathcal{O}\rangle_{\rm stat} + \omega_{\rm kin}a^4\sum_x\langle\mathcal{O}\mathcal{O}_{\rm kin}(x)\rangle_{\rm stat} + \omega_{\rm spin}a^4\sum_x\langle\mathcal{O}\mathcal{O}_{\rm spin}(x)\rangle_{\rm stat}\\ &\equiv \langle\mathcal{O}\rangle_{\rm stat} + \omega_{\rm kin}\langle\mathcal{O}\rangle_{\rm kin} + \omega_{\rm spin}\langle\mathcal{O}\rangle_{\rm spin}\,,\end{aligned} \tag{153}$$

where

$$\langle\mathcal{O}\rangle_{\rm stat} = \frac{1}{\mathcal{Z}}\int_{\rm fields}\mathcal{O}\exp(-a^4\sum_x[\mathcal{L}_{\rm light}(x) + \mathcal{L}_{\rm h}^{\rm stat}(x)]) \tag{154}$$

is defined with respect to the lowest order action, which is power counting renormalizable. The path integral defining the average extends over all fields and the normalization $\mathcal{Z}$ is fixed by $\langle 1\rangle_{\rm stat} = 1$.

In order to compute matrix elements or correlation functions in the effective theory, we also need the effective composite fields. At the classical level they can again be obtained from the Foldy-Wouthuysen rotation. In the quantum theory one adds all local fields with the proper quantum numbers and dimensions. For example the effective axial current (time component) is given by

$$A_0^{\rm HQET}(x) = Z_{\rm A}^{\rm HQET}\,[A_0^{\rm stat}(x) + c_{\rm A}^{\rm HQET}\,\delta A_0^{\rm stat}(x)]\,, \tag{155}$$

$$\delta A_0^{\rm stat}(x) = \overline{\psi}_{\rm l}(x)\frac{1}{2}(\overleftarrow{\nabla}_i + \overleftarrow{\nabla}_i^*)\gamma_i\gamma_5\psi_{\rm h}(x)\,. \tag{156}$$

Before entering into more details on the renormalization, we show some examples how the $1/m_{\rm b}$-expansion works.

III.2.2.1. $1/m_{\rm b}$-*expansion of correlation functions and matrix elements*

For now we assume that the coefficients

$$\begin{aligned} {\rm O}(1)\ &:\ \delta m\,,\ Z_{\rm A}^{\rm HQET}\,, \\ {\rm O}(1/m_{\rm b})\ &:\ \omega_{\rm kin}\,,\ \omega_{\rm spin}\,,\ c_{\rm A}^{\rm HQET}\,, \end{aligned} \tag{157}$$

are known as a function of the bare coupling g_0 and the quark mass $m_{\rm b}$. Their non-perturbative determination will be discussed later.

The rules of the $1/m_{\rm b}$-expansion are illustrated on the example of $C_{\rm AA}(x_0)$, Eq. (148). One uses Eq. (153) and the HQET representation of the composite field Eq. (155). Then the expectation value is expanded consistently in $1/m_{\rm b}$, counting powers of $1/m_{\rm b}$ as in Eq. (157). At order $1/m_{\rm b}$, terms proportional to $\omega_{\rm kin}\times c_{\rm A}^{\rm HQET}$ etc. are to be dropped. As a last step, we have to take the energy shift between HQET and QCD into account. Therefore the correlation function obtains an extra factor $\exp(-x_0 m_{\rm b})$, where the scheme dependence of $m_{\rm b}$ is compensated by a corresponding one in δm. One arrives at the expansion

$$\begin{aligned} C_{\rm AA}(x_0) = {\rm e}^{-m_{\rm b}x_0}(Z_{\rm A}^{\rm HQET})^2 \Big[& C_{\rm AA}^{\rm stat}(x_0) + c_{\rm A}^{\rm HQET}\, C_{\rm \delta AA}^{\rm stat}(x_0) \\ & + \omega_{\rm kin}\, C_{\rm AA}^{\rm kin}(x_0) + \omega_{\rm spin}\, C_{\rm AA}^{\rm spin}(x_0) \Big] \end{aligned} \tag{158}$$

with (remember the definitions in Eq. (153))

$$\begin{aligned} C_{\rm \delta AA}^{\rm stat}(x_0) &= \langle A_0^{\rm stat}(x)(\delta A_0^{\rm stat}(0))^\dagger\rangle_{\rm stat} + \langle \delta A_0^{\rm stat}(x)(A_0^{\rm stat}(0))^\dagger\rangle_{\rm stat}\,, \\ C_{\rm AA}^{\rm kin}(x_0) &= \langle A_0^{\rm stat}(x)(A_0^{\rm stat}(0))^\dagger\rangle_{\rm kin}\,, \\ C_{\rm AA}^{\rm spin}(x_0) &= \langle A_0^{\rm stat}(x)(A_0^{\rm stat}(0))^\dagger\rangle_{\rm spin}\,. \end{aligned} \tag{159}$$

It is now a straight forward exercise to obtain the expansion of the B-meson mass[u]

$$m_{\rm B} = m_{\rm b} + \widehat{\delta m} + E_{\rm stat} + \omega_{\rm kin} E_{\rm kin} + \omega_{\rm spin} E_{\rm spin}\,, \tag{160}$$

$$E_{\rm stat} = -\lim_{x_0\to\infty} \tilde\partial_0 \ln C_{\rm AA}^{\rm stat}(x_0)\Big|_{\delta m=0}\,, \tag{161}$$

$$E_{\rm kin} = -\lim_{x_0\to\infty} \tilde\partial_0\, \rho_{\rm kin}(x_0)\,, \quad \rho_{\rm kin}(x_0) = \frac{C_{\rm AA}^{\rm kin}(x_0)}{C_{\rm AA}^{\rm stat}(x_0)}\,, \tag{162}$$

$$E_{\rm spin} = -\lim_{x_0\to\infty} \tilde\partial_0\, \rho_{\rm spin}(x_0)\,, \quad \rho_{\rm spin}(x_0) = \frac{C_{\rm AA}^{\rm spin}(x_0)}{C_{\rm AA}^{\rm stat}(x_0)}\,, \tag{163}$$

---

[u] It follows from the simple form of the static propagator that there is no dependence on δm except for the explicitly shown energy shift $\widehat{\delta m}$.

and its decay constant

$$F_{\rm B}\sqrt{m_{\rm B}} = \lim_{x_0\to\infty} \left\{2\exp(m_{\rm B}x_0)\, C_{\rm AA}(x_0)\right\}^{1/2}, \tag{164}$$

$$\begin{aligned} &= Z_{\rm A}^{\rm HQET} \lim_{x_0\to\infty} \Phi^{\rm stat}\left\{1+\tfrac{1}{2}x_0\left[\omega_{\rm kin}E_{\rm kin}+\omega_{\rm spin}E_{\rm kin}\right]\right. \\ &\quad \left. +\tfrac{1}{2}c_{\rm A}^{\rm HQET}\rho_{\delta{\rm A}}(x_0)+\tfrac{1}{2}\omega_{\rm kin}\rho_{\rm kin}(x_0)+\tfrac{1}{2}\omega_{\rm spin}\rho_{\rm spin}(x_0)\right\}, \end{aligned} \tag{165}$$

$$\Phi^{\rm stat} = \lim_{x_0\to\infty} \left\{2\exp(E_{\rm stat}x_0)\, C_{\rm AA}^{\rm stat}(x_0)\right\}^{1/2}, \quad \rho_{\delta{\rm A}}(x_0) = \frac{C_{\delta{\rm AA}}^{\rm stat}(x_0)}{C_{\rm AA}^{\rm stat}(x_0)}\,.$$

Using the transfer matrix formalism (with normalization $\langle B|B\rangle = 1$), one further observes that

$$\begin{aligned} E_{\rm kin} &= -\langle B|a^3\sum_{\mathbf{z}} \mathcal{O}_{\rm kin}(0,\mathbf{z})|B\rangle_{\rm stat}\,, \\ E_{\rm spin} &= -\langle B|a^3\sum_{\mathbf{z}} \mathcal{O}_{\rm spin}(0,\mathbf{z})|B\rangle_{\rm stat}\,. \end{aligned} \tag{166}$$

As expected, only the parameters of the action are relevant in the expansion of hadron masses.

A correct split of the terms in Eq. (160) and Eq. (165) into leading order and next to leading order pieces which *separately have a continuum limit* requires more thought on the renormalization of the $1/m_{\rm b}$-expansion. We turn to this now.

III.2.2.2. *Renormalization and continuum limit*

For illustration we first check the self consistency of Eq. (158). The relevant question concerns renormalization, namely: are the "free" parameters $\delta m \ldots c_{\rm A}^{\rm HQET}$ sufficient to absorb all divergences on the r.h.s.? We consider the most difficult term, $C_{\rm AA}^{\rm kin}(x_0)$. According to the standard rules, it is renormalized as

$$\begin{aligned} \left(C_{\rm AA}^{\rm kin}\right)_{\rm R}(x_0) &= \left(Z_{\rm A}^{\rm stat}\right)^2 \times \\ &\quad a^7\sum_{\mathbf{x},\,z}\left\langle A_0^{\rm stat}(x)\,(A_0^{\rm stat}(0))^\dagger\,\left(\mathcal{O}_{\rm kin}\right)_{\rm R}(z)\right\rangle_{\rm stat} + {\rm C.T.}\,, \end{aligned} \tag{167}$$

where C.T. denotes contact terms to be discussed shortly. The operator $\left(\mathcal{O}_{\rm kin}\right)_{\rm R}(z)$ involves a subtraction of lower dimensional ones,

$$\left(\mathcal{O}_{\rm kin}\right)_{\rm R}(z) = Z_{\mathcal{O}_{\rm kin}}\left(\mathcal{O}_{\rm kin}(z)+\frac{c_1}{a}\,\overline{\psi}_{\rm h}(z)D_0\psi_{\rm h}(z)+\frac{c_2}{a^2}\,\overline{\psi}_{\rm h}(z)\psi_{\rm h}(z)\right), \tag{168}$$

written here in terms of dimensionless c_i. Since we are interested in on-shell observables ($x_0 > 0$ in Eq. (158)), we may use the equation of motion $D_0\psi_{\rm h}(z) = 0$ to eliminate the second term. The third one, $\frac{c_2}{a^2}\overline{\psi}_{\rm h}(z)\psi_{\rm h}(z)$, is equivalent to a mass shift and only changes δm, which is hence quadratically divergent [v]. Thus all terms which are needed for the renormalization of $\mathcal{O}_{\rm kin}$ are present in Eq. (158).

It remains to consider the contact terms in Eq. (167). They originate from singularities in the operator products $\mathcal{O}_{\rm kin}(z)A_0^{\rm stat}(x)$ as $z \to x$ (and $\mathcal{O}_{\rm kin}(z)\big(A_0^{\rm stat}\big)^\dagger(0)$ as $z \to 0$), in complete analogy to the discussion in Sect. I.2.2.3. Using the operator product expansion they can be represented as linear combinations of $A_0^{\rm stat}(x)$ and $\delta A_0^{\rm stat}(x)$. Such terms are contained in Eq. (158) in the form of $C_{\rm AA}^{\rm stat}$ and $C_{\delta \rm AA}^{\rm stat}$. Indeed $A_0^{\rm stat}(x)$ and $\delta A_0^{\rm stat}(x)$ are the only operators of dimension 3 and 4 with the correct quantum numbers. Higher dimensional operators contribute only terms of order a.

Note that the coefficient of $A_0^{\rm stat}(x)$ in the operator product expansion of $a^4\sum_z \mathcal{O}_{\rm kin}(z)A_0^{\rm stat}(x)$ is power divergent $\sim 1/a$, for simple dimensional reasons. This means that there is a power divergent contribution to $Z_{\rm A}^{\rm HQET}$. As this happens only at order $1/m_{\rm b}$, not at the lowest order, this contribution to $Z_{\rm A}^{\rm HQET}$ behaves like $\sim 1/(am_{\rm b})$ for small lattice spacing.

We conclude that all terms which are needed for the renormalization of $C_{\rm AA}^{\rm kin}(x_0)$ are present in Eq. (158); the parameters may thus be adjusted to absorb all infinities and with properly chosen coefficients the continuum limit of the r.h.s. is expected to exist. The basic assumption of the effective field theory is that once the finite parts of the coefficients have been determined by matching a set of observables to QCD, these coefficients are applicable to any other observables.

III.2.2.3. *The flavor currents in the effective theory*

For later use we here give the expressions for the heavy-light currents. They are relevant in weak B-meson decays. For better readability, we include again the time component of the axial current.

Following our general rules for finding the HQET fields which represent

[v]Using the explicit form of the static propagator, Eq. (147), one can check that indeed $a^3\sum_{\mathbf{x}}\left\langle A_0^{\rm stat}(x)\,(A_0^{\rm stat}(0))^\dagger a^4\sum_z \overline{\psi}_{\rm h}(z)\psi_{\rm h}(z)\right\rangle_{\rm stat} = x_0 C_{\rm AA}^{\rm stat}(x_0)$, which can be absorbed by a $1/m_{\rm b}$ correction to δm.

the QCD ones we find

$$A_0^{\rm HQET}(x) = Z_{\rm A}^{\rm HQET}\,[A_0^{\rm stat}(x) + c_{\rm A}^{\rm HQET}\delta A_0^{\rm stat}(x)]\,, \tag{169}$$
$$V_0^{\rm HQET}(x) = Z_{\rm V}^{\rm HQET}\,[V_0^{\rm stat}(x) + c_{\rm V}^{\rm HQET}\delta V_0^{\rm stat}(x)]\,, \tag{170}$$
$$V_k^{\rm HQET}(x) = Z_{\mathbf{V}}^{\rm HQET}\,[V_k^{\rm stat}(x) + c_{\mathbf{V}}^{\rm HQET}\delta V_k^{\rm stat}(x)]\,, \tag{171}$$
$$A_k^{\rm HQET}(x) = Z_{\mathbf{A}}^{\rm HQET}\,[A_k^{\rm stat}(x) + c_{\mathbf{A}}^{\rm HQET}\delta A_k^{\rm stat}(x)]\,. \tag{172}$$

Ignoring that one can in principle simplify, for instance $\overline{\psi}_{\rm l}(x)\gamma_0\gamma_5\psi_{\rm h}(x) = -\overline{\psi}_{\rm l}(x)\gamma_5\psi_{\rm h}(x)$ due to $P_+\psi_{\rm h} = \psi_{\rm h}$, the basis fields are written in full analogy to the ones in QCD:

$$V_0^{\rm stat} = \overline{\psi}_{\rm l}\gamma_0\psi_{\rm h}\,, \quad A_0^{\rm stat} = \overline{\psi}_{\rm l}\gamma_0\gamma_5\psi_{\rm h}\,, \tag{173}$$
$$V_k^{\rm stat} = \overline{\psi}_{\rm l}\gamma_k\psi_{\rm h}\,, \quad A_k^{\rm stat} = \overline{\psi}_{\rm l}\gamma_k\gamma_5\psi_{\rm h}\,, \tag{174}$$
$$\delta V_0^{\rm stat} = \overline{\psi}_{\rm l}\frac{\overleftarrow{\nabla}_i + \overleftarrow{\nabla}_i^*}{2}\gamma_i\psi_{\rm h}\,, \quad \delta A_0^{\rm stat} = \overline{\psi}_{\rm l}\frac{\overleftarrow{\nabla}_i + \overleftarrow{\nabla}_i^*}{2}\gamma_i\gamma_5\psi_{\rm h}\,, \tag{175}$$
$$\delta V_k^{\rm stat} = -\overline{\psi}_{\rm l}\frac{\overleftarrow{\nabla}_i + \overleftarrow{\nabla}_i^*}{2}\gamma_i\gamma_k\psi_{\rm h}\,, \quad \delta A_k^{\rm stat} = \overline{\psi}_{\rm l}\frac{\overleftarrow{\nabla}_i + \overleftarrow{\nabla}_i^*}{2}\gamma_i\gamma_k\gamma_5\psi_{\rm h}\,, \tag{176}$$

We have chosen the bare fields such that they are exactly related by spin rotations,[w]

$$\begin{aligned}\delta_\sigma^k\left[V_0^{\rm stat}\right] &= iA_k^{\rm stat}\,, \quad \delta_\sigma^k\left[\delta V_0^{\rm stat}\right] = i\delta A_k^{\rm stat}\,,\\ \delta_\sigma^k\left[A_0^{\rm stat}\right] &= iV_k^{\rm stat}\,, \quad \delta_\sigma^k\left[\delta A_0^{\rm stat}\right] = i\delta V_k^{\rm stat}\,.\end{aligned} \tag{177}$$

In general, all Z-factors and coefficients $c^{\rm HQET}$ are functions of g_0 and $am_{\rm b}$ which are to be determined by matching non-perturbatively to QCD. Eq. (177) will be needed in Sect. III.3.1 when we discuss the static limit of the currents.

III.2.3. *Schrödinger functional correlation functions*

For an understanding of the details of the tests of HQET (Sect. III.4) as well as the non-perturbative matching to QCD (Sect. III.5) we will also need some Schrödinger functional correlation functions and their HQET expansion, which have not been defined yet. We give these details[140] now. The reader who is only interested in the general concepts may skip this section.

---

[w]The unnatural $-$ sign in Eq. (176) is present because we remain with the definition of $\delta A_0^{\rm stat}$ in Ref. 9 and do not want to introduce signs in Eq. (177).
The other rotations look like $\delta_\sigma^j A_k^{\rm stat} = i\,\epsilon_{kjl}V_l^{\rm stat} - i\,\delta_{kj}A_0^{\rm stat}$.

In Ref. 9 static quarks in the Schrödinger functional were discussed including Symanzik O(a)-improvement. It turns out that there are no dimension four composite fields which involve static quarks fields and which are compatible with the symmetries of the static action and the Schrödinger functional boundary conditions and which do not vanish by the equations of motion. Thus there are no O(a) boundary counter terms with static quark fields. For the same reason there are also no O($1/m_{\rm b}$) boundary terms in HQET. This then means the HQET expansion of the boundary quark fields $\zeta, \bar{\zeta}$ is trivial up to and including $1/m_{\rm b}$ terms.

In the spatial boundary condition of the fermion fields,

$$\psi(x+\hat{k}L) = {\rm e}^{i\theta}\psi(x)\,, \quad \overline{\psi}(x+\hat{k}L) = {\rm e}^{-i\theta}\overline{\psi}(x)\,, \tag{178}$$

the same phase θ is taken for all quark fields, whether relativistic or described by HQET.[x] In QCD, relevant correlation functions in the pseudoscalar and vector channel are

$$f_{\rm A}(x_0,\theta) = -\frac{a^6}{2}\sum_{\mathbf{y},\mathbf{z}} \left\langle (A_{\rm I})_0(x)\,\overline{\zeta}_{\rm b}(\mathbf{y})\gamma_5\zeta_{\rm l}(\mathbf{z})\right\rangle \,, \tag{179}$$

$$k_{\rm V}(x_0,\theta) = -\frac{a^6}{6}\sum_{\mathbf{y},\mathbf{z},k} \left\langle (V_{\rm I})_k(x)\,\overline{\zeta}_{\rm b}(\mathbf{y})\gamma_k\zeta_{\rm l}(\mathbf{z})\right\rangle \,, \tag{180}$$

with the O(a) improved currents[139]

$$(A_{\rm I})_\mu = \overline{\psi}_{\rm l}\gamma_\mu\gamma_5\psi_{\rm b} + ac_{\rm A}\tilde{\partial}_\mu\overline{\psi}_{\rm l}\gamma_5\psi_{\rm b} \tag{181}$$

$$(V_{\rm I})_\mu = \overline{\psi}_{\rm l}\gamma_\mu\psi_{\rm b} + ac_{\rm V}\tilde{\partial}_\mu\overline{\psi}_{\rm l}i\sigma_{\mu\nu}\psi_{\rm b}\,. \tag{182}$$

Furthermore we consider boundary to boundary correlation functions

$$f_1(\theta) = -\frac{a^{12}}{2L^6}\sum_{\mathbf{u},\mathbf{v},\mathbf{y},\mathbf{z}} \left\langle \overline{\zeta}_{\rm l}{}'(\mathbf{u})\gamma_5\zeta'_{\rm h}(\mathbf{v})\,\overline{\zeta}_{\rm b}(\mathbf{y})\gamma_5\zeta_{\rm l}(\mathbf{z})\right\rangle \,, \tag{183}$$

$$k_1(\theta) = -\frac{a^{12}}{6L^6}\sum_{\mathbf{u},\mathbf{v},\mathbf{y},\mathbf{z},k} \left\langle \overline{\zeta}_{\rm l}{}'(\mathbf{u})\gamma_k\zeta'_{\rm b}(\mathbf{v})\,\overline{\zeta}_{\rm b}(\mathbf{y})\gamma_k\zeta_{\rm l}(\mathbf{z})\right\rangle \,. \tag{184}$$

Their renormalization is standard [46], for example at vanishing light quark masses

$$[f_{\rm A}]_{\rm R}\,(x_0,\theta) = Z_{\rm A}\,(1+\tfrac{1}{2}b_{\rm A}am_{\rm q,b})Z_\zeta^2\,(1+b_\zeta am_{\rm q,b})f_{\rm A}(x_0,\theta)\,, \tag{185}$$

$$[f_1]_{\rm R}\,(\theta) = Z_\zeta^4\,(1+b_\zeta am_{\rm q,b})^2 f_1(\theta)\,, \tag{186}$$

[x]In principle we can easily have different θ for different fields, as long as they are quenched, so in particular for the heavy field. Since this freedom has not yet been used, we do not discuss it here. In applications with dynamical fermions it may well become relevant.

with Z_ζ a renormalization factor of the relativistic boundary quark fields and b_ζ another improvement coefficient.

Their expansions to first order in $1/m_\mathrm{b}$ read

$$[f_\mathrm{A}]_\mathrm{R} = Z_\mathrm{A}^\mathrm{HQET} Z_{\zeta_\mathrm{h}} Z_\zeta \mathrm{e}^{-m_\mathrm{b} x_0} \left\{ f_\mathrm{A}^\mathrm{stat} + c_\mathrm{A}^\mathrm{HQET} f_{\delta\mathrm{A}}^\mathrm{stat} + \omega_\mathrm{kin} f_\mathrm{A}^\mathrm{kin} + \omega_\mathrm{spin} f_\mathrm{A}^\mathrm{spin} \right\} \tag{187}$$

$$[k_\mathrm{V}]_\mathrm{R} = Z_\mathrm{V}^\mathrm{HQET} Z_{\zeta_\mathrm{h}} Z_\zeta \mathrm{e}^{-m_\mathrm{b} x_0} \left\{ k_\mathrm{V}^\mathrm{stat} + c_\mathrm{V}^\mathrm{HQET} k_{\delta\mathrm{V}}^\mathrm{stat} + \omega_\mathrm{kin} k_\mathrm{V}^\mathrm{kin} + \omega_\mathrm{spin} k_\mathrm{V}^\mathrm{spin} \right\} \tag{188}$$

$$= -Z_\mathrm{V}^\mathrm{HQET} Z_{\zeta_\mathrm{h}} Z_\zeta \mathrm{e}^{-m_\mathrm{b} x_0} \{ f_\mathrm{A}^\mathrm{stat} + c_\mathrm{V}^\mathrm{HQET} f_{\delta\mathrm{A}}^\mathrm{stat} + \omega_\mathrm{kin} f_\mathrm{A}^\mathrm{kin} - \tfrac{1}{3}\omega_\mathrm{spin} f_\mathrm{A}^\mathrm{spin} \} ,$$

$$[f_1]_\mathrm{R} = Z_{\zeta_\mathrm{h}}^2 Z_\zeta^2 \mathrm{e}^{-m_\mathrm{b} T} \left\{ f_1^\mathrm{stat} + \omega_\mathrm{kin} f_1^\mathrm{kin} + \omega_\mathrm{spin} f_1^\mathrm{spin} \right\} , \tag{189}$$

$$[k_1]_\mathrm{R} = Z_{\zeta_\mathrm{h}}^2 Z_\zeta^2 \mathrm{e}^{-m_\mathrm{b} T} \left\{ f_1^\mathrm{stat} + \omega_\mathrm{kin} f_1^\mathrm{kin} - \tfrac{1}{3}\omega_\mathrm{spin} f_1^\mathrm{spin} \right\} . \tag{190}$$

Apart from

$$f_{\delta\mathrm{A}}^\mathrm{stat}(x_0, \theta) = -\frac{a^6}{2} \sum_{\mathbf{y},\mathbf{z}} \langle \delta A_0^\mathrm{stat}(x) \, \overline{\zeta}_\mathrm{h}(\mathbf{y}) \gamma_5 \zeta_1(\mathbf{z}) \rangle \tag{191}$$

the labeling of the different terms follows directly the one introduced in Eq. (153). We have used identities such as $f_\mathrm{A}^\mathrm{kin} = -k_\mathrm{V}^\mathrm{kin}$, $f_\mathrm{A}^\mathrm{spin} = 3k_\mathrm{V}^\mathrm{spin}$. As a simple consequence of the spin symmetry of the static action, these are valid at any lattice spacing.

III.3. The scope of the theory

We are now in the position to discuss what can be done in the effective theory. This concerns also the continuum effective theory and in particular the question, where perturbation theory is sufficient to give an answer and where it is not.

III.3.1. *A first example: the decay constant*

III.3.1.1. *Renormalization and matching in perturbation theory*

The matrix element Φ^QCD, Eq. (149), is scale independent, due to the chiral symmetry of QCD in the massless limit. But of course it depends on the mass of the b-quark. In the effective theory this symmetry is absent and, remaining a while at the lowest order in $1/m_\mathrm{b}$, $Z_\mathrm{A}^\mathrm{stat}$ depends on the renormalization scale, μ, used in the renormalization condition which defines the finite current. On the other hand, as long as the effective theory is considered by itself, $Z_{\rm A}^{\rm stat}$ does not depend on the mass of the b-quark in an obvious way. The mass-dependence comes in by choosing an appropriate finite renormalization such that the matrix elements of the current in the effective theory are equal to the ones of the QCD current (up to $\mathrm{O}(1/m_{\rm b})$). This step is called matching. Choosing to renormalize the current in the effective theory in the $\overline{\rm MS}$ scheme (any other scheme would of course be possible) we therefore have

$$\Phi^{\rm QCD} = C_{\rm match}(m_{\rm b},\mu) \times \Phi_{\overline{\rm MS}}(\mu) + \mathrm{O}(1/m_{\rm b}) \tag{192}$$

with

$$C_{\rm match}(m_{\rm b},\mu) = 1 + c_1(m_{\rm b}/\mu)\bar{g}^2_{\overline{\rm MS}}(\mu) + \ldots \tag{193}$$

The finite renormalization factor $C_{\rm match}$ is determined (usually in perturbation theory) such that Eq. (192) holds for some particular matrix element of the current and will then be valid *for all matrix elements.* Obviously, the μ-dependence in Eq. (192) is artificial, since we have a scale-independent quantity in QCD. Only the $m_{\rm b}$-dependence is for real.

The μ-dependence in Eq. (192) is removed explicitly by changing from $\Phi_{\overline{\rm MS}}(\mu) \equiv Z^{\rm stat}_{\rm A,\overline{MS}}\langle 0|A_0|B\rangle_{\rm stat}$ to the RGI matrix element

$$\Phi_{\rm RGI} = \lim_{\mu\to\infty} \left[\,2b_0\bar{g}^2(\mu)\,\right]^{-\gamma_0/2b_0} \Phi_{\overline{\rm MS}}(\mu)\,. \tag{194}$$

Here, the lowest order coefficient of the β-function, b_0 enters as well as $\gamma_0 = -1/(4\pi^2)$ defined by

$$\gamma(\bar{g}) \equiv \frac{\mu}{Z_{\rm A}^{\rm stat}}\frac{\mathrm{d}}{\mathrm{d}\mu}Z_{\rm A}^{\rm stat} = -\gamma_0\bar{g}^2 + \mathrm{O}(\bar{g}^4)\,. \tag{195}$$

We can now write down the HQET-expansion of the QCD matrix element

$$\Phi^{\rm QCD} = C_{\rm PS}(M_{\rm b}/\Lambda_{\overline{\rm MS}}) \times \Phi_{\rm RGI} + \mathrm{O}(1/m_{\rm b})\,. \tag{196}$$

The relation between Eq. (196) and Eq. (192) is easily seen by using

$$\frac{\Phi_{\rm RGI}}{\Phi_{\overline{\rm MS}}(\mu)} = \left[\,2b_0\bar{g}^2(\mu)\,\right]^{-\gamma_0/2b_0} \exp\left\{-\int_0^{\bar{g}(\mu)} \mathrm{d}g \left[\frac{\gamma_{\overline{\rm MS}}(g)}{\beta_{\overline{\rm MS}}(g)} - \frac{\gamma_0}{b_0 g}\right]\right\} \tag{197}$$

setting the arbitrary renormalization point μ to $m_{\rm b}$ and identifying

$$\begin{aligned} C_{\rm PS}\left(\frac{M_{\rm b}}{\Lambda_{\overline{\rm MS}}}\right) &= C_{\rm match}(1)\frac{\Phi_{\overline{\rm MS}}(m_{\rm b})}{\Phi_{\rm RGI}} \qquad (198)\\ &= \left[\,2b_0\bar{g}^2(m_{\rm b})\,\right]^{\gamma_0/2b_0} \exp\left\{\int_0^{\bar{g}(m_{\rm b})} \mathrm{d}g \left[\frac{\gamma_{\rm match}(g)}{\beta_{\overline{\rm MS}}(g)} - \frac{\gamma_0}{b_0 g}\right]\right\}, \end{aligned}$$

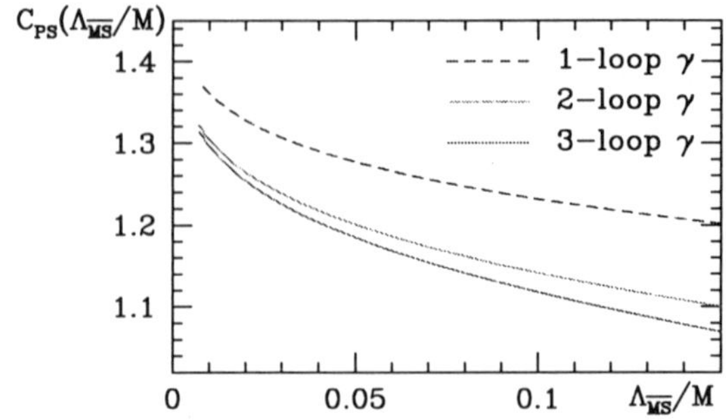

Fig. 24. $C_{\rm PS}$ estimated in perturbation theory. For B-physics we have $\Lambda_{\overline{\rm MS}}/M_{\rm b} \approx 0.04$.

where $\bar{g}$ is taken in the $\overline{\rm MS}$ scheme. The last equation provides a definition of the anomalous dimension $\gamma_{\rm match}$ in the "matching scheme". Perturbatively, it has contributions from $\gamma_{\overline{\rm MS}}$ as well as from $C_{\rm match}$, namely

$$\gamma_{\rm match}(\bar{g}) = -\gamma_0 \bar{g}^2 - [\gamma_1^{\overline{\rm MS}} + 2b_0 c_1(1)]\bar{g}^4 + \dots \quad (199)$$

Replacing the $\overline{\rm MS}$ coupling by a non-perturbative one, $\gamma_{\rm match}$ may also be defined beyond perturbation theory through Eqs. (198,196).[y] Another advantage of Eq. (198) (compared to Eq. (192)) is that $C_{\rm PS}$ is independent of the arbitrary choice of renormalization scheme for the composite operators in the effective theory. Apart from the choice of the QCD coupling, the "convergence" of the series Eq. (199) is dictated by the physics, nothing else.

Note further that (at leading order in $1/m_{\rm b}$) the conversion function $C_{\rm PS}$ contains the full (logarithmic) mass-dependence. The non-perturbative effective theory matrix elements, $\Phi_{\rm RGI}$, are mass independent numbers. Conversion functions such as $C_{\rm PS}$ are universal for all (low energy) matrix elements of their associated operator. Thus

$$C_{\rm AA}(x_0) \overset{x_0 \gg 1/m_{\rm b}}{\sim} [C_{\rm PS}(\tfrac{M_{\rm b}}{\Lambda_{\overline{\rm MS}}})\, Z^{\rm stat}_{\rm A,RGI}]^2 \langle A_0^{\rm stat}(x)^\dagger A_0^{\rm stat}(0)\rangle + {\rm O}(\tfrac{1}{m_{\rm b}}) \quad (200)$$

is a straight forward generalization of Eq. (196). Here, $Z^{\rm stat}_{\rm A,RGI} = \lim_{\mu\to\infty} \left[2b_0\bar{g}^2(\mu)\right]^{-\gamma_0/2b_0} Z^{\rm stat}_{\rm A,\overline{MS}}(\mu)$.

Analogous expressions for the conversion functions are valid for the time component of the axial current replaced by other composite fields,

[y]Clearly the r.h.s. of Eq. (198) is a function of $\bar{g}^2(m_{\rm b})$, i.e. a function of $m_{\rm b}/\Lambda_{\overline{\rm MS}}$. We prefer to write it as a function of the ratio of renormalization group invariants, $M_{\rm b}/\Lambda_{\overline{\rm MS}}$.

for example the space components of the vector current. Based on the work of Refs. 147, 148, 149 and recent efforts their perturbative expansion is known including the 3-loop anomalous dimension $\gamma_{\rm match}$ obtained from the 3-loop anomalous dimension $\gamma_{\overline{\rm MS}}$ [150] and the 2-loop matching function $C_{\rm match}$ [151–153]. Figure 24, taken from Ref. 154, illustrates that the remaining $\mathrm{O}(\bar{g}^6(m_{\rm b}))$ errors in $C_{\rm PS}$ seem to be relatively small.

We return to the full set of heavy-light flavor currents Sect. III.2.2.3. The bare fields satisfy the symmetry relations Eq. (177). The same is then true for the RGI fields in static approximation. Furthermore, the axial currents are related to the vector ones by a chiral rotation of the light quark fields[155]. It then follows that *in static approximation* the effective currents are given by

$$A_0^{\rm HQET} = C_{\rm PS}(M_{\rm b}/\Lambda_{\overline{\rm MS}})\, Z_{\rm A,RGI}^{\rm stat}(g_0)\, A_0^{\rm stat}\,, \tag{201}$$

$$V_k^{\rm HQET} = C_{\mathbf{V}}(M_{\rm b}/\Lambda_{\overline{\rm MS}})\, Z_{\rm A,RGI}^{\rm stat}(g_0)\, V_k^{\rm stat}\,, \tag{202}$$

$$V_0^{\rm HQET} = C_{\rm V}(M_{\rm b}/\Lambda_{\overline{\rm MS}})\, Z_{\rm A,RGI}^{\rm stat}(g_0)\, Z_{\rm V/A}^{\rm stat}(g_0)\, V_0^{\rm stat}\,, \tag{203}$$

$$A_k^{\rm HQET} = C_{\mathbf{A}}(M_{\rm b}/\Lambda_{\overline{\rm MS}})\, Z_{\rm A,RGI}^{\rm stat}(g_0) Z_{\rm V/A}^{\rm stat}(g_0)\, A_k^{\rm stat}\,. \tag{204}$$

The finite renormalization $Z_{\rm V/A}^{\rm stat}(g_0)$ can either be computed by a Ward identity [155] or, if one has a regularization with exact chiral symmetry, it is one. The factor $Z_{\rm A,RGI}^{\rm stat}(g_0)$ is analogous to $Z_{\rm M}(g_0)$ in Eq. (120) and has been computed in a similar way [10]. It can be split into

$$Z_{\rm A,RGI}^{\rm stat}(g_0) = \frac{\Phi_{\rm RGI}}{\Phi(\mu)} \times Z_{\rm A}^{\rm stat}(g_0, 2L_{\rm max}/a)\Big|_{\mu=1/(2L_{\rm max})}\,, \tag{205}$$

where the second factor depends on the lattice action, the first one does not. The first factor has been shown in Fig. 22 for a series of μ in the quenched approximation with a smallest scale of $\mu = 1/(2L_{\rm max})$.

We finally note that the bare pseudo scalar density, P, and the scalar density, S, are identical to $-A_0^{\rm stat}$ and $V_0^{\rm stat}$, respectively. This then also holds for the RGI ones; they are given by $-Z_{\rm A,RGI}^{\rm stat}(g_0)A_0^{\rm stat}$ and $Z_{\rm A,RGI}^{\rm stat}(g_0)\, Z_{\rm V/A}^{\rm stat}(g_0)V_0^{\rm stat}$.

For the renormalized $P^{\rm HQET}$ and $S^{\rm HQET}$, there is in principle an arbitrariness, since in QCD they depend on a scale and a scheme. This arbitrariness is fixed by considering the RGI fields in QCD. They satisfy the PCAC and PCVC relation with the RGI quark masses, for example ($M_{\rm l}$ is the mass of the light quark)

$$\partial_\mu (A_{\rm R})_\mu^{\rm QCD} = (M_{\rm b} + M_{\rm l}) P_{\rm RGI}\,. \tag{206}$$

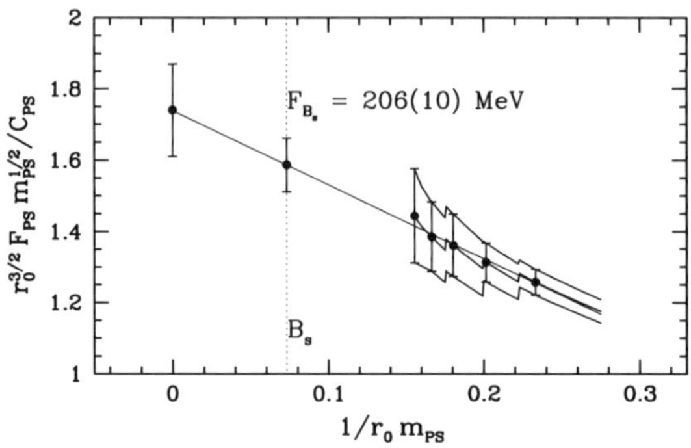

Fig. 25. Example of an interpolation between a static result and results with $m_h < m_b$. Continuum extrapolations are done before the interpolation [156]. The point at $1/r_0 m_{PS} = 0$ is given by $r_0^{3/2}\Phi_{RGI}$.

Taking the vacuum-to-B-meson matrix element of this relation (in static approximation) fixes

$$P^{\rm HQET} = -C_{\rm PS}(M_{\rm b}/\Lambda_{\overline{\rm MS}})\,\frac{m_{\rm B}}{M_{\rm b}}\,Z^{\rm stat}_{\rm A,RGI}(g_0)A_0^{\rm stat}\,, \tag{207}$$

$$S^{\rm HQET} = C_{\rm V}(M_{\rm b}/\Lambda_{\overline{\rm MS}})\,\frac{m_{\rm B}}{M_{\rm b}}\,Z^{\rm stat}_{\rm V/A}(g_0)\,Z^{\rm stat}_{\rm A,RGI}(g_0)\,V_0^{\rm stat} \tag{208}$$

in static approximation. (Choosing a different matrix element would change $m_{\rm B} \to m_{\rm B} + {\rm O}(\Lambda_{\rm QCD})$.) We leave it as an exercise to verify these relations.

All conversion functions C are known up to (presumably small) $\alpha(m_{\rm b})^3$ errors. At least when the conversion functions are known with such high precision, the knowledge of the leading term in expansions such as Eq. (196) is very useful to constrain the large mass behavior of QCD observables, computed on the lattice with unphysical quark masses $m_{\rm h} < m_{\rm b}$, typically $m_{\rm h} \approx m_{\rm charm}$. As illustrated in Fig. 25, one can then, with a reasonable smoothness assumption, interpolate to the physical point.

The relation between the RGI fields and the bare fields has so far been obtained for

- $Z^{\rm stat}_{\rm A,RGI}$ with both $N_{\rm f} = 0$ [10] and $N_{\rm f} = 2$ [157];
- the parity violating $\Delta B = 2$ four fermion operators [11,158] for $N_{\rm f} = 0$ while $N_{\rm f} = 2$ has been started.

In the second case, the matrix elements of the two operators, evaluated in twisted mass QCD will give the standard model B-parameter for B-$\overline{\rm B}$ mixing.

Thus, soon one will be able to do interpolations such as Fig. 25 also for semi-leptonic decays such as $B \to \pi\ l\ \nu$ (close to zero recoil) and for the B-parameter.

III.3.1.2. *Beyond the leading order: the need for non-perturbative conversion functions*

Still, getting the continuum extrapolations of the data at finite heavy quark masses in Fig. 25 under control may represent a challenge when dynamical fermions are included. Furthermore one should not forget that the functional form of the interpolation in that figure does essentially assume that the $1/m$ expansion remains reasonably accurate also significantly below $m_{\rm b}$.

It is therefore natural to try to compute the $1/m_{\rm b}$ correction directly in HQET. However, if one wants to do this consistently, the leading order conversion functions such as $C_{\rm PS}$ have to be known non-perturbatively. This general problem in the determination of power corrections in QCD is seen in the following way. Consider the error made in Eq. (198) (or Eq. (192)) when the anomalous dimension has been computed at l loops and $C_{\rm match}$ at $l-1$ loop order. The conversion function $C_{\rm PS}$ is then known up to an *error*

$$\Delta(C_{\rm PS}) \propto [\bar{g}^2(m_{\rm b})]^l \sim \left\{ \frac{1}{2b_0 \ln(m_{\rm b}/\Lambda_{\rm QCD})} \right\}^l \stackrel{m_{\rm b}\to\infty}{\gg} \frac{\Lambda_{\rm QCD}}{m_{\rm b}} . \qquad (209)$$

As $m_{\rm b}$ is made large, this perturbative error becomes dominant over the power correction one wants to determine. Taking a perturbative conversion function and adding power corrections to the leading order effective theory is thus a phenomenological approach, where one assumes that the coefficient of the $[\bar{g}^2(m_{\rm b})]^l$ term (as well as higher order ones) is small, such that the $\Lambda/m_{\rm b}$ corrections dominate over a certain mass interval. In such a phenomenological determination of a power correction, its size depends on the order of perturbation theory considered. A theoretically consistent evaluation of power corrections requires a fully non-perturbative formulation of the theory including a non-perturbative matching to QCD, see Sect. III.5.

Note for experts. Eq. (209) has little to do with renormalons. Rather it is due to the truncation of perturbation theory at a fixed order. Of course a renormalon-like growth of the perturbative coefficients does not help.

III.3.1.3. *Splitting leading order (LO) and next to leading order (NLO)*

We just learned that the very definition of a NLO correction to $F_{\rm B}$ means to take Eq. (165) with all coefficients $Z_{\rm A}^{\rm HQET} \ldots c_{\rm A}^{\rm HQET}$ determined non-perturbatively. We want to briefly explain that the split between LO and NLO is not unique. This is fully analogous to the case of standard perturbation theory in α, where the split depends on the renormalization scheme used, or better on the experimental observable used to determine α in the first place.

Consider the lowest order. The only coefficient needed in Eq. (165) is then $C_{\rm PS} Z_{\rm A,RGI}^{\rm stat} = \lim_{m_{\rm b}\to\infty} Z_{\rm A}^{\rm HQET}$. It has to be fixed by matching some matrix element of $A_0^{\rm stat}$ to the matrix element of A_0 in QCD. For example one may choose $\langle B'|A_0^\dagger|0\rangle$, with $|B'\rangle$ denoting the first pseudo-scalar excited state. Or one may take a finite volume matrix element $\langle B(L)|A_0^\dagger|\Omega(L)\rangle$, see Sect. III.4. Since the matching involves the QCD matrix element, there are higher order in $1/m_{\rm b}$ "pieces" in these equations. There is no reason for them to be independent of the particular matrix element. So from matching condition to matching condition, $C_{\rm PS} Z_{\rm A,RGI}^{\rm stat}$ differs by $\mathrm{O}(\Lambda_{\rm QCD}/m_{\rm b})$ terms.

The matrix element $F_{\rm B}$ in static approximation inherits this $\mathrm{O}(\Lambda_{\rm QCD}/m_{\rm b})$ ambiguity. The $\Lambda_{\rm QCD}/m_{\rm b}$ corrections are hence not unique. Fixing a matching condition, the leading order $F_{\rm B}$ as well as the one including the corrections can be computed and has a continuum limit. Their difference can be defined as the $1/m_{\rm b}$ correction. However, what matters is not the ambiguous NLO term, but the fact that the uncertainty is reduced from $\mathrm{O}(\Lambda_{\rm QCD}/m_{\rm b})$ in the LO term to $\mathrm{O}(\Lambda_{\rm QCD}^2/m_{\rm b}^2)$ in the sum.

III.3.2. *A second example: mass formulae*

Let us represent the formula for the meson mass, Eq. (160), including also the vector meson as

$$m_{\rm B}^{\rm av} \equiv \frac{1}{4}[m_{\rm B} + 3m_{\rm B^*}] = m_{\rm b} + \widehat{\delta m} + E_{\rm stat} + \omega_{\rm kin} E_{\rm kin} + \mathrm{O}(1/m_{\rm b}^2) \quad (210)$$

and

$$\Delta m_{\rm B} \equiv m_{\rm B^*} - m_{\rm B} = -\tfrac{4}{3}\omega_{\rm spin} E_{\rm spin} + \mathrm{O}(1/m_{\rm b}^2)\,, \quad (211)$$

where the fact that $\mathcal{O}_{\rm spin}$ does not contribute in Eq. (210) is a consequence of the exact spin symmetry of the lowest order. It is a simple exercise to verify the above equations explicitly by considering correlation functions of $V_k = \overline{\psi}_{\rm l}\gamma_k\psi_{\rm b}$ instead of A_0.

Usually these equations are written as [159]

$$m_{\rm B}^{\rm av} \sim m_{\rm b} + \overline{\Lambda} + \frac{1}{2m_{\rm b}}\lambda_1\,, \quad \Delta m_{\rm B} \sim -\frac{2}{m_{\rm b}}\lambda_2\,. \tag{212}$$

We arrive at a similar form by defining power divergent subtractions for $E_{\rm kin}$ and $E_{\rm stat}$ vs.

$$m_{\rm B}^{\rm av} = m_{\rm B}^{(0a)} + m_{\rm B}^{(0b)} + m_{\rm B}^{(1a)} + m_{\rm B}^{(1b)} + {\rm O}(\Lambda_{\rm QCD}^3/m_{\rm b}^2)\,, \tag{213}$$

where we have split up as

$$\begin{aligned} {\rm O}(m_{\rm b}):\ m_{\rm B}^{(0a)} &= m_{\rm b} + \widehat{\delta m}^{\rm stat} + E_{\rm stat}^{\rm sub}\,, & (214)\\ {\rm O}(\Lambda_{\rm QCD}):\ m_{\rm B}^{(0b)} &= E_{\rm stat} - E_{\rm stat}^{\rm sub}\,, & (215)\\ {\rm O}(\Lambda_{\rm QCD}^2/m_{\rm b}):\ m_{\rm B}^{(1a)} &= \widehat{\delta m}^{(1)} + \omega_{\rm kin}E_{\rm kin}^{\rm sub}\,, \quad m_{\rm B}^{(1b)} = \omega_{\rm kin}[E_{\rm kin} - E_{\rm kin}^{\rm sub}]\,. & (216)\end{aligned}$$

The subtraction terms (the Hamiltonian should be defined as $\mathbb{H} = -\frac{1}{a}\ln(\mathbb{T})$ in terms of the transfer matrix $\mathbb{T}$)

$$E_{\rm stat}^{\rm sub} = \langle\beta|\mathbb{H}|\beta\rangle_{\rm stat}|_{\delta m=0}\,, \quad E_{\rm kin}^{\rm sub} = \langle\beta| - a^3\sum_{\mathbf{x}} \mathcal{O}_{\rm kin}(x)|\beta\rangle_{\rm stat}\,. \tag{217}$$

are chosen such that $m_{\rm B}^{(0a)} \ldots m_{\rm B}^{(1b)}$ are finite and have a continuum limit. This follows from our discussion of renormalization. By $\widehat{\delta m}^{\rm stat}$ we denote $\widehat{\delta m}$ in static approximation and by $\widehat{\delta m}^{(1)}$ the piece that is to be added when $\omega_{\rm kin} \neq 0$; it accounts for the term c_2/a^2 in Eq. (168).

This rewriting exposes that the split up of the meson formulae into various orders of $1/m_{\rm b}$ is not unique. In the chosen form it depends on the arbitrary state $|\beta\rangle$. In other words: non-perturbatively, $\overline{\Lambda}$ and λ_1 can be defined exactly, but in many ways.[z]

Toward the end of this lecture we will see how the subtraction terms and $m_{\rm b} + \widehat{\delta m}$ can be defined such that they are computable in practice and the remaining error is reduced to ${\rm O}(\Lambda_{\rm QCD}^3/m_{\rm b}^2)$.

---

[z] In phenomenological applications[160], one works in dimensional regularization where the subtraction terms can be omitted. One defines a perturbative scheme for $m_{\rm b}$. It is then clear that (for example) $\overline{\Lambda}$ depends on the order of perturbation theory as $[\bar{g}^2(m_{\rm b})]^{l+1}m_{\rm b}] \overset{m_{\rm b}\to\infty}{\gg} \Lambda_{\rm QCD}$ as in Eq. (209). Quantities such as $\overline{\Lambda}$ are then *effective parameters*, not directly related to the asymptotic $1/m_{\rm b}$ expansion.

III.4. Non-perturbative tests of HQET

Although it is generally accepted that HQET is an effective theory of QCD, tests of this equivalence are rare and mostly based on phenomenological analysis of experimental results. A pure theory test can be performed if QCD including a heavy enough quark can be simulated on the lattice at lattice spacings which are small enough to be able to take the continuum limit. This has recently been achieved [154] and will be summarized below.

We start with the QCD side of such a test. Lattice spacings such that $am_{\rm b} \ll 1$ can be reached if one puts the theory in a finite volume, $L^3 \times T$ with L, T not too large. We shall use $T = L$. For various practical reasons, Schrödinger functional boundary conditions are chosen. Equivalent boundary conditions are imposed in the effective theory. We then consider correlation functions such as $f_{\rm A}$ and f_1. The first one is the correlator of boundary quark fields ζ (located at $x_0 = 0$) and the time component of the axial current in the bulk ($0 < x_0 < T$). The second one describes the propagation of a quark-antiquark pair from the $x_0 = 0$ boundary to the $x_0 = T$ boundary. See Sect. III.2.3 for details.

We then take a ratio for which the renormalization factors of the boundary fields cancel,

$$Y_{\rm PS}(L, M_{\rm b}) \equiv Z_{\rm A} \left. \frac{f_{\rm A}(L/2)}{\sqrt{f_1}} \right|_{T=L} = \frac{\langle \Omega(L)|A_0|B(L)\rangle}{||\,|\Omega(L)\rangle\,||\;||\,|B(L)\rangle\,||}, \tag{218}$$

$$|B(L)\rangle = {\rm e}^{-L\mathbb{H}/2}|\varphi_{\rm B}(L)\rangle\,, \; |\Omega(L)\rangle = {\rm e}^{-L\mathbb{H}/2}|\varphi_0(L)\rangle\,.$$

As shown in the above equations, $Y_{\rm PS}$ can be represented as a matrix element of the axial current between a normalized state $|B(L)\rangle$ with the quantum numbers of a B-meson and $|\Omega(L)\rangle$ which has vacuum quantum numbers. The time evolution ${\rm e}^{-L\mathbb{H}/2}$ ensures that both of these states are dominated by energy eigenstates with energies around $2/L$ and less (above $m_{\rm b}$). In other words, HQET is applicable if $1/L \ll m_{\rm b}$ (and of course $\Lambda_{\rm QCD} \ll m_{\rm b}$).

One then expects (for fixed $L\Lambda_{\rm QCD}$)

$$Y_{\rm PS}(L, M_{\rm b})/C_{\rm PS}(M_{\rm b}/\Lambda) = X_{\rm RGI} + {\rm O}(1/z)\,, \quad z = M_{\rm b}L\,, \tag{219}$$

where the $1/m_{\rm b}$ corrections are written in the dimensionless variable $1/z$ and $X_{\rm RGI}$ is defined as $Y_{\rm PS}$ but at lowest order in the effective theory and normalized as in Eq. (194). Of course such relations are expected after the continuum limit of both sides has been taken separately. For the case of $Y_{\rm PS}(L, M_{\rm b})$, this is done by the following steps:

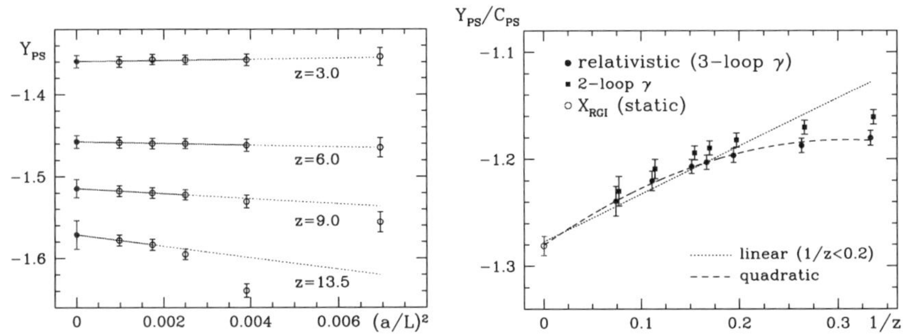

Fig. 26. Testing Eq. (219) through numerical simulations in the quenched approximation and for $L \approx 0.2\,\mathrm{fm}$ [154]. The physical mass of the b-quark corresponds to $z \approx 5$.

- Fix a value u_0 for the renormalized coupling $\bar{g}^2(L)$ (in the Schrödinger functional scheme) at vanishing quark mass. In Ref. 154 u_0 is chosen such that $L \approx 0.2\,\mathrm{fm}$.
- For a given resolution L/a, determine the bare coupling from the condition $\bar{g}^2(L) = u_0$. This can easily be done since the relation between bare and renormalized coupling is known [98].
- Fix the bare quark mass m_q of the heavy quark such that $LM = z$ using the known renormalization factors Z_M, Z in $M = Z_\mathrm{M} Z\,(1 + ab_\mathrm{m} m_\mathrm{q})\, m_\mathrm{q}$, where Z was introduced in Eq. (83) and Z_M in Eq. (120).
- Evaluate Y_PS and repeat for better resolution a/L.
- Extrapolate to the continuum as shown in Fig. 26, left.

As can be seen in the figure, the continuum extrapolation becomes more difficult as the mass of the heavy quark is increased and $\mathrm{O}((a\,m)^2)$ discretization errors become more and more important. In contrast the continuum extrapolation in the static effective theory (Fig. 27) is much easier (once the renormalization factor relating bare current and RGI current is known [10]). After the continuum limit has been taken, the finite mass QCD observable $Y_\mathrm{PS}(L, M)$ turns smoothly into the prediction from the effective theory as illustrated in the r.h.s. of Fig. 26. Several such successful tests were performed in Ref. 154, two of them with the static ($M \to \infty$) limit known from the spin symmetry of HQET. For lack of space we do not show more examples but only note that the coefficient of the $1/z^n$ terms in fits to the finite mass results together with the static limit are roughly of order unity.

Of course, finite mass lattice QCD results have been compared to the static limit over the years, see for example Refs. 161, 162, 163, 164, 165,

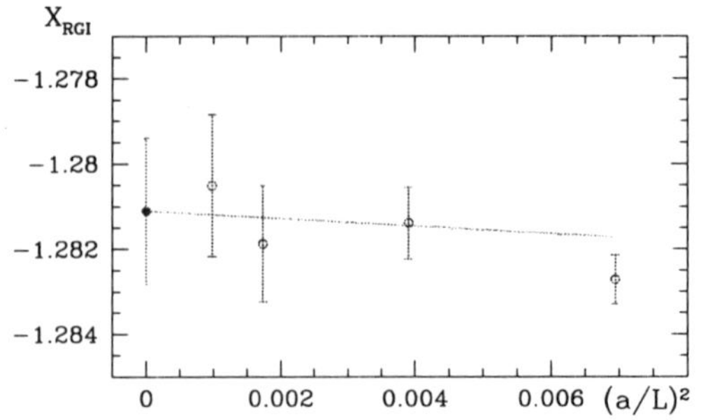

Fig. 27. Continuum extrapolation of $X_{\rm RGI}$ [154].

166, 167, 168, 169, 170, 171, 172, 173 and references therein. So what is new in the tests just discussed? The composite fields were renormalized non-perturbatively throughout and, by considering a small volume, the continuum limit could be taken at large quark masses.

III.5. Strategy for non-perturbative matching

Following Sect. III.3.1.2, the missing piece for a general computation including $1/m_{\rm b}$ corrections is a practical strategy for determining the parameters in the Lagrangian and in the effective fields beyond perturbation theory. Let us denote the number of parameters which have to be determined, not counting the parameters in the light sector of QCD, by $N_{\rm eff}$. For instance, including $1/m_{\rm b}$ terms but not considering matrix elements of any composite fields, we have $N_{\rm eff} = 3$, namely $\omega_{\rm kin}, \omega_{\rm spin}$ and $m_{\rm b} + \widehat{\delta m}$. Given these three parameters, all masses can be computed. If in addition we want to compute matrix elements of A_0, such as $F_{\rm B}$, we have $N_{\rm eff} = 5$, since also $Z_{\rm A}^{\rm HQET}, c_{\rm A}^{\rm HQET}$ are parameters of the theory.

III.5.1. *Matching in small volume*

Observables, e.g. dimensionless renormalized correlation functions or energies are denoted by $\Phi^{\rm HQET}$ in the effective theory and by $\Phi^{\rm QCD}$ in QCD. The $N_{\rm eff}$ unknown parameters can be determined from $\Phi_k^{\rm HQET} = \Phi_k^{\rm QCD}, k = 1 \ldots N_{\rm eff}$, provided the sensitivity of these conditions to the desired parameters is sufficient. In general the determination of $\Phi_k^{\rm QCD}$ will be very difficult because a b-quark has to be simulated and $\mathrm{O}((am_{\rm b})^2)$ cutoff effects will be large. The way around is once again to consider a finite volume, where small lattice spacings are accessible. In practice using furthermore Schrödinger functional boundary conditions is a good idea,

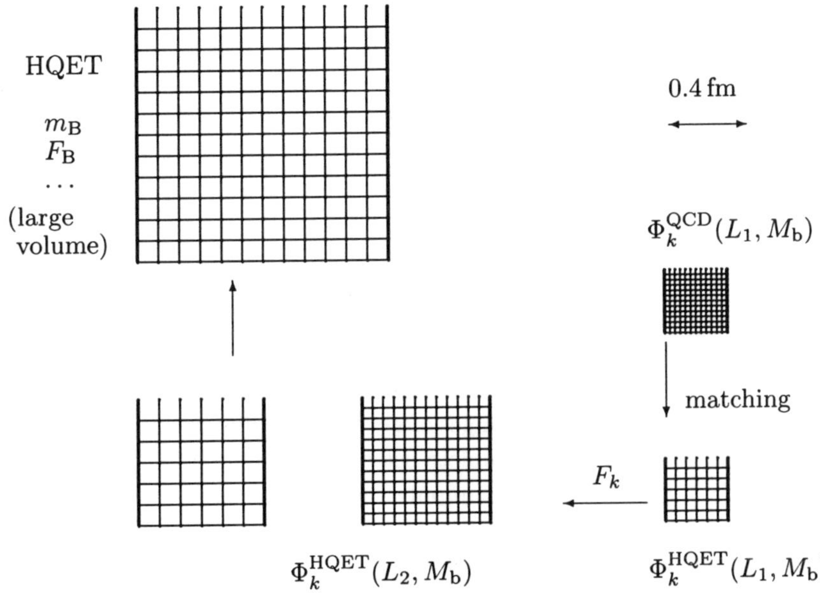

Fig. 28. The strategy for a non-perturbative determination of the HQET-parameters from QCD simulations in a small volume. Steps indicated by arrows are to be repeated at smaller lattice spacings to reach a continuum limit.

since then the simulations can easily be done also with dynamical fermions. Given the experience of the tests of HQET (Sect. III.4), a good choice is $L = L_1 \approx 0.4\,\mathrm{fm}$. Then $1/z = 1/(LM_b) \approx 1/10$ and the HQET expansion is very accurate. So for the matching step we impose

$$\Phi_k^{HQET}(L_1, M_b) = \Phi_k^{QCD}(L_1, M_b)\,, \quad k = 1, \ldots, N_{\text{eff}}\,. \tag{220}$$

to determine the N_{eff} parameters in the effective theory (right hand side of Fig. 28). We assume that the observables $\Phi_k(L, M_b)$ have been made dimensionless by multiplication with appropriate powers of L. They should be chosen with care (e.g. no large momenta should appear) but the effect of variations in the matching conditions on the final results is in any case of a higher order in the $1/m_b$ expansion.

III.5.2. *Step scaling functions*

The matching conditions, Eq. (220), define the HQET parameters for any value of the lattice spacing (or equivalently bare coupling). In practice, for

$L_1 \approx 0.4\,\text{fm}$, the parameters of the effective theory are then determined at rather small lattice spacings in a range of $a \approx 0.02\,\text{fm}$ to $a \approx 0.05\,\text{fm}$. Large volumes as they are needed to compute the physical mass spectrum or matrix elements then require very large lattices ($L/a > 50$). A further step is needed to bridge the gap to practicable lattice spacings. A well-defined procedure is as follows (bottom part of Fig. 28). We define step scaling functions [37], F_k, by

$$\Phi_k^{\text{HQET}}(sL, M) = F_k(\{\Phi_j^{\text{HQET}}(L, M)\,, j = 1 \ldots N_{\text{eff}}\})\,, \\ k = 1 \ldots N_{\text{eff}}\,, \tag{221}$$

where usually one uses scale changes of $s = 2$. These dimensionless functions describe the change of the complete set of observables $\{\Phi_k^{\text{HQET}}\}$ under a scaling of $L \to sL$. In order to compute them one

i selects a lattice with a certain resolution a/L.
ii The specification of $\Phi_j^{\text{HQET}}(L, M)$, $j = 1, \ldots, N_n$, then fixes all (bare) parameters of the theory.
iii The l.h.s. of Eq. (221) is now computed, keeping the bare parameters fixed while changing $L/a \to L'/a = sL/a$.
iv The values for the continuum F_k are reached by extrapolating the resulting lattice numbers to $a/L \to 0$.

Starting from $L = L_1$ it turns out that a single step going to $L_2 = 2L_1$ is sufficient [140]. Then one can switch at fixed bare parameters to a large volume where finite size effects are negligible (left part of Fig. 28). This is done in full analogy to the steps i…iv above. One only computes the large volume quantities for the bare parameters fixed in step ii (with $L = L_2$). Again a continuum extrapolation can be carried out.

III.5.3. *Example: The mass of the b quark*

For illustration purposes we consider a simple example, the computation of the b-quark mass, starting from the observed (spin averaged) B-meson mass. For this calculation one obviously has to consider a range of masses M_{b} in Eq. (220) and determine the physical point from the requirement $m_{\text{B}}^{\text{av}} = m_{\text{B}}^{\text{av}}|_{\text{experiment}}$. It is thus the first computation to be carried out. Subsequently, one may directly choose the physical point in Eq. (220).

III.5.3.1. *Static approximation*

Remembering Eq. (214), we are after a precise definition and calculation of $E_{\rm stat}^{\rm sub}$ and $m_{\rm B}^{(0a)} = \widehat{\delta m}^{\rm stat} + E_{\rm stat}^{\rm sub}$. Already in the static approximation, a non-perturbative matching is required, if one wants to take the continuum limit. (Perturbatively one determines $\widehat{\delta m}^{\rm stat}$ with an in the continuum limit divergent error term of order $\Delta(\widehat{\delta m}^{\rm stat}) = c_{l+1} g_0^{2l+2}/a$.)

In Eq. (220) we have the simple case $N_{\rm eff} = 1$. We omit the discussion of fixing the bare light quark masses and coupling. Obviously any finite volume energy in the b-sector, denoted by Γ, will do to fix $\widehat{\delta m}$. Two precise definitions are given in the following four equations. The reader who is only interested in the general concept may skip this detail.

The first choice which comes to mind is[139,60,140]

$$\Gamma^{\rm av}(L,\theta_0) = -\frac{\partial_0 + \partial_0^*}{2} F_{\rm av}(x_0,\theta_0) \text{ at } x_0 = L/2\,,\; T = L\,, \tag{222}$$

with (see Sect. III.2.3)

$$F_{\rm av}(x_0,\theta) = \frac{1}{4}\ln\left[-f_{\rm A}(x_0,\theta)\,(k_{\rm V}(x_0,\theta))^3\right]. \tag{223}$$

However, at order $1/m_{\rm b}$ the energy $\Gamma^{\rm av}$ depends on $c_{\rm A}^{\rm HQET}, c_{\rm V}^{\rm HQET}$. This is clearly inconvenient and can be avoided by choosing instead

$$\Gamma_1(L,\theta_0) = -\frac{\partial_T + \partial_T^*}{2} F_1(L,\theta_0) \text{ at } T = L/2\,, \tag{224}$$

with

$$F_1(L,\theta) = \frac{1}{4}\ln\left[f_1(\theta)\,(k_1(\theta))^3\right]. \tag{225}$$

A spin average is taken which will be relevant when we include the first order in $1/m_{\rm b}$. Both $\Gamma^{\rm av}$ and Γ_1 turn into $m_{\rm B}^{\rm av}$ when L and T are large.

In the numerical evaluation we chose $\theta_0 = 0$. We require matching, $\Phi_1^{\rm QCD}(L_1,M_{\rm b}) = \Phi_1^{\rm HQET}(L_1,M_{\rm b})$, with

$$\Phi_1^{\rm QCD}(L_1,M_{\rm b}) \equiv L_1\,\Gamma_1(L_1,\theta_0)\,, \tag{226}$$

$$\Phi_1^{\rm HQET}(L_1,M_{\rm b}) \equiv L_1\,(\Gamma_1^{\rm stat}(L_1,\theta_0) + m_{\rm b})\,. \tag{227}$$

Here $\Gamma_1^{\rm stat}$ refers to Eq. (224) at the lowest order in $1/m_{\rm b}$. Eq. (221) can then be written in the simple form,

$$\Phi_1^{\rm HQET}(2L,M_{\rm b}) = 2\Phi_1^{\rm HQET}(L,M_{\rm b}) + \sigma_{\rm m}\left(\bar g^2(L)\right)\,, \tag{228}$$

$$\sigma_{\rm m}\left(\bar g^2(L)\right) \equiv 2L\left[\,\Gamma_1^{\rm stat}(2L,\theta_0) - \Gamma_1^{\rm stat}(L,\theta_0)\,\right]. \tag{229}$$

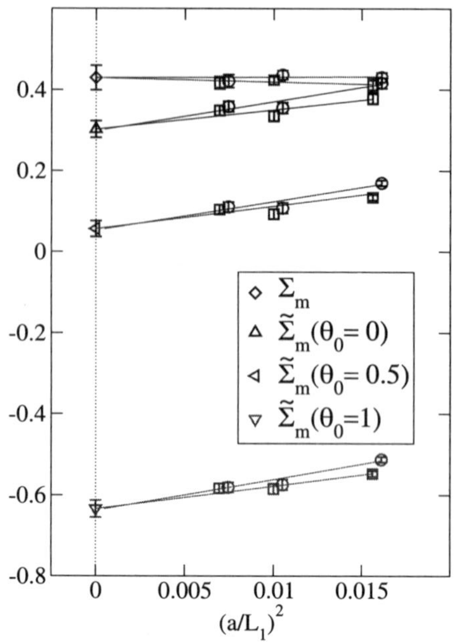

Fig. 29. Continuum extrapolation of $\sigma_{\rm m} = \lim_{a/L\to 0} \Sigma_{\rm m}$ and $\tilde{\sigma}_{\rm m}$. Two different discretizations for the static quark action are used and extrapolated to a common continuum limit[140]. Dividing by L_2, the y-axes covers about an energy range of 300 MeV.

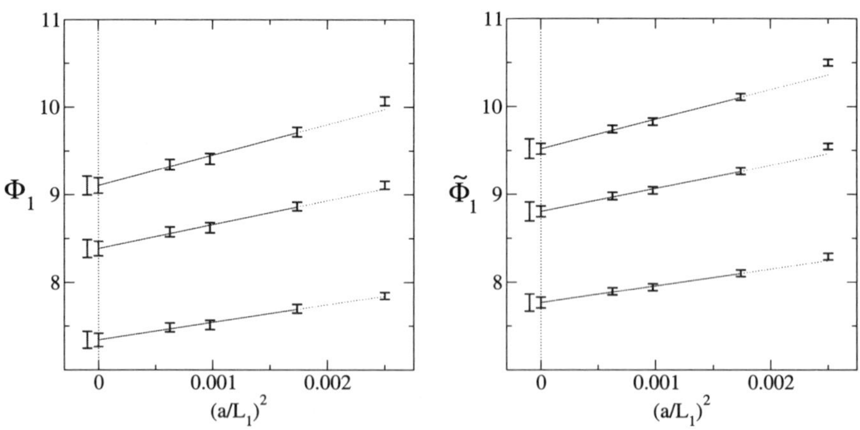

Fig. 30. Continuum extrapolation of $\Phi_1(L_1, M_{\rm b})$, for $z = 10.4\,,\ 12.1\,,\ 13.3$ from bottom to top[140]. On the right, the equivalent in the alternative strategy is shown with $\theta_0 = 1/2$. Dividing by L_1, the y-axes covers about a range of 2 GeV.

In $\sigma_{\rm m}$ the divergent δm as well as the mass shift $m_{\rm b}$ cancel. Its continuum extrapolation is illustrated in Fig. 29.

We now see immediately that

$$m_{\rm B}^{\rm av} = m_{\rm B}^{(0a)} + m_{\rm B}^{(0b)} + {\rm O}(\Lambda_{\rm QCD}^2/m_{\rm b}) \tag{230}$$

$$m_{\rm B}^{(0b)} = \underbrace{E_{\rm stat} - \Gamma_1^{\rm stat}(L_2,\theta_0)}_{a\to 0 \text{ in HQET}} \tag{231}$$

$$m_{\rm B}^{(0a)} = \underbrace{\Gamma_1^{\rm stat}(L_2,\theta_0) - \Gamma_1^{\rm stat}(L_1,\theta_0)}_{a\to 0 \text{ in HQET}} + \tfrac{1}{L_1}\underbrace{\Phi_1^{\rm QCD}(L_1,M_{\rm b})}_{a\to 0 \text{ in QCD}}, \tag{232}$$

where the first term in Eq. (232) (in dimensionless form) is given by Eq. (229) and $E_{\rm stat}$ is the infinite volume energy of a B-meson in static approximation introduced earlier. It is often called the static binding energy. As indicated, the continuum limit can be taken in each individual step; a numerical example for the last term is shown in Fig. 30.

After obtaining all pieces in Eq. (230), the equation is numerically solved for $z_{\rm b} = M_{\rm b}L_1$, see Fig. 31. Since also the size of L_1 in units of r_0 [47] is known, one can quote

$$r_0 M_{\rm b}^{(0)} = 17.25(20) \;\to\; M_{\rm b}^{(0)} = 6.806(79)\,{\rm GeV}, \tag{233}$$

where $r_0 = 0.5\,{\rm fm}$ is used. This result is in the quenched approximation but includes the lowest non-trivial order in $1/m_{\rm b}$ and a continuum limit. In the form of Eq. (213), we have defined the subtraction $E_{\rm stat}^{\rm sub} = \Gamma_1^{\rm stat}(L_2)$. The term $m_{\rm B}^{(0a)}$ is then given entirely by the finite volume computations while $m_{\rm B}^{(0b)}$ results from the large volume $E_{\rm stat}$ and $\Gamma_1^{\rm stat}(L_2)$. We proceed to discuss the $1/m_{\rm b}$ corrections.

III.5.3.2. *Including $1/m_{\rm b}$ corrections.*

Both the spin average $m_{\rm B}^{\rm av}$ and the finite volume energy Γ_1 are constructed such that $\omega_{\rm spin}$ drops out of their $1/m_{\rm b}$ expansion. Furthermore, Γ_1 does not obtain any $1/m_{\rm b}$-terms due to the expansion of composite fields, a convenient property of the Schrödinger functional boundary fields. Thus, for the particular problem of relating $m_{\rm B}^{\rm av}$ to the mass of the quark, only one more matching observable has to be defined, to determine $\omega_{\rm kin}$.

The choice in Ref. 140 was

$$\Phi_2(L,M_{\rm b}) = R_1(L,\theta_1,\theta_2) - R_1^{\rm stat}(L,\theta_1,\theta_2), \tag{234}$$

$$R_1(L,\theta_1,\theta_2) = F_1(L,\theta_1) - F_1(L,\theta_2) \;\text{ at } T = L/2, \tag{235}$$

$$R_1^{\rm stat}(L,\theta_1,\theta_2) = \ln\left[f_1^{\rm stat}(\theta_1)/f_1^{\rm stat}(\theta_2)\right], \;\text{ at } T = L/2 \tag{236}$$

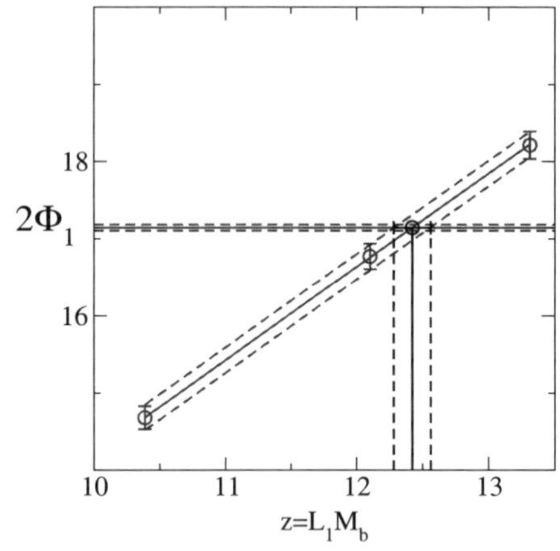

Fig. 31. Graphical solution of Eq. (230). Data points are $2\Phi_1$ and the horizontal error band is $L_2 m_{\rm B}^{\rm av} - \sigma_{\rm m} - L_2[E^{\rm stat} - \Gamma_1^{\rm stat}(L_2, \theta_0)]$.

where the static piece $R_1^{\rm stat}$ is subtracted such that Φ_2 is proportional to $\omega_{\rm kin}$. It is now a matter of simple algebra to compute the step scaling functions and finally the $1/m_{\rm b}$-correction to Eq. (233). In the numerical evaluation three combinations $(\theta_1, \theta_2) = (0, 1/2)\,, (1/2, 1)\,, (0, 1)$ were chosen. The $1/m_{\rm b}$-correction, split as in Eq. (213), but for the quark mass, turned out to be about independent of θ_1, θ_2 with (after continuum extrapolation)

$$r_0\, M_{\rm b}^{(1a)} = -0.06(3)\,, \quad r_0\, M_{\rm b}^{(1b)} = -0.06(8) \;\rightarrow\; r_0 M_{\rm b} = 17.12(22). \tag{237}$$

Indeed since for our choice of matching condition the static approximation is independent of θ_1, θ_2, the values for $r_0\, M_{\rm b}^{(1)}$ have to be independent up to small $r_0\, \Lambda_{\rm QCD}^3/m_{\rm b}^2$ terms. This is so because the result including all terms has this precision.

With $\Lambda_{\overline{\rm MS}} r_0 = 0.602(48)$ [17,93], the 4-loop β function and the mass anomalous dimension [174–177], we translate $M_{\rm b} = M_{\rm b}^{(0)} + M_{\rm b}^{(1)}$ to the mass in the $\overline{\rm MS}$ scheme,

$$\overline{m}_{\rm b}(\overline{m}_{\rm b}) = 4.347(48){\rm GeV}\,; \tag{238}$$

the associated perturbative uncertainty can safely be neglected. In the $\overline{\rm MS}$ scheme the $1/m_{\rm b}$ term amounts to $-27(22)$MeV.

III.5.3.3. *An alternative strategy.*

Also an alternative strategy has been tested. It is based on Γ^{av}, Eq. (222). It hence also involves the $1/m_{\mathrm{b}}$-corrections to the currents, but only in the combination $c_{\mathrm{av}}^{\mathrm{HQET}} = (c_{\mathrm{A}}^{\mathrm{HQET}} + 3c_{\mathrm{V}}^{\mathrm{HQET}})/4$. Three matching observables are necessary. They can be found in Ref. 140. Here we only mention that again three angles $\theta_0, \theta_1, \theta_2$ appear, but now in nine different combinations. Together with the above discussed strategy, there are 12 different matching conditions.

At lowest order in $1/m_{\mathrm{b}}$, the equations relating quark mass and the spin-averaged B-meson mass remain unchanged. Only the step scaling function σ_{m} is replaced by $\tilde{\sigma}_{\mathrm{m}}$, and Φ_1 is replaced by $\tilde{\Phi}_1$. Both depend on the angle θ_0, but not on θ_1, θ_2. A comparison is shown in Fig. 29, Fig. 30. Dividing by L_1 and L_2 respectively, one observes differences of about 250 MeV in these quantities. According to the HQET counting they are expected to be of order Λ_{QCD}. In the combination, which leads to $M_{\mathrm{b}}^{(0)}$, these differences are reduced to $\mathrm{O}(\Lambda_{\mathrm{QCD}}^2/m_{\mathrm{b}})$.

In the table shown, such differences are barely visible in the lowest order results, where one would estimate $\mathrm{O}(\Lambda_{\mathrm{QCD}}^2/m_{\mathrm{b}}^2) \sim 1\%$ effects. After including the $1/m_{\mathrm{b}}$ corrections no signs of differences remain. Indeed one would estimate them to be $\mathrm{O}((\Lambda_{\mathrm{QCD}}/m_{\mathrm{b}})^3) \sim 0.1\%$, quite below our statistical precision. We conclude that the expansion works very well, as expected.

θ_0	$r_0\, M_{\mathrm{b}}^{(0)}$	$r_0\, M_{\mathrm{b}} = r_0\,(M_{\mathrm{b}}^{(0)} + M_{\mathrm{b}}^{(1a)} + M_{\mathrm{b}}^{(1b)})$		
		$\theta_1 = 0$, $\theta_2 = 1/2$	$\theta_1 = 1/2$, $\theta_2 = 1$	$\theta_1 = 1$, $\theta_2 = 0$
		Main strategy		
0	17.25(20)	17.12(22)	17.12(22)	17.12(22)
		Alternative strategy		
0	17.05(25)	17.25(28)	17.23(27)	17.24(27)
1/2	17.01(22)	17.23(28)	17.21(27)	17.22(28)
1	16.78(28)	17.17(32)	17.14(30)	17.15(30)

We finally note that the overall uncertainty of the b-quark mass is dominated by the renormalization factors in relativistic QCD, i.e. by the relation between the bare mass and the RGI mass in QCD. The second most important contribution to the errors is the large volume E_{kin}. This quantity contains a quadratic divergence. Once $E_{\mathrm{kin}}^{\mathrm{sub}}$ is subtracted, the remaining finite piece has errors which grow fast as one approaches the continuum limit. However, the methods of Ref. 178 have not yet been applied and we expect to reduce this source of error.

III.6. More literature

Before summarizing and discussing the perspectives for HQET on the lattice, let us attempt to give a brief guide to literature on subjects which we did not cover.

Early numerical computations have been summarized in reviews [162,179,180]. The power divergent mixing of operators of different dimensions was seen in an explicit perturbative computation [181]. The $\mathrm{B}^* - \mathrm{B}$ mass splitting, which is a $1/m_\mathrm{b}$ effect given at lowest order by E_spin has been investigated with perturbative ω_spin by Refs. 182, 183, 184, 185.

In static approximation, the four fermion operators responsible for $\mathrm{B}-\overline{\mathrm{B}}$ mixing with Wilson fermions require the subtraction of operators with different chiralities, which was attempted with perturbatively estimated mixing coefficients [186,184]. Later it was realized that the mixing is completely avoided with an action with exact chiral symmetry (and spin symmetry) [187] and a calculation of $\mathrm{B} - \overline{\mathrm{B}}$ mixing was carried out with perturbative renormalization [188].

The dependence of heavy-light meson properties on the mass of the light quark can be described by a suitable chiral effective theory [189,190]. Its Lagrangian involves the $\mathrm{B}^*\mathrm{B}\pi$ coupling as a low energy constant. This was estimated in a static computation [191]. Also an exploratory computation of the Isgur-Wise functions $\tau_{1/2}, \tau_{3/2}$ at zero recoil has been carried out [192,193]. These functions give the form factors for transitions between heavy-light mesons of different orbital angular momentum.

Finally, the formulation of HQET at a finite velocity has been investigated by Aglietti et al. We refer to Refs. 138, 194 and references therein. Both perturbative investigations of the renormalization [195,193] and numerical simulations [196,192] have been carried out.

III.7. Summary and perspectives

Non-perturbative HQET at the leading order in $1/m_\mathrm{b}$ has reached a satisfactory status. To underline this statement, we comment briefly on the progress made in recent years. The ground breaking work of Refs. 132, 197, 198 was followed by intense activity leading to a computation where (within the quenched approximation) all error sources were investigated and controlled as well as possible in the middle of the nineties [199]. For the B decay constant, the authors estimated three sources of errors between 7% and 12% each, one of them due to an estimated precision of the perturbative renormalization.

It took almost a decade until a non-perturbative method for the renormalization in the effective theory was fully developed [9,10], but meanwhile a total error of 4% in the B_s-meson decay constant has been reached [200], including the continuum extrapolation. In reaching this accuracy also the reduction of statistical errors [145,146] was relevant.

Also the standard model B-parameter for B-$\overline{\rm B}$ mixing is well on its way [11,158,201]. In these cases, bare perturbation theory is now avoided by non-perturbative renormalization of the lattice operators. As explained in Sect. III.3.1, all static-light bilinears require only one common renormalization factor $Z^{\rm stat}_{\rm A,RGI}$. As this is known [10], also semi-leptonic decay form factors can be computed with non-perturbative renormalization. For completeness we note that a source of a perturbative error remains in $C_{\rm PS}$, Fig. 24, and its relatives for other bilinears, but here high order continuum perturbation theory is available [151–153,150] and is applied at the b-scale. This error is under reasonable control. Applying these methods to the theory with dynamical fermions is straight forward; "only" the usual problems of simulations with light quarks have to be solved.

By themselves such lowest order (in $1/m_{\rm b}$) results are not expected to have an interesting precision for phenomenological applications, but certainly they can constrain the large mass behavior computed with other methods [161–173]. An interesting application has been the combination of the approach of Refs. 172, 173 with the large mass behavior computed via HQET [200].

The $1/m_{\rm b}$ corrections can also be computed directly in the effective theory. Here, the necessary steps have been carried out in detail for the mass of the b-quark in the quenched approximation. The resulting precision is rather satisfactory and higher order $1/m_{\rm b}$-corrections can be neglected. The latter has also been verified explicitly by comparing the results following from a number of different matching conditions. No significant differences were found.

An extension of this computation to full QCD has been started by the ALPHA collaboration. This is of particular interest because for full QCD it is of course more difficult to reach the small lattice spacings needed for computations with relativistic quarks of masses around $m_{\rm charm}$ – the basis of the alternatives mentioned above. It is time to address other observables such as the B-meson decay constant in this direct approach!

Acknowledgements

I am grateful to the organizers of this school for composing a very interesting programme and a pleasant atmosphere In particular I thank Y. Kuramashi for his efforts and his patience. I would like to thank my friends in the ALPHA-collaboration for the enjoyable and fruitful collaboration and all I learned from them. In particular it is worth emphasizing that the lecture on HQET is based on joint work with M. Della Morte, N. Garron, J. Heitger and M. Papinutto. I am grateful to N. Garron, J. Heitger and S. Takeda for sending figures or data to be used in this writeup and to M. Della Morte, D. Guazzini and H. Simma and U. Wolff for comments on the manuscript. I finally thank NIC/DESY for allocating computer time for the ALPHA-projects, which was essential for most of the numerical investigations that were discussed in these lectures.

References

1. K. Symanzik, *Nucl. Phys.* **B226**, 187 (1983).
2. H. Neuberger, *Phys. Lett.* **B417**, 141 (1998), hep-lat/9707022.
3. P. Hasenfratz, V. Laliena and F. Niedermayer, *Phys. Lett.* **B427**, 125 (1998), hep-lat/9801021.
4. M. Lüscher, *Phys. Lett.* **B428**, 342 (1998), hep-lat/9802011.
5. ALPHA, R. Frezzotti et al., *JHEP* **08**, 058 (2001), hep-lat/0101001.
6. ALPHA, R. Frezzotti, S. Sint and P. Weisz, *JHEP* **07**, 048 (2001), hep-lat/0104014.
7. R. Frezzotti and G.C. Rossi, *JHEP* **08**, 007 (2004), hep-lat/0306014.
8. R. Frezzotti et al., *JHEP* **04**, 038 (2006), hep-lat/0503034.
9. ALPHA, M. Kurth and R. Sommer, *Nucl. Phys.* **B597**, 488 (2001), hep-lat/0007002.
10. ALPHA, J. Heitger, M. Kurth and R. Sommer, *Nucl. Phys.* **B669**, 173 (2003), hep-lat/0302019.
11. F. Palombi et al., *JHEP* **08**, 017 (2006), hep-lat/0604014.
12. M. Lüscher et al., *Nucl. Phys.* **B384**, 168 (1992), hep-lat/9207009.
13. K. Symanzik, *Nucl. Phys.* **B190**, 1 (1981).
14. M. Lüscher, *Nucl. Phys.* **B254**, 52 (1985).
15. R. Narayanan and U. Wolff, *Nucl. Phys.* **B444**, 425 (1995), hep-lat/9502021.
16. ALPHA, A. Bode, P. Weisz and U. Wolff, *Nucl. Phys.* **B576**, 517 (2000), Erratum, *ibid.* **B600**, 453 (2001), Erratum, *ibid.* **B608**, 481 (2001), hep-lat/9911018.
17. M. Lüscher et al., *Nucl. Phys.* **B413**, 481 (1994), hep-lat/9309005.
18. S. Sint, *Nucl. Phys.* **B421**, 135 (1994), hep-lat/9312079.
19. M. Lüscher et al., *Nucl. Phys.* **B478**, 365 (1996), hep-lat/9605038.
20. M. Bochicchio et al., *Nucl. Phys.* **B262**, 331 (1985).
21. M. Lüscher et al., *Nucl. Phys.* **B491**, 344 (1997), hep-lat/9611015.
22. M. Della Morte et al., *JHEP* **07**, 007 (2005), hep-lat/0505026.

23. ALPHA, M. Della Morte et al., *Nucl. Phys.* **B729**, 117 (2005), hep-lat/0507035.
24. G.C. Rossi and M. Testa, *Nucl. Phys.* **B163**, 109 (1980).
25. G.C. Rossi and M. Testa, *Nucl. Phys.* **B176**, 477 (1980).
26. Y. Taniguchi, *JHEP* **12**, 037 (2005), hep-lat/0412024.
27. Y. Taniguchi, *JHEP* **10**, 027 (2006), hep-lat/0604002.
28. S. Sint, *PoS* **LAT2005**, 235 (2006), hep-lat/0511034.
29. R. Frezzotti and G.C. Rossi, *JHEP* **10**, 070 (2004), hep-lat/0407002.
30. L. Maiani and G. Martinelli, *Phys. Lett.* **B178**, 265 (1986).
31. R. Sommer, in *Schladming 1997, Computing particle properties 65-113* (1997), hep-ph/9711243.
32. K. Jansen et al., *Phys. Lett.* **B372**, 275 (1996), hep-lat/9512009.
33. ALPHA, J. Garden et al., *Nucl. Phys.* **B571**, 237 (2000), hep-lat/9906013.
34. R. Sommer, *Nucl. Phys. Proc. Suppl.* **42**, 186 (1995), hep-lat/9411024.
35. M. Lüscher, (1998), hep-lat/9802029.
36. M. Hasenbusch et al., *Nucl. Phys. Proc. Suppl.* **106**, 911 (2002), hep-lat/0110202.
37. M. Lüscher, P. Weisz and U. Wolff, *Nucl. Phys.* **B359**, 221 (1991).
38. P. Hasenfratz and F. Niedermayer, *Nucl. Phys.* **B596**, 481 (2001), hep-lat/0006021.
39. F. Knechtli, B. Leder and U. Wolff, *Nucl. Phys.* **B726**, 421 (2005), hep-lat/0506010.
40. U. Wolff et al., *PoS* **LAT2005**, 253 (2006), hep-lat/0509043.
41. K.G. Wilson, *Phys. Rev.* **D10**, 2445 (1974).
42. B. Sheikholeslami and R. Wohlert, *Nucl. Phys.* **B259**, 572 (1985).
43. M. Lüscher and P. Weisz, *Nucl. Phys.* **B479**, 429 (1996), hep-lat/9606016.
44. R. Wohlert, *Improved continuum limit lattice action for quarks*, DESY 87/069 .
45. G. Heatlie et al., *Nucl. Phys.* **B352**, 266 (1991).
46. S. Sint and P. Weisz, *Nucl. Phys.* **B502**, 251 (1997), hep-lat/9704001.
47. R. Sommer, *Nucl. Phys.* **B411**, 839 (1994), hep-lat/9310022.
48. ALPHA, J. Heitger, *Nucl. Phys.* **B557**, 309 (1999), hep-lat/9903016.
49. C.M. de Divitiis and R. Petronzio, *Phys. Lett.* **B419**, 311 (1998), hep-lat/9710071.
50. T. Bhattacharya et al., *Phys. Rev.* **D63**, 074505 (2001), hep-lat/0009038.
51. ALPHA, M. Guagnelli et al., *Nucl. Phys.* **B595**, 44 (2001), hep-lat/0009021.
52. M. Lüscher et al., *Nucl. Phys.* **B491**, 323 (1997), hep-lat/9609035.
53. ALPHA, K. Jansen and R. Sommer, *Nucl. Phys.* **B530**, 185 (1998), hep-lat/9803017.
54. UKQCD, S. Collins et al., *Phys. Rev.* **D67**, 014504 (2003), hep-lat/0110159.
55. S. Dürr and M. Della Morte, *Nucl. Phys. Proc. Suppl.* **129**, 417 (2004), hep-lat/0309169.
56. M. Della Morte, R. Hoffmann and R. Sommer, *JHEP* **03**, 029 (2005), hep-lat/0503003.
57. S. Sint and R. Sommer, *Nucl. Phys.* **B465**, 71 (1996), hep-lat/9508012.
58. Y. Taniguchi and A. Ukawa, *Phys. Rev.* **D58**, 114503 (1998), hep-

lat/9806015.
59. S. Aoki and Y. Kuramashi, *Phys. Rev.* **D68**, 094019 (2003), hep-lat/0306015.
60. ALPHA, J. Heitger and J. Wennekers, *JHEP* **02**, 064 (2004), hep-lat/0312016.
61. JLQCD, N. Yamada et al., *Phys. Rev.* **D71**, 054505 (2005), hep-lat/0406028.
62. S. Aoki, R. Frezzotti and P. Weisz, *Nucl. Phys.* **B540**, 501 (1999), hep-lat/9808007.
63. CP-PACS, S. Aoki et al., *Phys. Rev.* **D73**, 034501 (2006), hep-lat/0508031.
64. K.H. Mutter et al., in *Brookhaven 1986, proceedings, Lattice Gauge Theory '86* 257-267.
65. T.A. DeGrand, A. Hasenfratz and T.G. Kovacs, *Nucl. Phys.* **B547**, 259 (1999), hep-lat/9810061.
66. JLQCD, S. Aoki et al., *Phys. Rev.* **D72**, 054510 (2005), hep-lat/0409016.
67. L. Del Debbio et al., *JHEP* **02**, 011 (2006), hep-lat/0512021.
68. ALPHA, M. Della Morte et al., *Comput. Phys. Commun.* **165**, 49 (2005), hep-lat/0405017.
69. ALPHA, A. Bode et al., *Phys. Lett.* **B515**, 49 (2001), hep-lat/0105003.
70. ALPHA, M. Della Morte et al., *Nucl. Phys.* **B713**, 378 (2005), hep-lat/0411025.
71. ALPHA, JLQCD and CPPACS, R. Sommer et al., *Nucl. Phys. Proc. Suppl.* **129**, 405 (2004), hep-lat/0309171.
72. S.R. Sharpe, (2006), hep-lat/0607016.
73. G. Martinelli et al., *Nucl. Phys.* **B445**, 81 (1995), hep-lat/9411010.
74. G. Martinelli et al., *Nucl. Phys.* **B611**, 311 (2001), hep-lat/0106003.
75. S. Capitani et al., *Nucl. Phys.* **B593**, 183 (2001), hep-lat/0007004.
76. S.R. Sharpe, *Nucl. Phys. Proc. Suppl.* **106**, 817 (2002), hep-lat/0110021.
77. T. Bhattacharya et al., *Phys. Rev.* **D73**, 034504 (2006), hep-lat/0511014.
78. O. Bar, G. Rupak and N. Shoresh, *Phys. Rev.* **D67**, 114505 (2003), hep-lat/0210050.
79. M. Lüscher, *JHEP* **05**, 052 (2003), hep-lat/0304007.
80. M. Lüscher, *Comput. Phys. Commun.* **156**, 209 (2004), hep-lat/0310048.
81. M. Lüscher, *Comput. Phys. Commun.* **165**, 199 (2005), hep-lat/0409106.
82. M. Hasenbusch, *Phys. Lett.* **B519**, 177 (2001).
83. C. Urbach et al., *Comput. Phys. Commun.* **174**, 87 (2006), hep-lat/0506011.
84. M.A. Clark and A.D. Kennedy, (2006), hep-lat/0608015.
85. A.D. Kennedy, (2006), hep-lat/0607038.
86. W.M. Yao et al., *Journal of Physics* **G33**, 1 (2006).
87. G. Grunberg, *Phys. Lett.* **B95**, 70 (1980).
88. G. Grunberg, *Phys. Rev.* **D29**, 2315 (1984).
89. I. Montvay and G. Munster, "*Quantum fields on a lattice*", (Cambridge, 1994).
90. C. Michael, *Phys. Lett.* **B283**, 103 (1992), hep-lat/9205010.
91. UKQCD, S.P. Booth et al., *Nucl. Phys.* **B394**, 509 (1993), hep-lat/9209007.
92. G.S. Bali and K. Schilling, *Phys. Rev.* **D47**, 661 (1993), hep-lat/9208028.
93. S. Necco and R. Sommer, *Nucl. Phys.* **B622**, 328 (2002), hep-lat/0108008.

94. S. Necco and R. Sommer, *Phys. Lett.* **B523**, 135 (2001), hep-ph/0109093.
95. M. Lüscher et al., *Nucl. Phys.* **B389**, 247 (1993), hep-lat/9207010.
96. G.M. de Divitiis et al., *Nucl. Phys.* **B433**, 390 (1995), hep-lat/9407028.
97. ALPHA, G. de Divitiis et al., *Nucl. Phys.* **B437**, 447 (1995), hep-lat/9411017.
98. ALPHA, S. Capitani et al., *Nucl. Phys.* **B544**, 669 (1999), hep-lat/9810063.
99. F. Palombi, C. Pena and S. Sint, *JHEP* **03**, 089 (2006), hep-lat/0505003.
100. ALPHA, M. Guagnelli et al., *JHEP* **03**, 088 (2006), hep-lat/0505002.
101. ALPHA, P. Dimopoulos et al., (2006), hep-lat/0610077.
102. G. Martinelli et al., *Nucl. Phys.* **B445**, 81 (1995), hep-lat/9411010.
103. A. Donini et al., *Eur. Phys. J.* **C10**, 121 (1999), hep-lat/9902030.
104. D. Becirevic et al., **JHEP 08**, 022 (2004), hep-lat/0401033.
105. SPQcdR, D. Becirevic et al., *Nucl. Phys. Proc. Suppl.* **140**, 246 (2005), hep-lat/0409110.
106. ALPHA, M. Guagnelli, R. Sommer and H. Wittig, *Nucl. Phys.* **B535**, 389 (1998), hep-lat/9806005.
107. K. Symanzik, in *Mathematical problems in theoretical physics*, eds. R. Schrader et al., Lecture Notes in Physics Vol. 153 (Springer, New York, 1982).
108. M. Lüscher and P. Weisz, *Commun. Math. Phys.* **97**, 59 (1985).
109. Y. Iwasaki, *Nucl. Phys.* **B258**, 141 (1985).
110. QCD-TARO, P. de Forcrand et al., *Nucl. Phys.* **B577**, 263 (2000), hep-lat/9911033.
111. S. Aoki, R. Frezzotti and P. Weisz, *Nucl. Phys.* **B540**, 501 (1999), hep-lat/9808007.
112. S. Takeda, S. Aoki and K. Ide, *Phys. Rev.* **D68**, 014505 (2003), hep-lat/0304013.
113. ALPHA, J. Heitger et al., *Nucl. Phys. Proc. Suppl.* **106**, 859 (2002), hep-lat/0110201.
114. S. Takeda et al., *Phys. Rev.* **D70**, 074510 (2004), hep-lat/0408010.
115. JLQCD, S. Aoki et al., *Phys. Rev.* **D68**, 054502 (2003), hep-lat/0212039.
116. QCDSF, M. Göckeler et al., (2004), hep-ph/0409312.
117. M. Peter, *Nucl. Phys.* **B501**, 471 (1997), hep-ph/9702245.
118. Y. Schröder, *Phys. Lett.* **B447**, 321 (1999), hep-ph/9812205.
119. S. Bethke, (2004), hep-ex/0407021.
120. J. Blümlein, H. Böttcher and A. Guffanti, *Nucl. Phys. Proc. Suppl.* **135**, 152 (2004), hep-ph/0407089.
121. W. Bernreuther and W. Wetzel, *Nucl. Phys.* **B197**, 228 (1982).
122. HPQCD, Q. Mason et al., *Phys. Rev. Lett.* **95**, 052002 (2005), hep-lat/0503005.
123. ALPHA, J. Rolf and S. Sint, *JHEP* **12**, 007 (2002), hep-ph/0209255.
124. M. Della Morte et al., *PoS* **LAT2005**, 224 (2005), hep-lat/0509173.
125. A. Bucarelli et al., *Nucl. Phys.* **B552**, 379 (1999), hep-lat/9808005.
126. F. Palombi, R. Petronzio and A. Shindler, *Nucl. Phys.* **B637**, 243 (2002), hep-lat/0203002.
127. Zeuthen-Rome / ZeRo, M. Guagnelli et al., *Nucl. Phys.* **B664**, 276 (2003),

hep-lat/0303012.
128. ALPHA, M. Kurth and R. Sommer, *Nucl. Phys.* **B623**, 271 (2002), hep-lat/0108018.
129. A.S. Kronfeld, *Nucl. Phys. Proc. Suppl.* **129**, 46 (2004), hep-lat/0310063.
130. S. Hashimoto and T. Onogi, *Ann. Rev. Nucl. Part. Sci.* **54**, 451 (2004), hep-ph/0407221.
131. T. Onogi, *PoS* **LAT2006**, 017 (2006), hep-lat/0610115.
132. E. Eichten, *Nucl. Phys. Proc. Suppl.* **4**, 170 (1988), talk delivered at *the Int. Sympos. of Field Theory on the Lattice*, Seillac, France, Sep 28 - Oct 2, 1987.
133. E. Eichten and B. Hill, *Phys. Lett.* **B234**, 511 (1990).
134. B. Grinstein, *Nucl. Phys.* **B339**, 253 (1990).
135. H. Georgi, *Phys. Lett.* **B240**, 447 (1990).
136. N. Isgur and M.B. Wise, *Phys. Lett.* **B232**, 113 (1989).
137. N. Isgur and M.B. Wise, *Phys. Lett.* **B237**, 527 (1990).
138. U. Aglietti, *Nucl. Phys.* **B421**, 191 (1994), hep-ph/9304274.
139. ALPHA, J. Heitger and R. Sommer, *JHEP* **02**, 022 (2004), hep-lat/0310035.
140. M. Della Morte et al., (2006), hep-ph/0609294.
141. J.G. Korner and G. Thompson, *Phys. Lett.* **B264**, 185 (1991).
142. S. Weinberg, *Physica* **A96**, 327 (1979).
143. J. Polchinski, *Nucl. Phys.* **B231**, 269 (1984).
144. H. Georgi, *Nucl. Phys.* **B361**, 339 (1991).
145. ALPHA, M. Della Morte et al., *Phys. Lett.* **B581**, 93 (2004), hep-lat/0307021.
146. M. Della Morte, A. Shindler and R. Sommer, *JHEP* **08**, 051 (2005), hep-lat/0506008.
147. D.J. Broadhurst and A.G. Grozin, *Phys. Lett.* **B267**, 105 (1991).
148. M.A. Shifman and M.B. Voloshin, *Sov. J. Nucl. Phys.* **45**, 292 (1987).
149. H.D. Politzer and M.B. Wise, *Phys. Lett.* **B206**, 681 (1988).
150. K.G. Chetyrkin and A.G. Grozin, *Nucl. Phys.* **B666**, 289 (2003), hep-ph/0303113.
151. X. Ji and M.J. Musolf, *Phys. Lett.* **B257**, 409 (1991).
152. D.J. Broadhurst and A.G. Grozin, *Phys. Rev.* **D52**, 4082 (1995), hep-ph/9410240.
153. V. Gimenez, *Nucl. Phys.* **B375**, 582 (1992).
154. ALPHA, J. Heitger et al., *JHEP* **11**, 048 (2004), hep-ph/0407227.
155. S. Hashimoto, T. Ishikawa and T. Onogi, *Nucl. Phys. Proc. Suppl.* **106**, 352 (2002).
156. ALPHA, J. Rolf et al., *Nucl. Phys. Proc. Suppl.* **129**, 322 (2004), hep-lat/0309072.
157. ALPHA, M. Della Morte, P. Fritzsch and J. Heitger, in preparation .
158. ALPHA, F. Palombi et al., in preparation .
159. A.F. Falk and M. Neubert, *Phys. Rev.* **D47**, 2965 (1993), hep-ph/9209268.
160. M. Battaglia et al., (2003), hep-ph/0304132.
161. C. Alexandrou et al., *Z. Phys.* **C62**, 659 (1994), hep-lat/9312051.
162. R. Sommer, *Phys. Rept.* **275**, 1 (1996), hep-lat/9401037.

163. H. Wittig, *Int. J. Mod. Phys.* **A12**, 4477 (1997), hep-lat/9705034.
164. A.X. El-Khadra et al., *Phys. Rev.* **D58**, 014506 (1998), hep-ph/9711426.
165. JLQCD, S. Aoki et al., *Phys. Rev. Lett.* **80**, 5711 (1998).
166. C.W. Bernard et al., *Phys. Rev. Lett.* **81**, 4812 (1998), hep-ph/9806412.
167. D. Becirevic et al., *Phys. Rev.* **D60**, 074501 (1999), hep-lat/9811003.
168. CP-PACS, A. Ali Khan et al., *Phys. Rev.* **D64**, 034505 (2001), hep-lat/0010009.
169. UKQCD, K.C. Bowler et al., *Nucl. Phys.* **B619**, 507 (2001), hep-lat/0007020.
170. UKQCD, L. Lellouch and C.J.D. Lin, *Phys. Rev.* **D64**, 094501 (2001), hep-ph/0011086.
171. S.M. Ryan, *Nucl. Phys. Proc. Suppl.* **106**, 86 (2002), hep-lat/0111010.
172. G.M. de Divitiis et al., *Nucl. Phys.* **B672**, 372 (2003), hep-lat/0307005.
173. G.M. de Divitiis et al., *Nucl. Phys.* **B675**, 309 (2003), hep-lat/0305018.
174. T. van Ritbergen, J.A.M. Vermaseren and S.A. Larin, *Phys. Lett.* **B400**, 379 (1997), hep-ph/9701390.
175. K.G. Chetyrkin, *Phys. Lett.* **B404**, 161 (1997), hep-ph/9703278.
176. J.A.M. Vermaseren, S.A. Larin and T. van Ritbergen, *Phys. Lett.* **B405**, 327 (1997), hep-ph/9703284.
177. M. Czakon, *Nucl. Phys.* **B710**, 485 (2005), hep-ph/0411261.
178. J. Foley et al., *Comput. Phys. Commun.* **172**, 145 (2005), hep-lat/0505023.
179. H. Wittig, *Int. J. Mod. Phys.* **A12**, 4477 (1997), hep-lat/9705034.
180. J.M. Flynn and C.T. Sachrajda, *Adv. Ser. Direct. High Energy Phys.* **15**, 402 (1998), hep-lat/9710057.
181. L. Maiani, G. Martinelli and C.T. Sachrajda, *Nucl. Phys.* **B368**, 281 (1992).
182. M. Bochicchio et al., *Nucl. Phys.* **B372**, 403 (1992).
183. J.M. Flynn and B.R. Hill, *Phys. Lett.* **B264**, 173 (1991).
184. UKQCD, A.K. Ewing et al., *Phys. Rev.* **D54**, 3526 (1996), hep-lat/9508030.
185. V. Gimenez, G. Martinelli and C.T. Sachrajda, *Nucl. Phys.* **B486**, 227 (1997), hep-lat/9607055.
186. D. Becirevic et al., *JHEP* **04**, 025 (2002), hep-lat/0110091.
187. D. Becirevic and J. Reyes, *Nucl. Phys. Proc. Suppl.* **129**, 435 (2004), hep-lat/0309131.
188. D. Becirevic et al., *PoS* **LAT2005**, 218 (2006), hep-lat/0509165.
189. M.B. Wise, *Phys. Rev.* **D45**, 2188 (1992).
190. B. Grinstein et al., *Nucl. Phys.* **B380**, 369 (1992), hep-ph/9204207.
191. D. Becirevic et al., *PoS* **LAT2005**, 212 (2006), hep-lat/0510017.
192. D. Becirevic et al., *Phys. Lett.* **B609**, 298 (2005), hep-lat/0406031.
193. B. Blossier et al., *Phys. Lett.* **B632**, 319 (2006), hep-lat/0507024.
194. U. Aglietti and V. Gimenez, *Nucl. Phys.* **B439**, 91 (1995), hep-lat/9503001.
195. J.C. Christensen, T. Draper and C. McNeile, *Phys. Rev.* **D62**, 114006 (2000), hep-lat/9912046.
196. J.E. Mandula and M.C. Ogilvie, *Phys. Rev.* **D57**, 1397 (1998), hep-lat/9703020.
197. C.R. Allton et al., *Nucl. Phys.* **B349**, 598 (1991).
198. C. Alexandrou et al., *Phys. Lett.* **B256**, 60 (1991).

199. A. Duncan et al., *Phys. Rev.* **D51**, 5101 (1995), hep-lat/9407025.
200. D. Guazzini, R. Sommer and N. Tantalo, *PoS* **LAT2006**, 084 (2006).
201. C. Pena, *PoS* **LAT2006**, 019 (2006), hep-lat/0610109.